## Other Titles in This Series

23 **P. M. Pardalos, D. Shalloway, and G. Xue, Editors,** Global Minimization of Nonconvex Energy Functions: Molecular Conformation and Protein Folding

22 **Panos M. Pardalos, Mauricio G. C. Resende, and K. G. Ramakrishnan, Editors,** Parallel Processing of Discrete Optimization Problems

21 **D. Frank Hsu, Arnold L. Rosenberg, and Dominique Sotteau, Editors,** Interconnection Networks and Mapping and Scheduling Parallel Computations

20 **William Cook, László Lovász, and Paul Seymour, Editors,** Combinatorial Optimization

19 **Ingemar J. Cox, Pierre Hansen, and Bela Julesz, Editors,** Partitioning Data Sets

18 **Guy E. Blelloch, K. Mani Chandy, and Suresh Jagannathan, Editors,** Specification of Parallel Algorithms

17 **Eric Sven Ristad, Editor,** Language Computations

16 **Panos M. Pardalos and Henry Wolkowicz, Editors,** Quadratic Assignment and Related Problems

15 **Nathaniel Dean and Gregory E. Shannon, Editors,** Computational Support for Discrete Mathematics

14 **Robert Calderbank, G. David Forney, Jr., and Nader Moayeri, Editors,** Coding and Quantization: DIMACS/IEEE Workshop

13 **Jin-Yi Cai, Editor,** Advances in Computational Complexity Theory

12 **David S. Johnson and Catherine C. McGeoch, Editors,** Network Flows and Matching: First DIMACS Implementation Challenge

11 **Larry Finkelstein and William M. Kantor, Editors,** Groups and Computation

10 **Joel Friedman, Editor,** Expanding Graphs

9 **William T. Trotter, Editor,** Planar Graphs

8 **Simon Gindikin, Editor,** Mathematical Methods of Analysis of Biopolymer Sequences

7 **Lyle A. McGeoch and Daniel D. Sleator, Editors,** On-Line Algorithms

6 **Jacob E. Goodman, Richard Pollack, and William Steiger, Editors,** Discrete and Computational Geometry: Papers from the DIMACS Special Year

5 **Frank Hwang, Fred Roberts, and Clyde Monma, Editors,** Reliability of Computer and Communication Networks

4 **Peter Gritzmann and Bernd Sturmfels, Editors,** Applied Geometry and Discrete Mathematics, The Victor Klee Festschrift

3 **E. M. Clarke and R. P. Kurshan, Editors,** Computer-Aided Verification '90

2 **Joan Feigenbaum and Michael Merritt, Editors,** Distributed Computing and Cryptography

1 **William Cook and Paul D. Seymour, Editors,** Polyhedral Combinatorics

# Global Minimization of Nonconvex Energy Functions: Molecular Conformation and Protein Folding

# DIMACS

Series in Discrete Mathematics
and Theoretical Computer Science

Volume 23

# Global Minimization of Nonconvex Energy Functions: Molecular Conformation and Protein Folding

DIMACS Workshop
March 20–21, 1995

P. M. Pardalos
D. Shalloway
G. Xue
Editors

NSF Science and Technology Center
in Discrete Mathematics and Theoretical Computer Science
A consortium of Rutgers University, Princeton University,
AT&T Bell Labs, Bellcore

**American Mathematical Society**

This DIMACS volume on Global Minimization of Nonconvex Energy Functions contains refereed papers selected from a workshop of the same title held on March 20–21, 1995, that was part of the DIMACS Special Year program on Mathematical Support for Molecular Biology.

1991 *Mathematics Subject Classification*. Primary 65K10, 90C30, 90C27, 92E10, 90C05, 90C25.

---

**Library of Congress Cataloging-in-Publication Data**

Global minimization of nonconvex energy functions : molecular conformation and protein folding : DIMACS workshop, March 20–21, 1995 / P. M. Pardalos, D. Shalloway, G. Xue, editors.

p. cm. — (DIMACS series in discrete mathematics and theoretical computer science, ISSN 1052-1798; v. 23)

Includes bibliographical references.

ISBN 0-8218-0471-5 (alk. paper)

1. Protein folding—Mathematical models—Congresses. I. Pardalos, P. M. (Panos M.), 1954– II. Shalloway, D. (David), 1948– . III. Hsüeh, Kuo-liang. IV. DIMACS (Group) V. Series.

QP551.G63 1995

574.19′245—dc20 95-25197

CIP

---



∞ The paper used in this book is acid-free and falls within the guidelines
established to ensure permanence and durability.
♻ Printed on recycled paper.

10 9 8 7 6 5 4 3 2 1 01 00 99 98 97 96

*"Innumerable secret signs, with which the universe is studded and which constitute so many syllables of an unknown language, urge us to compose words, and with words, phrases whose deciphering puts us at the threshold of the deepest truth. "*

- ODYSSEUS ELYTIS, Nobel Lecture (December 8, 1979)

# Contents

# Foreword

This DIMACS volume on *Global minimization of nonconvex energy functions* contains refereed papers selected from a workshop of the same title that was part of the DIMACS Special Year program on Mathematical Support for Molecular Biology.

We would like to particularly thank the organizers of the Special Year program: Joachim Messing and Fred Roberts (chairs), Lawrence Shepp and Michael Waterman (co-chairs), and Martin Farach (assistant chair) for their tremendous efforts initiating this workshop, along with many others, organizing seminars, postdoctoral fellowship programs, tutorials, and distinguished lecturer talks, and attending to many other tasks large and small.

It is a pleasure also to thank Panos M. Pardalos, David Shalloway, and Guoliang Xue for their dedicated work in organizing the workshop and producing this volume.

Finally, we would like to acknowledge the generous support of the National Science Foundation, AT&T Bell Laboratories, Bellcore, the New Jersey Commission on Science and Technology, the National Center for Human Genome Research, the National Security Agency, the Department of Energy, the Centers for Disease Control, the National Institute of Allergy and Infectious Diseases, Roche Molecular Systems, Rutgers University, Sandia Laboratories, and Smith-Kline Beecham for the entire Special Year program and this workshop in particular.

Andras Hajnal, Director<br>
Andrew Yao, Co-Director<br>
Stephen R. Mahaney, Associate Director

# Preface

In the context of the 1994-1995 DIMACS special year on Mathematical Support for Molecular Biology, a two-day workshop entitled "Global Minimization of Nonconvex Energy Functions: Molecular Conformation and Protein Folding" was held on March 20-21, 1995. One of the most significant and challenging problems in molecular biophysics and biochemistry is that of computing the native 3-dimensional conformation (folded state) of a globular protein given its amino acid sequence, possibly in the presence of additional agents (e.g., drugs). The solution will probably require the use of experimental or knowledge-based methods to provide initial guesses for subsequent numerical refinement. The numerical methods require: (a) determination of suitable energy functions that can reliably model protein behavior and (b) optimization methods that can find global minima of energy functions given the many degrees of freedom available to protein structures. This workshop focused on global optimization methods and related problems through an interdisciplinary effort that was represented from computational groups in global optimization, computer science and biochemistry.

The workshop featured about twenty invited speakers who talked about protein folding, simulation and modeling techniques, heuristics and exact algorithms for globally optimizing nonconvex energy functions. This volume contains a collection of refereed papers from the workshop.

The workshop was sponsored by DIMACS, through grants from the National Science Foundation, the New Jersey Commission on Science and Technology, the National Center for Human Genome Research, the Department of Energy, the National Security Agency, the Centers for Disease Control, the National Institute of Allergy and Infectious Diseases, Rutgers University, and other sources. We would like to take this opportunity to thank the sponsors, the DIMACS staff, the participants, the authors, the referees, and the American Mathematical Society, for making the workshop successful and the publication of this volume possible.

Panos M. Pardalos, David Shalloway, and Guoliang Xue
June 1995

DIMACS Series in Discrete Mathematics
and Theoretical Computer Science
Volume **23**, 1996

# Global minimization on rugged energy landscapes

Patricia Amara, Jianpeng Ma and John E. Straub

ABSTRACT. Three algorithms for global energy minimization based on the simulated annealing of the classical density distribution are presented. Two algorithms are based on the approximate solution of the classical Liouville equation for the dynamics of a system coupled to a heat bath using Fokker-Planck and Smoluchowski dynamics. A third algorithm is based on annealing the classical density distribution directly in temperature and is the classical analog of imaginary time quantum dynamics. These three methods are compared with standard simulated annealing based on molecular dynamics. The results for a model potential, Lennard-Jones clusters, water clusters, a model homopolymer, and a model heteropolymer demonstrate that by annealing the continuous density distribution (representing a volume of phase or configuration space) the likelihood of finding the global minimum is dramatically enhanced.

## 1. Introduction

When a system must be described quantum mechanically, except for the most simple systems, approximate basis set methods and perturbation theory must be employed. However, once one assumes that a system (for example, a liquid, crystal or protein) is well described by classical mechanics, one has the luxury of calculating the classical dynamics exactly by the numerical solution of Newton's equations of motion. To some extent, this explains why relatively little work has been done to develop approximate classical dynamical simulation algorithms.

When one wishes to find the global energy minimum of a complex molecular system, or search for binding sites on a protein for a ligand or potential drug, one finds that exact molecular dynamics simulation does not provide the most effective means to sample the potential energy hypersurface [**1**]. As a result, approximate classical molecular dynamics methods have been developed to address these hard problems; many are based on the dynamics of a continuous classical density distribution (representing an ensemble of systems) which is defined by the Liouville equation. Due to the similarity between the Liouville equation of classical statistical mechanics and the quantum mechanical von Neumann (quantum Liouville) equation[**2, 3**] and Schrödinger equation [**4, 5, 6**], many of the methods developed

1991 *Mathematics Subject Classification.* Primary 65K10, 90C30; Secondary 92C05, 92E10.
JES gratefully acknowledges the support of the National Science Foundation (CHE-9306375).

to treat quantum mechanical dynamics have been extended as approximate, coarse grained classical dynamics methods [**7, 8, 9, 10, 11, 12**].

In this paper, we summarize the development of methods based on the dynamical and thermodynamical evolution of the classical density distribution. We present general derivations of moment equations for the classical density distribution. Just as methods based on Gaussian packet dynamics have been particularly effective in providing insights and useful algorithms in quantum dynamics, we have found that approximate Gaussian density distribution based methods are effective in providing both (1) effective "black box" optimization algorithms and (2) a profitable way of thinking about global optimization problems where many potentially useful areas of research remain unexplored. Results from the application of these methods to a number of optimization problems including atomic clusters, water clusters, model homopolymer collapse and model protein folding are described. These applications help to provide an understanding of the mechanism used by these optimization methods as well as their limitations and long suits.

## 2. Evolution of the classical density distribution in time and temperature

In this section we present simulated annealing algorithms based on the evolution of the classical density distribution defined by three different equations of motion. (1) The Fokker-Planck equation is used to derive equations of motion in time for the classical phase space density distribution $\rho(\mathbf{x}, \mathbf{p}, t)$. (2) The Smoluchowski equation is used to derive equations of motion in time for a reduced configurational space density distribution $\hat{\rho}(\mathbf{x}, t)$ where the inertial terms dependent on the momenta are removed. (3) The Bloch equation is used to derive equations of motion in *temperature* for the configurational density distribution $\hat{\rho}(\mathbf{x}, \beta)$. In each case, we first derive the general moment equations which provide an exact description of the evolution of the distribution in time or temperature. We then focus on the special case of the Gaussian density approximation.

### 2.1. Fokker-Planck packet dynamics of $\rho(\mathbf{x}, \mathbf{p}, t)$.

2.1.1. *Time evolution of the total position and momentum moments* $\mathbf{M}_{n,k}$.

The total position and momentum moment $\mathbf{M}_{n,k}$ is defined in general as a symmetrized tensor of the form

$$\mathbf{M}_{n,k} = \int d\mathbf{x}d\mathbf{p}\,\mathcal{P}\left(\mathcal{M}^{n,k}\right)\rho = \int d\mathbf{x}d\mathbf{p}\,\mathcal{P}\left(\Delta\mathbf{x}^{n}\Delta\mathbf{p}^{k}\right)\rho$$

where $\mathcal{P}(.)$ stands for the average over all possible permutations of the individual tensor indices and $n+k$ is the rank of the tensor $\mathbf{M}_{n,k}$. The increment of phase space is $d\mathbf{x}d\mathbf{p} = dx_1dx_2dx_3dp_1dp_2dp_3$ (in three dimensions), $\Delta\mathbf{x} = \mathbf{x} - \mathbf{x}_0 = \Delta x_i\hat{e}_i$ and $\Delta\mathbf{p} = \mathbf{p} - \mathbf{p}_0 = \Delta p_i\hat{e}_i$ where $\mathbf{x}_0 = \langle x\rangle$ and $\mathbf{p}_0 = \langle p\rangle$ and $\langle . \rangle$ indicates an average over the normalized density distribution.

The Fokker-Planck equation for the time evolution of the phase space density distribution $\rho(\mathbf{x}, \mathbf{p}, t)$ is

$$\frac{\partial}{\partial t}\rho = -\left[\frac{p_i}{m}\frac{\partial}{\partial x_i} + F_i(\mathbf{x})\frac{\partial}{\partial p_i} - \gamma\left[\frac{\partial}{\partial p_i}\left[p_i + mk_BT\frac{\partial}{\partial p_i}\right]\right]\right]\rho. \tag{2.1}$$

From this equation of motion, the time derivative of the moments $\mathbf{M}_{n,k}$ are defined

$$\frac{d}{dt}\mathbf{M}_{n,k} = \int d\mathbf{x}d\mathbf{p}\left[\mathcal{P}\left(\frac{\partial \mathcal{M}^{n,k}}{\partial t}\right)\rho + \mathcal{P}\left(\mathcal{M}^{n,k}\frac{\partial \rho}{\partial t}\right)\right]. \tag{2.2}$$

The details of evaluation of this equation are given in Appendix A. The general result for the exact Fokker-Planck dynamics is

$$\frac{d}{dt}\mathbf{x}_0 = \frac{\mathbf{p}_0}{m} \qquad\qquad \frac{d}{dt}\mathbf{p}_0 = \mathbf{F}_0 - \gamma\mathbf{p}_0$$

and

$$\frac{d}{dt}\mathbf{M}_{n,k} = \frac{n}{m}\mathbf{M}_{n-1,k+1} + k\mathbf{W}_{n,k-1} - \gamma\left[k\mathbf{M}_{n,k} - k(k-1)mk_BT\mathbf{L}_{n,k-2}\right] \tag{2.3}$$

where we have defined the $n+k$ rank tensors $\mathbf{W}_{n,k} = \left\langle \mathcal{P}\left(\mathcal{M}^{n,k}(F_i(\mathbf{x}) - F_{0i})\hat{e}_i\right)\right\rangle$ and $\mathbf{L}_{n,k-2} = \left\langle \mathcal{P}\left(\mathcal{M}^{n,k-2}\hat{e}_i\hat{e}_i\right)\right\rangle$. This is our final result for the general moment equations which provide an exact description for the time evolution of the phase space density distribution in Fokker-Planck dynamics. Note that the equations of motion for the distribution center is similar to the Langevin equation for a single point in phase space. The center responds to the average force on the distribution

$$\mathbf{F}_0 = \int d\mathbf{x}d\mathbf{p}\,\mathbf{F}(\mathbf{x})\rho.$$

In general, one cannot integrate the unbounded hierarchy of equations. One possible way to proceed is to define a basis for $\rho$ and derive equations of motion for the parameters of that basis. An example is the Gaussian phase packet approximation.

2.1.2. *Gaussian phase packet (GPP) Fokker-Planck dynamics.*
The phase space density distribution for each degree of freedom may be approximated as a spherically symmetric GPP in the form

$$\rho(\mathbf{r},\mathbf{p},t) = \left(\frac{\sqrt{1-\alpha^2}}{2\pi\sigma_a\sigma_b}\right)^d \exp\left[-\frac{1}{2}\left(\frac{r-r_0}{\sigma_a}\right)^2 - \frac{1}{2}\left(\frac{p-p_0}{\sigma_b}\right)^2 - \alpha\left(\frac{r-r_0}{\sigma_a}\right)\left(\frac{p-p_0}{\sigma_b}\right)\right].$$

This approximation leads to the following time evolution for the center and second-order moments [**13**]

$$\begin{aligned} &\dot{\mathbf{r}}_0 = \frac{\mathbf{p}_0}{m} \qquad \dot{\mathbf{p}}_0 = -\nabla_{\mathbf{r}_0}\langle V\rangle - \gamma\mathbf{p}_0 \\ &\dot{M}_{0,2} = -\frac{2}{d}M_{1,1}\nabla^2_{r_0}\langle V\rangle - 2\gamma\left[M_{0,2} - dmk_BT\right] \qquad \dot{M}_{2,0} = \frac{2}{m}M_{1,1} \\ &\dot{M}_{1,1} = \frac{1}{m}M_{0,2} - \frac{1}{d}M_{2,0}\nabla^2_{r_0}\langle V\rangle - \gamma M_{1,1}. \end{aligned} \tag{2.4}$$

For a spherically symmetric Gaussian, the second rank tensor moments are isotropic so the scalars $\mathrm{M}_{2,0}$, $\mathrm{M}_{1,1}$ and $\mathrm{M}_{0,2}$ are the trace of the second rank tensors $\mathbf{M}_{2,0}$, $\mathbf{M}_{1,1}$ and $\mathbf{M}_{0,2}$ respectively [**14, 15**]. These equations of motion have been used, along with the simulated annealing protocol, in the global optimization of a variety of systems discussed in the next section. In practice, one may control the external bath temperature $T$ using a cooling schedule. The friction constant $\gamma$ acts as an effective force constant to constrain the temperature of the system about the bath temperature through the equation of motion for $M_{0,2}$. The center of the distribution follows a Langevin dynamics on a fluctuating effective potential energy surface

$$\langle V\rangle = \int d\mathbf{r}d\mathbf{p}\, V(\mathbf{r})\rho.$$

This averaged potential may be thought of as a coarse grained potential function, averaged over the extent of the distribution. The effect of a broad distribution of $\langle V\rangle$ is to raise minima, lower barriers, and smooth the potential surface dramatically reducing the number of local minima. This effect of potential smoothing, which arises naturally from the dynamical equations for the density distribution, added to the simulated annealing protocol, effectively combines two of the paradigms of molecular energy minimization [**16**].

**2.2. Smoluchowski packet dynamics of $\hat{\rho}(\mathbf{x}, t)$.**

2.2.1. *Time evolution of the total position moments $\mathbf{M}_n$.*

In many situations, the momenta of a system relax quickly compared to the coordinates and it is possible to eliminate the momenta from the density distribution adiabatically. The resulting reduced density distribution function is $\hat{\rho}(\mathbf{x}, t)$ whose dynamics may be defined in terms of symmetrized n$^{th}$ rank moment tensors in configuration space

$$\mathbf{M}_n = \int d\mathbf{x}\, \mathcal{P}\left(\mathcal{M}^n\right)\rho = \int d\mathbf{x}\, \mathcal{P}\left(\Delta\mathbf{x}^n\right)\rho.$$

The dynamics of the reduced configurational space distribution function $\hat{\rho}(\mathbf{x}, t)$ is defined by the Smoluchowski equation

$$\frac{\partial}{\partial t}\rho = \frac{1}{m\gamma}\left[\frac{\partial}{\partial x_i}\left[-F_i(\mathbf{x}) + k_B T\frac{\partial}{\partial x_i}\right]\rho\right]. \tag{2.5}$$

The time derivative of the moments $\mathbf{M}_n$ is given by

$$\frac{d}{dt}\mathbf{M}_n = \int d\mathbf{x}\mathcal{P}\left(\frac{\partial \mathcal{M}^n}{\partial t}\right)\rho + \mathcal{P}\left(\mathcal{M}^n\frac{\partial\rho}{\partial t}\right). \tag{2.6}$$

Evaluation of the integrals is described in Appendix B. Combined with the equation of motion for the center of the distribution, we arrive at

$$\frac{d}{dt}\mathbf{x_0} = \frac{1}{m\gamma}\mathbf{F}_0$$

and

$$\frac{d}{dt}\mathbf{M}_n = \frac{1}{m\gamma}\left[n\mathbf{W}_{n-1} + k_B T n(n-1)\mathbf{L}_{n-2}\right] \tag{2.7}$$

where the n$^{th}$ rank tensors are $\mathbf{M}_n = \langle\mathcal{P}\left(\mathcal{M}^n\right)\rangle$, $\mathbf{W}_n = \langle\mathcal{P}\left(\mathcal{M}^n(F_i(\mathbf{x}) - F_0)\right)\rangle$ and $\mathbf{L}_n = \langle\mathcal{P}\left(\mathcal{M}^{n-2}\hat{e}_i\hat{e}_i\right)\rangle$. These equations of motion describe the Smoluchowski dynamics of the density distribution exactly. Note that the equation of motion for the center of the distribution has the form of a steepest descent equation employing the force averaged over the density distribution.

2.2.2. *GPP Smoluchowski dynamics.*
The configurational distribution may be approximated as a spherical Gaussian

$$\hat{\rho}(\mathbf{r},t) = \left(\frac{2\pi M_2}{d}\right)^{-d/2} \exp\left[-\frac{d}{2M_2}(\mathbf{r}-\mathbf{r}_0)^2\right].$$

$\mathbf{r}_0$ is the center of the distribution and $\mathrm{M}_2$ is the trace of the isotropic second rank tensor $\mathbf{M}_2$. The resulting equations of motion for the center and second-order moment are

$$\begin{aligned}\frac{d\mathbf{r}_0}{dt} &= -\frac{1}{m\gamma}\nabla_{r_0}\langle V\rangle \\ \frac{dM_2}{dt} &= \frac{1}{m\gamma}\left[2dk_BT - \frac{2}{d}M_2\nabla^2_{r_0}\langle V\rangle\right].\end{aligned} \tag{2.8}$$

The center of the distribution follows a steepest descent path on a dynamic effective potential. When the packets are wide and the distribution is delocalized the effective potential is smoothed. At low temperatures, the packet narrows and the exact physical potential surface is recovered. As in the case of Fokker-Planck dynamics, a simulated annealing protocol can be followed by controlling the bath temperature $T$. In Fokker-Planck dynamics, the system temperature fluctuates about the bath temperature T, while in Smoluchowski dynamics, the system temperature equals the bath temperature at all times.

**2.3. Thermodynamical annealing in temperature of $\hat{\rho}(\mathbf{x},\beta)$.**
With all dynamical annealing algorithms, the choice of the cooling schedule is important. Typically, a realizable cooling schedule is far from optimal. Therefore, we have explored a different form of annealing where the density distribution is integrated in temperature directly and there is no real time dynamics.

2.3.1. *Temperature evolution of the total position moments* $\mathbf{M}_n$.
The temperature dependence of the reduced configurational distribution function $\hat{\rho}(\mathbf{x},\beta)$ is described by the reduced Bloch equation

$$\frac{\partial}{\partial\beta}\hat{\rho}(\mathbf{x},\beta) = -(V(\mathbf{x}) - \langle V(\mathbf{x})\rangle)\hat{\rho}(\mathbf{x},\beta). \tag{2.9}$$

The symmetrized tensor representing the position moment $\mathbf{M}_n$ is

$$\mathbf{M}_n = \int d\mathbf{x}\,\mathcal{P}\left(\mathcal{M}^n\right)\rho = \int d\mathbf{x}\,\mathcal{P}\left(\Delta\mathbf{x}^n\right)\rho.$$

It follows that the temperature dependence of the moments $\mathbf{M}_n$ is given by

$$\frac{d}{d\beta}\mathbf{M}_n = -\left\langle\mathcal{P}\left(V(\mathbf{x})\mathcal{M}^n\right)\right\rangle + \left\langle\mathcal{P}(\mathcal{M}^n)\right\rangle\left\langle V(\mathbf{x})\right\rangle. \tag{2.10}$$

These moment equations provide an exact description of the temperature evolution of the density distribution.

2.3.2. *Gaussian Density Annealing (GDA).*
In the GDA method we approximate the density as a Gaussian distribution which is fully defined by the center of the distribution $\mathbf{r}_0$ and variance $M_2$ or

$$\hat{\rho}(\mathbf{r},\beta) = \left(\frac{2\pi M_2}{d}\right)^{-d/2} \exp\left[-\frac{d}{2M_2}(\mathbf{r}-\mathbf{r}_0)^2\right]$$

as for Smoluchowski dynamics. The temperature evolution for the center and second-order moment is given by

$$\begin{aligned} \frac{d\mathbf{r}_0}{d\beta} &= -\frac{1}{d} M_2 \nabla_{r_0} \langle V \rangle \\ \frac{dM_2}{d\beta} &= -\frac{1}{d^2} M_2^2 \nabla_{r_0}^2 \langle V \rangle. \end{aligned} \tag{2.11}$$

Again, the effective potential $\langle V \rangle$ appears. The center follows the gradient of the effective potential while the variance adjusts to the Laplacian of the effective potential. As with all the GPP based algorithms, the annealing dynamics is informed by higher order derivatives of the potential, rather than the force alone as is the case with classical molecular dynamics.

## 3. Applications

The methods described above have been applied to a series of model and molecular potential energy minimization problems. In this section, we provide a summary of the results.

### 3.1. One-dimensional rough potentials.

One might assume that exact classical mechanical annealing should be superior to an approximate, coarse grained annealing method. However, application of exact MD and the approximate GPP based annealing methods has shown that the GPP dynamics, while providing an approximate thermodynamics and dynamics for the system, can be superior in locating low energy regions of the potential energy hypersurface.

To understand how this is accomplished, we have studied the one-dimensional rough potential [**17**]

$$V(x) = \frac{1}{2}\kappa x^2 + \epsilon \cos[qx + \pi]. \tag{3.1}$$

At high temperature, the density distribution is delocalized over a broad region of the potential surface. As the temperature is lowered, the distribution can collapse and become localized in a potential energy minimum. For this simple potential, it is possible to isolate two critical temperatures for the localization-delocalization transition (above which the packet is delocalized and below which the packet collapses). (1) For an *annealing* run, where the temperature is monotonically decreased, the critical temperature for the packet collapse transition is $T_{\mathrm{a}}$. It is possible to derive an upper bound for this critical annealing temperature in terms of the potential parameters [**18**]

$$T_{\mathrm{a}} = -\frac{4\kappa}{q^2 k_B} \ln \frac{\kappa}{\epsilon q^2}. \tag{3.2}$$

(2) For a *heating* run, the initially collapsed distribution is delocalized above a critical temperature $T_{\mathrm{h}}$. In general, for the GPP representation of the density distribution $T_{\mathrm{a}} \ll T_{\mathrm{h}}$. This hysteresis effectively reduces the temperature of the collapse transition for an annealing run. A similar observation has been made by Orešič and Shalloway in a study of Lennard-Jones cluster optimization [**19**]. By lowering the collapse transition temperature, the distribution remains delocalized over a broader region of temperature, possibly leading to further enhanced sampling.

This analysis implies that thermodynamic critical temperatures are dependent on the form of the approximation to the density distribution and the details of the annealing run. Coarse-grained dynamics might therefore be designed to move a critical temperature in such a way as to improve the probability that the algorithm will locate the global energy minimum. This is discussed below in the context of a homopolymer collapse transition.

**3.2. Lennard-Jones clusters.**
The benchmark for molecular optimization algorithms is the potential energy optimization of Lennard-Jones clusters. We have demonstrated that the GPP based Fokker-Planck and GDA algorithms are very effective in isolating the global energy minimum for clusters as large as $N = 55$ [**11, 13**]. The GPP based dynamical annealing algorithms are significantly more effective than standard molecular dynamics methods for the same cooling schedule. The GDA algorithm, which does not depend on a cooling schedule, was found to be approximately as effective as the GPP based Fokker-Planck method for the exponential cooling schedule $T(t) = 2\exp(-t/10)$. Of course, while the GDA algorithm may be improved on only by modifying the representation of the density distribution, the GPP method may be improved by optimizing the cooling schedule. We have done some work in this direction through the use of a cooling schedule where the rate of change in temperature with time is proportional to the heat capacity $C_v(T)$ at constant volume

$$dT/dt = -\text{constant}\frac{1}{C_v(T)}. \tag{3.3}$$

Similar ideas have been explored previously [**20**]. Scaling the cooling rate in this way allows the system to spend approximately the same period of time $dt$ in each energy increment $dE$ as the temperature is being reduced. This simple technique leads to a significant improvement in the results for the same investment of computational time [**13**].

Ideally, one would like to understand the relative success or failure of optimization for a particular system in terms of the details of the potential energy hypersurface. We found that the probability of finding the global energy minimum for clusters of $N = 8$ to 19 was strongly correlated with the size of the energy gap between the global minimum and lowest energy local minimum. "Magic number" clusters, which have particularly low lying global energy minima and large gaps, were easier optimization problems than clusters where the energy gap was small [**13**]. This implies that for molecular optimization problems, the difficulty does not scale in a simple way with N but depends on the characteristics of the potential energy hypersurface.

**3.3. Water clusters.**
Tsoo and Brooks have applied a variant of the GDA algorithm to study the optimization of water clusters [**21**]. Water clusters provide notoriously difficult molecular optimization problems (empirically more difficult than Lennard-Jones clusters and small peptides). In their study, they recognized that a straightforward integration of the GDA equations of motion in reciprocal temperature $\beta$ can result in a too rapid decrease in the variances ($M_2$) of the packets with relatively little movement of the centers. They found that it was useful to provide bounds on the variances while updating the position of the distribution center. The results of

Tsoo and Brooks show that their variation on the GDA algorithm performs slightly better than the Diffusion Equation Method (DEM) of Scheraga and coworkers [**22**] but is not successful in isolating the global energy minimum in certain cases.

This insight led to our development of the Adiabatic GDA method whereby one replaces the equation of motion for the center

$$d\mathbf{r}_0/d\beta = -(1/d)M_2\nabla_{r_0}\langle V\rangle = 0 \tag{3.4}$$

with a local energy minimization routine. The algorithm follows the protocol (1) solve Eq. (3.4) using a minimization routine, (2) integrate $dM_2/d\beta$ for a given step in $\beta$, (3) return to (1). This method allows the centers to respond instantaneously to changes in the variances. The resulting algorithm is a significant improvement over the standard GDA method [**16**].

### 3.4. Collapse transitions in Lennard-Jones homopolymers.

The GPP dynamical algorithm has been used to study the collapse of Lennard-Jones homopolymers [**23**]. The homopolymer was modeled as a freely jointed chain of Lennard-Jones atoms connected by harmonic bonds. An interesting point regarding the approximate thermodynamics generated by the GPP representation of the system emerged. In examining, for example, the radius of gyration as a function of temperature for the GPP and exact MD simulations, the homopolymer collapse transition temperature $T_\theta$ was found to occur at a significantly lower temperature in the GPP simulation than in the MD simulation. We believe that this is a manifestation of the property observed previously for the one-dimensional rough potential [**18**] where the critical annealing temperature $T_\mathrm{a}$ was suppressed. For the model one-dimensional potential and the Lennard-Jones homopolymer, as the temperature is lowered the GPP dynamics leads to a more extended or delocalized distribution at a given temperature. Therefore, above $T_\theta$ the GPP representation of the system has an effectively altered thermodynamics.

The time scale for nucleation or homopolymer collapse can depend on the difference between a thermodynamic transition temperature and a quench temperature. For example, de Gennes [**24**] has proposed that the characteristic time for the homopolymer collapse transition following a quench to a temperature $T_Q < T_\theta$ increases monotonically with the temperature ratio

$$\frac{T_\theta - T_Q}{T_\theta}. \tag{3.5}$$

Altering the thermodynamic transition temperature can lead to effectively faster (or slower) transitions. Therefore, if an approximate representation of the system leads to an altered value of $T_\theta$, the dynamics may also be affected in an important way. Potentially, this could be used as a criterion to develop more effective coarse grained dynamics for global optimization.

### 3.5. Folding model proteins.

We have completed an extensive study of global optimization for 22-mer and 46-mer model proteins [**25**] based on the three letter code potential of Honeycutt and Thirumalai [**26**]. The three letters correspond to hydrophobic-type beads (B), hydrophilic-type beads (L), and neutral-type beads (N). The potential consists of a linear polymer chain of harmonic bonds

$$V_{\mathrm{bond}}(r) = \frac{\kappa_b}{2}(r-\sigma)^2 \tag{3.6}$$

and angles

$$V_{\text{angle}}(\theta) = \frac{\kappa_\theta}{2}(\theta - \theta_0)^2 \tag{3.7}$$

where the equilibrium angle is $\theta_0 = 105°$. The dihedral angle potential is

$$E_{\text{torsion}}(\phi) = A(1 + \cos[\phi]) + B(1 + \cos[3\phi]) \tag{3.8}$$

where the dihedral constants are defined to be $A = 0$ and $B = 0.2$ if two or more of the defining atoms $i, j, k$ and $l$ are type N, and $A = B = 1.2$ otherwise.

The nonbonded potential is made of soft-sphere type repulsions and Lennard-Jones terms defined as

$$\begin{aligned} V_{BB}(r) &= 4\epsilon\left[(\sigma/r)^{12} - (\sigma/r)^6\right] \\ V_{LB,LL}(r) &= (8\epsilon/3)\left[(\sigma/r)^{12} + (\sigma/r)^6\right] \\ V_{NN,NL,NB}(r) &= 4\epsilon\,(\sigma/r)^{12}. \end{aligned} \tag{3.9}$$

The potential models the basic properties of attractive interactions between hydrophobic residues (B) and encourages the formation of a hydrophobic core. The repulsions between hydrophilic sites (L) and all other residues model their tendency to be found at the protein surface.

We arrived at a number of conclusions relating to the relative effectiveness of the GPP based algorithms, simulated annealing using MD, and the DEM [**22**] in a study of the 22-mer $(LB)_5N_2$ $(LB)_5$, a $\beta$-turn, and 46-mer $B_9$ $N_3$ $(LB)_4$ $N_3$ $B_9$ $N_3$ $(LB)_5$ L, a $\beta$-turn sandwich [**25**].

(1) Comparative analysis of the success in locating the global energy minimum for the 22-mer model showed that the GPP based methods, along with the DEM, are significantly more effective in isolating the global energy minimum than simulated annealing with MD. In particular, the GDA, DEM and GPP based Fokker-Planck dynamics algorithms were most effective.

(2) The probability of finding the global energy minimum for three sequences of the 22-mer, (a) the hetero sequence $(LB)_5$ $N_2$ $(LB)_5$, (b) the hybrid sequence $B_9$ $N_3$ $(LB)_5$ and the (c) the homo sequence $B_{22}$, (which uses the torsion potential of the hybrid sequence), was studied. The results showed that the hetero sequence (a) was easiest to fold (for all algorithms studied). We found that the energy gap between the global energy minimum and the next highest energy, incorrectly folded state was well correlated with algorithmic success. This observation was made previously in the context of Lennard-Jones clusters [**13**].

(3) In comparing the results for folding the 22-mer and 46-mer, we found that phase space based dynamical annealing methods (such as the GPP based Fokker-Planck dynamics) had a distinct advantage over configurational space based methods (such as the GDA) in finding the global energy minimum of the 46-mer. It appears that the configurational space based methods, such as the GPP based Smoluchowski algorithms and the GDA algorithm, which are based on the evolution of $\hat{\rho}(\mathbf{r})$ in the absence of ballistic motion, are very sensitive to the initial conditions. The phase space based dynamical methods, which maintain inertial terms in their dynamics, were better able to overcome an unfortunate initial placement.

## 4. Conclusions

Results for a variety of systems indicate that when the simulated annealing protocol is followed using an approximate GPP based classical dynamics, it is possible to combine the best aspects of the simulated annealing and potential smoothing paradigms for molecular optimization. Regarding the relative success of these methods, it has been emphasized that the effect of potential smoothing is to reduce the number of local minima and thereby ease the search of conformational space [**22, 27**]. We have shown that a related change in the thermodynamic transition temperatures can be used, in a compatible way, to account for the improved ability of the GPP based methods to locate the global energy minimum for a variety of molecular optimization problems.

## Appendix A. Derivation of $d\mathbf{M}_{n,k}/dt$ for Fokker-Planck dynamics

The right hand side of Eq. (2.2) may be divided into two integrals

$$I_1 = \int d\mathbf{x}d\mathbf{p}\,\mathcal{P}\left(\frac{\partial \mathcal{M}^{n,k}}{\partial t}\right)\rho \qquad I_2 = \int d\mathbf{x}d\mathbf{p}\,\mathcal{P}\left(\mathcal{M}^{n,k}\frac{\partial \rho}{\partial t}\right).$$

Given that

$$\frac{\partial}{\partial t}\Delta x_i\hat{e}_i = -\dot{x}_{0i}\hat{e}_i \qquad \frac{\partial}{\partial t}\Delta p_i\hat{e}_i = -\dot{p}_{0i}\hat{e}_i \tag{A.1}$$

and that for the Fokker-Planck dynamics

$$\dot{x}_{0i}\hat{e}_i = \frac{p_{0i}}{m}\hat{e}_i \qquad \dot{p}_{0i}\hat{e}_i = (F_{0i} - \gamma p_{0i})\hat{e}_i$$

we find

$$\frac{\partial}{\partial t}\mathcal{P}\left(\mathcal{M}^{n,k}\right) = \mathcal{P}\left(-\frac{n}{m}(p_{0i}\hat{e}_i)\mathcal{M}^{n-1,k} - k(F_{0i} - \gamma p_{0i})\hat{e}_i\mathcal{M}^{n,k-1}\right). \tag{A.2}$$

We advise that one performs the time derivatives using the Einstein summation convention. The permutation operator insures the simple final result. Therefore, $\mathrm{I}_1$ can be written as $I_1 = I_{11} + I_{12} + I_{13}$

$$\begin{aligned} I_1 = &-\frac{n}{m}\int d\mathbf{x}d\mathbf{p}\,\mathcal{P}\left(\mathcal{M}^{n-1,k}(p_{0i}\hat{e}_i)\right)\rho - k\int d\mathbf{x}d\mathbf{p}\,\mathcal{P}\left(\mathcal{M}^{n,k-1}(F_{0i}\hat{e}_i)\right)\rho \\ &+ \gamma k\int d\mathbf{x}d\mathbf{p}\,\mathcal{P}\left(\mathcal{M}^{n,k-1}(p_{0i}\hat{e}_i)\right)\rho. \end{aligned} \tag{A.3}$$

Substituting $\partial\rho/\partial t$ by its expression given in Eq. (2.1), $\mathrm{I}_2$ can be written as $I_2 = I_{21} + I_{22} + I_{23} + I_{24} + I_{25}$

$$\begin{aligned} I_2 = &-\frac{1}{m}\int d\mathbf{x}d\mathbf{p}\,\mathcal{P}\left(\mathcal{M}^{n,k}p_i\frac{\partial \rho}{\partial x_i}\right) - \int d\mathbf{x}d\mathbf{p}\,\mathcal{P}\left(\mathcal{M}^{n,k}F_i(\mathbf{x})\frac{\partial \rho}{\partial p_i}\right) \\ &+ \gamma\int d\mathbf{x}d\mathbf{p}\,\mathcal{P}\left(\mathcal{M}^{n,k}p_i\frac{\partial \rho}{\partial p_i}\right) + \gamma\int d\mathbf{x}d\mathbf{p}\,\mathcal{P}\left(\mathcal{M}^{n,k}\right)\rho \\ &+ \gamma m k_B T\int d\mathbf{x}d\mathbf{p}\,\mathcal{P}\left(\mathcal{M}^{n,k}\frac{\partial^2 \rho}{\partial p_i^2}\right). \end{aligned} \tag{A.4}$$

Holding the variable $\mathbf{p}$ constant, $\mathrm{I}_{21}$ may be integrated by parts with respect to $\mathbf{x}$. For all following integrations by parts we divide the integrand into two parts;

the integrated term is always a differential of $\rho$. We assume that the surface terms vanish since the normalized density distribution goes to 0 at $\infty$. This leads to

$$I_{21} = \frac{n}{m} \int d\mathbf{x} d\mathbf{p}\, \mathcal{P} \left( \mathcal{M}^{n-1,k}(p_i \hat{e}_i) \right) \rho.$$

In the same way, holding $\mathbf{x}$ constant, $I_{22}$ and $I_{23}$ may be integrated by parts with respect to $\mathbf{p}$ leading to

$$I_{22} = k \int d\mathbf{x} d\mathbf{p}\, \mathcal{P} \left( \mathcal{M}^{n,k-1} F_i(\mathbf{x}) \hat{e}_i \right) \rho$$

and

$$I_{23} = -\gamma k \int d\mathbf{x} d\mathbf{p}\, \mathcal{P} \left( \mathcal{M}^{n,k-1} p_i \hat{e}_i \right) \rho - \gamma \int d\mathbf{x} d\mathbf{p}\, \mathcal{P} \left( \mathcal{M}^{n,k} \right) \rho.$$

$I_{25}$ may be integrated by parts twice with respect to $\mathbf{p}$ leading to

$$I_{25} = \gamma m k_B T k(k-1) \int d\mathbf{x} d\mathbf{p}\, \mathcal{P} \left( \mathcal{M}^{n,k-2}(\hat{e}_i \hat{e}_i) \right) \rho. \tag{A.5}$$

Grouping all terms from Eq. (A.3) and Eq. (A.4) we find a final expression for the total time derivative of the moment $d\mathbf{M}_{n,k}/dt = (I_{11} + I_{21}) + (I_{12} + I_{22}) + (I_{13} + I_{23} + I_{24}) + I_{25}$ which is

(A.6)

$$\begin{aligned} \frac{d}{dt}\mathbf{M}_{n,k} = \frac{n}{m} \int d\mathbf{x} d\mathbf{p}\, \mathcal{P} \left( \mathcal{M}^{n-1,k+1} \right) \rho + k \int d\mathbf{x} d\mathbf{p}\, \mathcal{P} \left( \mathcal{M}^{n,k-1} (F_i(\mathbf{x}) - F_{0i}) \hat{e}_i \right) \rho \\ - \gamma k \int d\mathbf{x} d\mathbf{p}\, \mathcal{P} \left( \mathcal{M}^{n,k} \right) \rho + \gamma k(k-1) m k_B T \int d\mathbf{x} d\mathbf{p}\, \mathcal{P} \left( \mathcal{M}^{n,k-2} \hat{e}_i \hat{e}_i \right) \rho. \end{aligned}$$

## Appendix B. Derivation of $d\mathbf{M}_n/dt$ for Smoluchowski dynamics

The right hand side of Eq. (2.6) may be divided into two integrals

$$I_1 = \int d\mathbf{x}\, \mathcal{P} \left( \frac{\partial \mathcal{M}^n}{\partial t} \right) \rho \qquad I_2 = \int d\mathbf{x}\, \mathcal{P} \left( \mathcal{M}^n \frac{\partial \rho}{\partial t} \right).$$

Using the result of Eq. (A.1) and the fact that for the Smoluchowski dynamics

$$\dot{x}_{0i} \hat{e}_i = \frac{1}{m\gamma} F_{0i} \hat{e}_i$$

$I_1$ can be written as

$$I_1 = -\frac{n}{m\gamma} \int d\mathbf{x}\, \mathcal{P} \left( \mathcal{M}^{n-1} F_{0i} \hat{e}_i \right) \rho. \tag{B.1}$$

Substituting the expression of $\partial\rho/\partial t$ defined in Eq. (2.5) we find

$$I_2 = \frac{1}{m\gamma} \int d\mathbf{x}\, \mathcal{P} \left( \mathcal{M}^n \left[ \frac{\partial}{\partial x_i} \left[ -F_i(\mathbf{x}) + k_B T \frac{\partial}{\partial x_i} \right] \right] \right) \rho.$$

Recognizing that $-F_i(\mathbf{x}) = \partial V(x_i)/\partial x_i$

$$\begin{aligned} I_2 = & \frac{1}{m\gamma} \int d\mathbf{x}\, \mathcal{P} \left( \mathcal{M}^n \frac{\partial^2 V(x_i)}{\partial x_i^2} \right) \rho + \frac{1}{m\gamma} \int d\mathbf{x}\, \mathcal{P} \left( \mathcal{M}^n \frac{\partial V(x_i)}{\partial x_i} \frac{\partial \rho}{\partial x_i} \right) \\ & + \frac{k_B T}{m\gamma} \int d\mathbf{x}\, \mathcal{P} \left( \mathcal{M}^n \frac{\partial^2 \rho}{\partial x_i^2} \right) \end{aligned}$$

and integrating the second term in $I_2$ by parts we find

$$I_2 = -\frac{n}{m\gamma}\int d\mathbf{x}\mathcal{P}\left(\mathcal{M}^{n-1}\frac{\partial V(x_i)}{\partial x_i}\hat{e}_i\right)\rho + \frac{k_BT}{m\gamma}\int d\mathbf{x}\mathcal{P}\left(\mathcal{M}^n\frac{\partial^2\rho}{\partial x_i^2}\right). \tag{B.2}$$

Integrating the second term of Eq. (B.2) by parts twice we find

$$\begin{aligned} I_2 =& \frac{n}{m\gamma}\int d\mathbf{x}\mathcal{P}\left(\mathcal{M}^{n-1}\left(-\frac{\partial V(x_i)}{\partial x_i}\right)\hat{e}_i\right)\rho \\ &+\frac{n(n-1)k_BT}{m\gamma}\int d\mathbf{x}\mathcal{P}\left(\mathcal{M}^{n-2}\hat{e}_i\hat{e}_i\right)\rho. \end{aligned} \tag{B.3}$$

Combining Eq. (B.1) and Eq. (B.3) to form $d\mathbf{M}_n/dt = I_1 + I_2$ we arrive at

$$\begin{aligned} \frac{d}{dt}\mathbf{M}_n \quad = \quad & \int d\mathbf{x}\,\frac{n}{m\gamma}\mathcal{P}\left(\mathcal{M}^{n-1}\left(F_i(\mathbf{x})\hat{e}_i - F_{0i}\hat{e}_i\right)\right)\rho \\ &+\frac{k_BTn(n-1)}{m\gamma}\mathcal{P}\left(\mathcal{M}^{n-2}\hat{e}_i\hat{e}_i\right)\rho \end{aligned}$$

leading to the final expression for $d\mathbf{M}_n/dt$ given in Eq. (2.7).

## References


[1] J. E. Straub and D. Thirumalai. *Exploring the energy landscape in proteins.* *Proc. Natl. Acad. Sci. USA* **90**, 809–813 (1993). J. E. Straub and D. Thirumalai. *Theoretical Probes of conformational fluctuations in S-Peptide and RNase A/3'_UMP enzyme product complex.* Proteins **15**, 360-373 (1993). J. E. Straub, A. Rashkin and D. Thirumalai. *Dynamics in rugged energy landscapes with applications to the S-Peptide and Ribonuclease A.* J. Am. Chem. Soc. **116**, 2049-2063 (1994).

[2] J. Grad, Y. J. Yan, and S. Mukamel. *Time-dependent self-consistent field approximation for semiclassical dynamics using gaussian wavepackets in phase space.* Chem. Phys. Lett. **134**, 291-295(1987); J. Grad, Y. J. Yan, A. Haque, and S. Mukamel. *Reduced equations of motion for semiclassical dynamics in phase space.* *J. Chem. Phys.* **86**, 3441-3454 (1987).

[3] R. D. Coalson and M. Karplus. *Generalized quantum Liouville equation: Its solution by wave packet dynamics.* *J. Chem. Phys.* **79**, 6150–6161 (1983).

[4] E. J. Heller. *Time-Dependent approach to semiclassical dynamics.* *J. Chem. Phys.* **62**,1544-1555 (1975); E. J. Heller. *The semiclassical way to molecular spectroscopy.* Acc. Chem. Res. **14**, 368-375 (1981).

[5] N. Corbin and K. Singer. *Semiclassical molecular dynamics of wave packets.* Mol. Phys. **46**, 671-677 (1982); K. Singer and W. Smith. *Semiclassical many-particle dynamics with gaussian wave packets.* *Mol. Phys.* **57**, 761-775 (1986).

[6] R. B. Gerber and M. A. Ratner. *Self-consistent-field methods for vibrational excitations in polyatomic systems.* *Adv. Chem. Phys.* **70**, 97–132 (1988).

[7] R. Elber and M. Karplus. *Enhanced sampling in molecular dynamics: Use of the time-dependent Hartree approximation for a simulation of carbon monoxide diffusion through myoglobin.* *J. Am. Chem. Soc* **112**, 9161–9175 (1990).

[8] R. Czerminski and R. Elber. *Computational studies of ligand diffusion in globins: I. Leghemoglobin.* *Proteins* **10**, 70–80 (1991).

[9] P. Amara, D. Hsu, and J. E. Straub. *Global energy minimum searches using an approximate solution of the imaginary time Schrödinger equation.* *J. Phys. Chem.* **97**, 6715–6721 (1993).

[10] A. B. Finnila, M.A. Gomez, C. Sebenik, C. Stenson, and J.D. Doll. *Quantum Annealing: A new method for minimizing multidimensional functions.* Chem. Phys. Lett **219**, 343–348 (1994).

[11] J. Ma, D. Hsu, and J. E. Straub. *Approximate solution of the classical Liouville equation using Gaussian phase packet dynamics: Application to enhanced equilibrium averaging and global optimization.* *J. Chem. Phys.* **99**, 4024–4035 (1993).

[12] R. E. Walkup. *A local-Gaussian approximation for the propagation of a classical distribution function. J. Chem. Phys.* **95**, 6440–6448 (1991).

[13] J. Ma and John E. Straub. *Simulated annealing using the classical density distribution. J. Chem. Phys.* **101**, 533–541 (1994).

[14] H. Jeffreys. *Cartesian Tensors. Cambridge University Press*, London, (1974).

[15] These equations are identical to those derived from the Dirac-Frenkel variational principle. J. Frenkel, *Wave Mechanics: Advanced General Theory* (Clarendon Press, Oxford, 1934). In the Dover edition (1950), see p. 253. Frenkel cites an appendix in the Russian edition of Dirac's *Principles of Quantum Mechanics* (1930).

[16] J. E. Straub. In R. Elber, editor, *New Developments in Theoretical Studies of Proteins.* World Scientific, Singapore, (in press).

[17] R. Zwanzig. *Diffusion in a rough potential. Proc. Natl Acad. Sci.* **85**, 2029–2030 (1988).

[18] J. E. Straub J. Ma and P. Amara. *Simulated annealing using coarse grained classical dynamics: Fokker-Planck and Smoluchowski dynamics in the Gaussian density approximation. submitted to J. Chem. Phys.*

[19] M. Orešič and D. Shalloway. *Hierarchical characterization of energy landscapes using Gaussian packet states. J. Chem. Phys.* **101**, 9844–9857 (1994).

[20] P. Salamon, J. D. Nulton, J. R. Harland, J. Pedersen, G. Ruppeiner, and L. Liao. *Simulated annealing with constant thermodynamic speed. Comp. Phys. Comm.* **49**, 423–428 (1988).

[21] C. Tsoo and C. L. Brooks III. *Cluster stucture determination using Gaussian density distribution global minimization methods. J. Chem. Phys.* **101**, 6405–6411 (1994).

[22] L. Piela, J. Kostrowicki, and H. A. Scheraga. *The multiple-minima problem in the conformational analysis of molecules. Deformation of the potential energy hypersurface by the Diffusion Equation Method. J. Phys. Chem.* **93**, 3339–3346 (1989). J. Kostrowicki and L. Piela and B. J. Cherayil and H. A. Scheraga. *Performance of the Diffusion Equation Method in searches for optimum structures of clusters of Lennard-Jones atoms.i*bid., **95** 4113-4119 (1991); R. J. Wawak and M. M. Wimmer and H. A. Scheraga. *Application of the Diffusion Equation Method of Global Optimization to Water Clusters. i*bid., **96**, 5138-5145 (1992);J. Kostrowicki and H. A. Scheraga.*Application of the Diffusion Equation Method for Global Optimization to Oligopeptides i*bid. **96**, 7442-7449 (1992).

[23] J. Ma, J. E. Straub, and E. I. Shakhnovich. *Simulation study of the collapse of Lennard-Jones homopolymers. submitted to J. Chem. Phys.*

[24] P. G. de Gennes. *Kinetics of collapse for a flexible coil. J. Phys. Lett.* **46**, L–639_L–642 (1985).

[25] P. Amara and J. E. Straub. *Folding model proteins using kinetic and thermodynamic annealing of the classical density distribution. submitted to J. Phys. Chem.*

[26] J. D. Honeycutt and D. Thirumalai. *The Nature of Folded States of Globular Proteins. Biopolymers* **32**, 695–709 (1992).

[27] D. Shalloway, in: *Recent advances in global optimization*, eds. A. Floudas and P. M. Pardalos (Princeton University Press, Princeton, 1992) p. 433.

Department of Chemistry, Boston University, Boston, Massachusetts 02215
*E-mail address*: `amara@chem.bu.edu`
*E-mail address*: `jma@chem.bu.edu`
*E-mail address*: `straub@chem.bu.edu`

DIMACS Series in Discrete Mathematics
and Theoretical Computer Science
Volume **23**, 1996

# Energy Directed Conformational Search of Protein Loops and Segments

Robert E. Bruccoleri

ABSTRACT. Systematic conformational search is a powerful method for modeling segments within proteins, because it enumerates all the stereochemically possible conformations the segment may adopt. However, the fact that computation time grows exponentially with peptide length has limited its utility to short segments. The computational burden can be significantly reduced by a directed conformational search, which evaluates the partial conformations generated during the search process, and uses these evaluations to direct the search process towards the lowest energy conformation. Previously, it had been demonstrated that the directed conformational search could be used to fit the protein backbone to the alpha carbon coordinates (D. Bassolino-Klimas and R. E. Bruccoleri, *Proteins: Struc. Funct. Gen.* **14,** 465-474 (1992)) [**2**]. In this paper, the use of energy directed conformational search is tested on loops within a set of proteins, and it is shown that the method is effective in ordering the resulting conformations by their energy. Therefore, use of this method can significantly enhance the efficiency of loop modeling by conformational search.

## 1. Introduction

The protein folding problem remains one of the most significant unsolved problems in modern biology[**12**]. At the genetic level, the three dimensional structure of protein is encoded as a one dimensional sequence of amino acids, but after synthesis, most proteins spontaneously fold into their native, compact structure.

Determination and analysis of this three dimensional structure is essential for understanding the mechanisms of protein function and, ultimately, to modify the function by either protein engineering or rational drug design. However, at the present time, experimental techniques for determining protein sequence are much faster than those for determining protein structure. As a result, the ability to predict protein structure from sequence is of great scientific and practical interest.

1991 *Mathematics Subject Classification.* Primary 92C05; Secondary 65K10.

*Key words and phrases.* Conformational search, protein modeling, homology modeling, directed search, protein folding.

I thank Keith Constantine and Jiri Novotny for careful reading and critique of the manuscript, and Malcolm Davis for help with $\mathcal{A}\mathcal{M}\mathcal{S}$-LaTeX. I am grateful to John Wiley and Sons, Inc. for permission to use copyrighted material.

The prediction of protein structure is difficult for several fundamental reasons. First, since the chemical structure of proteins is quite flexible, a typical protein can theoretically adopt a cosmologically large number of conformational states[**11, 16, 24**]. Second, the calculation of the Gibbs free energy of proteins is still problematic. Approximate, empirical methods do exist to differentiate native structures from many compact, misfolded structures[**9, 13, 15, 19, 23**], but not from all possible compact, misfolded structures. Third, it is possible that in some instances, the folded form of the protein is a kinetically trapped non-global minimum[**10**].

The strategy which has been selected for solving the problem begins with the realization that the energy function is largely independent of exploration of the conformational space, and thus, these two problems can be separately addressed. The initial predictive efforts have been confined to short protein segments or loops within proteins. In such problems, the number of possible conformational states is limited to a size that can be enumerated on modern day computers. Using relatively small systems, search algorithms and free energy functions can be tested for the entire conformational space, and the accuracy of these methods can be determined unequivocally. Progress toward solving the complete protein folding problem can be made both by independently improving the methods for searching the conformational space as well as improving the calculation of the free energy.

With regard to the issue of global energy minima versus kinetic trapping, the above strategy may not prove complete. However, it is very unlikely that the conformation of individual loops within a protein represent kinetically trapped intermediates, and therefore, a Boltzmann average of the global energy minima is the native structure. Regardless of the actual mechanism of folding, accurate free energy functions will be useful, and it may be possible to use a search methodology which mimics the biological process.

This paper describes an improvement to the conformational search methodology which increases the speed with which the lowest energy conformation can be found. The basic concept assumes that the lowest energy structure is composed of low energy segments, and that the calculated energy of a segment gives a good indication of the overall energy of the constructed loop. Thus, energy calculations of the partial conformations generated during the search process are used to guide the traversal of the search tree so as to generate the lowest energy conformations early in the search[**2**]. This concept was inspired from state space search methods used in artificial intelligence[**20**]. Thus, the search could be terminated prior to a complete exploration of the search tree.

Previously, the directed search methodology was successfully applied to the problem of generating backbone coordinates from $\alpha$-carbon coordinates[**2**], where the direction of the search was guided by the root mean square (RMS) deviations of the $\alpha$-carbons. In this paper, this methodology is applied to a set of test loops in order to characterize its performance when directed by energy calculations. This paper demonstrates the ability of the search method to accelerate the appearance of the lowest energy conformation. The accuracy of the energy function is not addressed.

## 2. Methods

The directed conformational search methodology has been described previously [**2**]. Briefly, a particular conformational search is defined by a list of degrees of freedom which are successively sampled in an iterative, nested order. The search itself can be represented as a tree search where the root of the search tree (Figure 1) represents the initial state of the search when nothing has been constructed. Each level in the search tree represents the sampling of a particular degree of freedom, and each node in the search tree represents partial conformations resulting from the sampling of degrees of freedom at the level of the node and all previous degrees of freedom. Node expansion means the sampling of the degree of freedom at the level of the node. A node is called a "leaf" node if it has been generated from the sampling of all degrees of freedom. A node is called an "open" node if it is not a leaf node and if it has not been expanded.

The search is directed by the values associated with each open node. In the previous paper[**2**], the value of each open node was set to the RMS deviation between the $\alpha$-carbon coordinates of the generated conformations and the X-ray structure. In this work, the value of each open node was set to the CHARMM energy function[**4**] of the partial conformation that the node represents. All calculations were made using the program, CONGEN[**5, 6**].

There are several different search methods available in CONGEN. The simplest exhaustive search method is a depth first search. Here, CONGEN expands the deepest and leftmost open node in the search tree at any given time. In this context, "left" means the first in the ordering of samples for the degrees of freedom. Likewise, "right" means the last in the ordering. Depth first search explores the search tree in left to right order.

There are currently three directed search methods in CONGEN; best first (EVAL), deepening evaluation (DEVAL), and a mixture of EVAL and DEVAL (MIX). In the best first search, an open node is selected for expansion if it is the lowest energy open node in the tree. In the deepening evaluation method, CONGEN proceeds sequentially through each level in the search tree, and expands the lowest energy open node at each level. If the terminal level of the tree or a level with no open nodes is reached, the search method restarts at the first degree of freedom. In the MIX strategy, CONGEN alternates between the EVAL and DEVAL strategies as if the two strategies were executed as coroutines. Since both strategies operate using the same search tree, the results can be a substantial improvement over either method individually. In the backbone reconstruction paper[**2**], the MIX strategy was the only effective method.

Because the number of open nodes grows rapidly with all of these strategies, it is generally necessary to prune the search tree by removing open nodes with high energy. However, in the runs described here, only the 26-10 antibody run and the ANO2 antibody using alternating backbone residues required tree pruning. In those runs, the tree pruning limit was set to 900000 nodes, and the first six levels in the search tree were not pruned. When the tree grew to the 900000 node limit, only the highest energy 20% of the open nodes were deleted.

**2.1. Test Systems.** For these tests, a set of segments was used from several different proteins which had a modest number of conformations under the search parameters described below. In this context, "modest" means between 1000 and 100000 conformations. Such runs completed within two days elapsed time on a 20 processor SGI Onyx system with 150 mHZ R4400 processors and 2 gigabytes of RAM, the machine used for all calculations described here. The test systems are given in Table I. All the protein coordinates except for ANO2 were taken from the

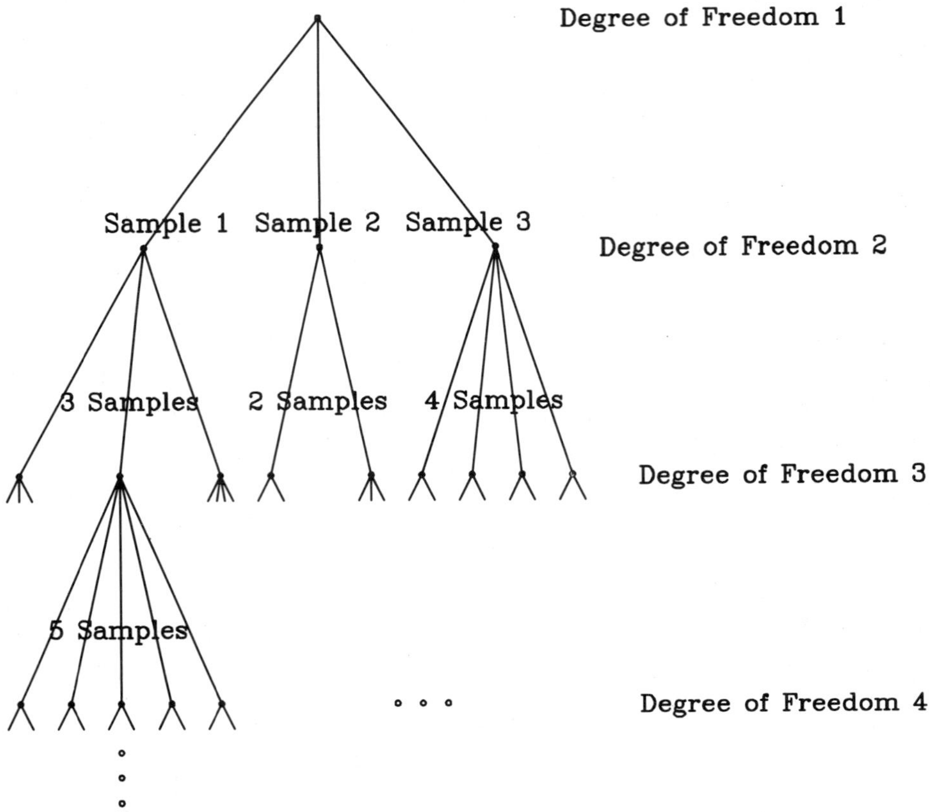


FIGURE 1. Presented is the top portion of a search tree showing how the sampling process can be represented. Each node in the tree represents a conformation. The root node of the tree, which is at the top of the figure, represents the system before the sampling of degrees of freedom has begin. Each successively deeper (meaning lower) level in the tree corresponds to a sampling of further degrees of freedom. The lines connecting nodes, or edges, represent the specific samples of degrees of freedom. The leaf nodes, which have no nodes emanating from them and which are not shown, are either completely constructed conformations or partially constructed conformations which were blocked because of the various constraints on the sampling. The completely constructed conformations will be as deep in the tree as are there are degrees of freedom to be sampled. The blocked conformations will be higher up. Intermediate nodes represent partially constructed conformations with a particular sampling of the degrees of freedom above it. Constraints applied early serve to delete many more possible nodes than pruning later on in the generation process. Adapted from ref. [**7**] and used by permission. ©1987 John Wiley & Sons, Inc.

TABLE I. Search Results

| Protein[a] | Residues[b] | Number[c] | Appearance Ratio[d] Depth First[e] | Directed Mix[f] | Directed Eval[g] |
|---|---|---|---|---|---|
| T4 Endopeptidase V[**18**] | 39-44 | 1147 | 0.119 | 0.105 | 0.008 |
| Antibody Kol H[**17**] | 41-47 | 1451 | 0.498 | 0.038 | 0.014 |
| Antibody AN02 H[**8**] | 49-57[h] | 2131 | 0.881 | 0.026 | 0.002 |
| Antibody 17-9 H[**22**] | 100-107 | 2620 | 0.521 | 0.073 | 0.116 |
| Antibody ANO2 H[**8**] | 49-57[hi] | 11033 | 0.189 | 0.370 | 0.001 |
| Carboxypeptidase A[**21**] | 53-59 | 27489 | 0.822 | 0.052 | 0.013 |
| Antibody 26-10 H[**14**] | 26-34 | 65627 | 0.545 | 0.151 | 0.140 |

[a] For the antibodies, the last letter gives the chain from which the segment was taken.
[b] Residue numbers of the test segment.
[c] The total number of conformations found in an exhaustive depth first search.
[d] Conformation number of the lowest energy conformer divided by the total number of conformations.
[e] Depth first search.
[f] Directed search using MIX strategy.
[g] Best first directed search.
[h] Includes the sidechain from heavy chain residue 33 in the sidechain degree of freedom
[i] The backbone degrees of freedom were searched in the order; 49, 57, 50, 56, 51, 55.

---

Brookhaven Data Bank[**3**]. Coordinates for the anti-dinitrophenyl antibody ANO2 [**8**] were taken from a previous antibody modeling effort[**1**]. All the segments are loops, with the exception of the T4 Endonuclease V segment, which is an exposed helix.

The parameters for all searches were as follows: Let $n$ symbolize the number of residues in the loop. With the exception of the alternating backbone ANO2 test, the first $n-3$ degrees of freedom were backbone constructions starting at the amino terminal side of the loop. In the alternating backbone ANO2 test, the backbone degrees of freedom alternated from each terminus until three residues (52-54) remained. The next degree of freedom was the chain closure needed to complete the backbone. An iterative sidechain degree of freedom was used to construct the sidechains. At this point, the conformations consisted of all the atoms in the loops, and these conformations were energy evaluated and written to a disk file. Energies were calculated using the CHARMM potential energy function[**4**] using the PARAM5 parameter set[**5**]. A non-bonded cutoff of 8.0 Å was used with distance dependent dielectric constant of $4r$. Hydrogen bonds were cutoff at 4.5 Å.

## 3. Results and Discussion

For each of the test systems, three searches were run: depth first, best first, and mixed strategy search methods. The results are summarized in Table I, and are illustrated in Figures 2–8.

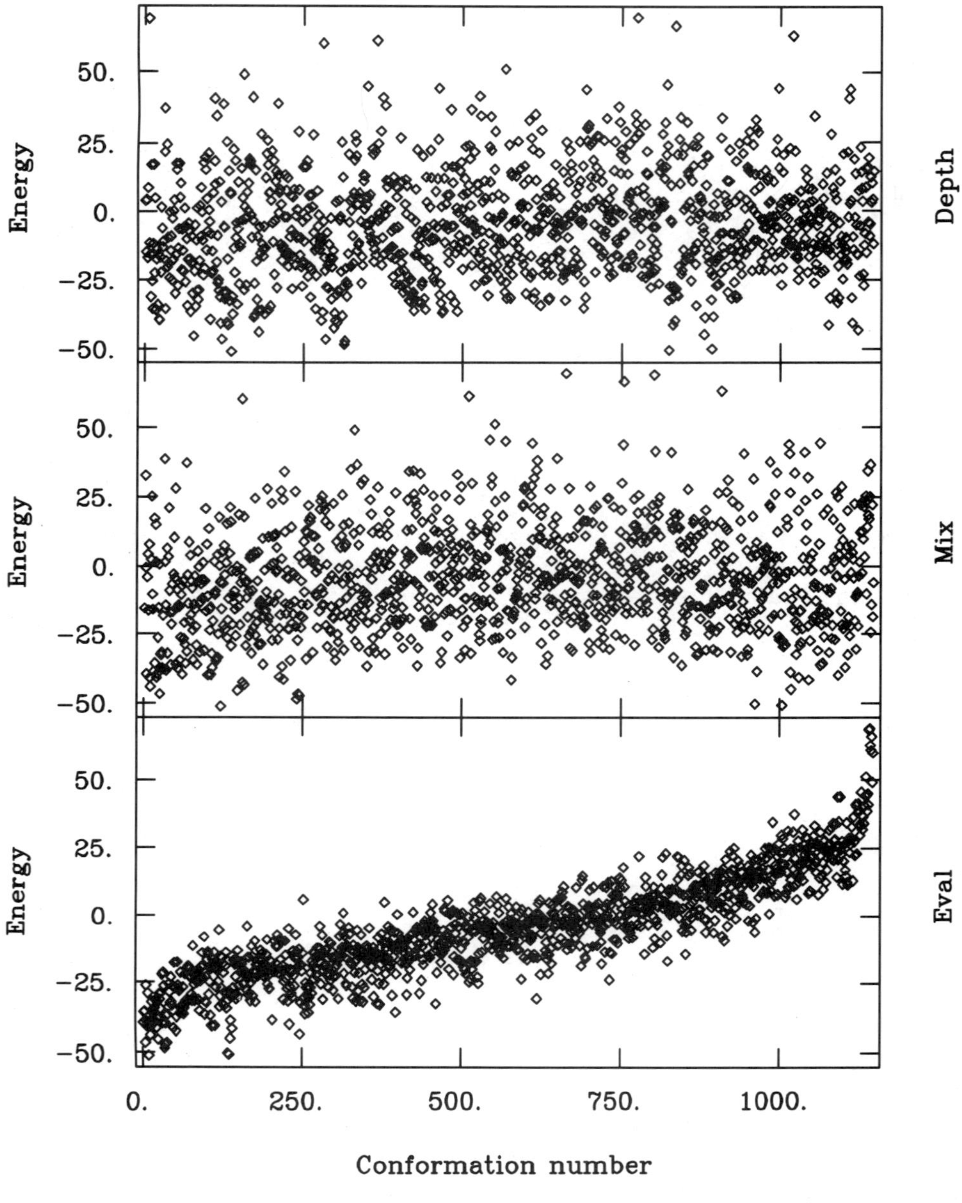


FIGURE 2. Endopeptidase V 39-44 Segment Conformation Energies

Each panel plots the energy of the conformations against their position in the order as they appear in the conformation file. Effectively, these are plots of conformation energy versus elapsed time in the search. The top panel shows the plot for a depth first search, which has no energy ordering. The middle panel shows the plot for a directed search using the mixed strategy, and the bottom panel shows the plot for a best first search.

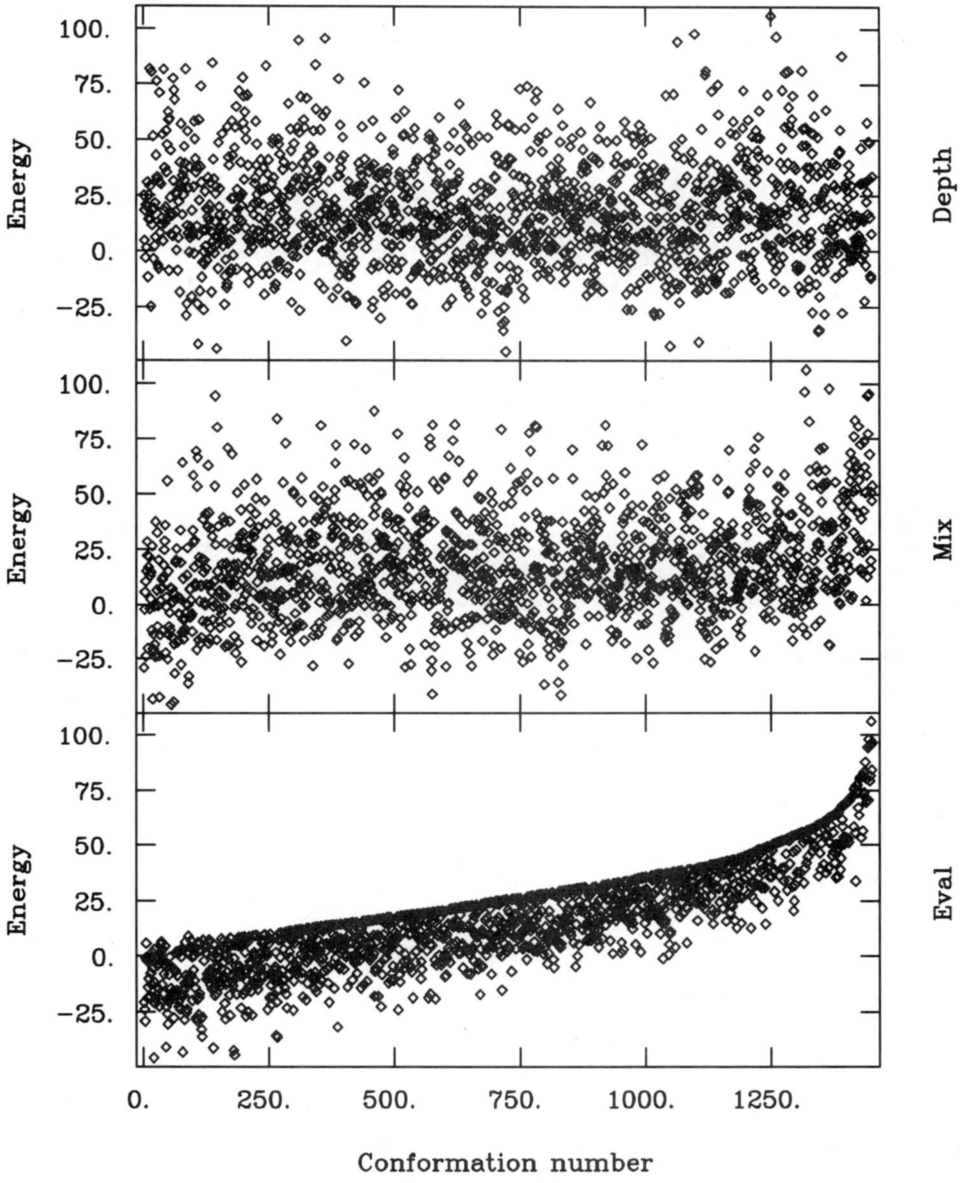


FIGURE 3. Antibody Kol H 41-47 Segment Conformation Energies. See the legend for figure 2 for details.

---

The key attribute of the directed search methodology is the speed with which it generates the lowest energy conformation. This speed is measured by the ratio of the conformation number of the lowest conformation to the total number of possible conformations, which shall be referred to as the appearance ratio. The expectation value for the appearance ratio in a random or depth first search is 0.5. As seen in Table I, the appearance ratio for the directed search methods is generally much smaller. In particular, the best first search has very low appearance ratios

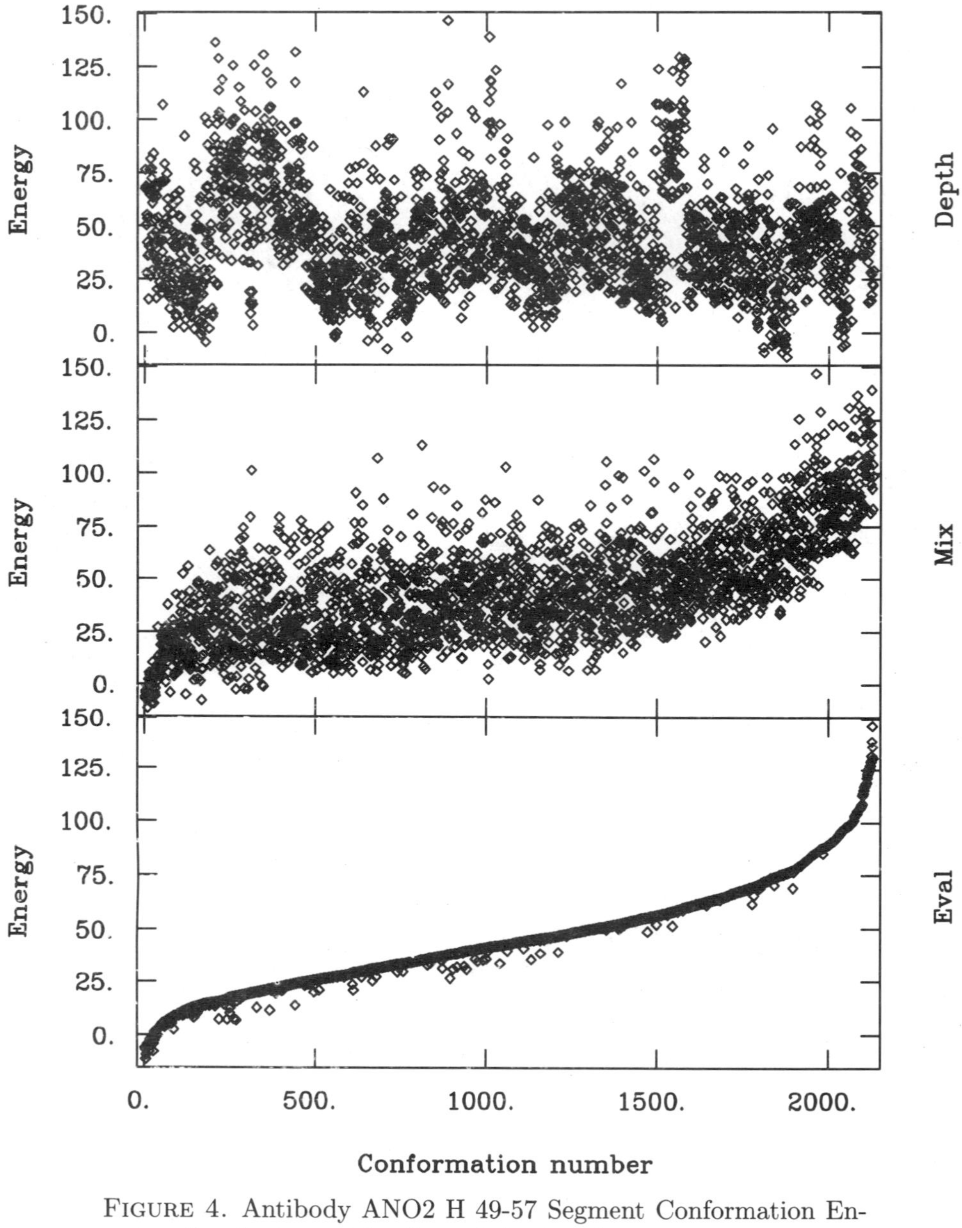


FIGURE 4. Antibody ANO2 H 49-57 Segment Conformation Energies. See the legend for figure 2 for details.

---

(< 0.015) in five of the seven test cases. The mixed directed search also has an ordering effect, but it is very weak compared to the best first search. This contrasts to the previous experience with the reconstruction of the peptide backbone from the $\alpha$-carbon coordinates, where the mixed strategy was the only successful one.

Another important characteristic of the directed search methodology is the identification of the lowest energy conformer prior to the completion of the search. Unfortunately, it is not possible to do this with certainty. Each of the plots of

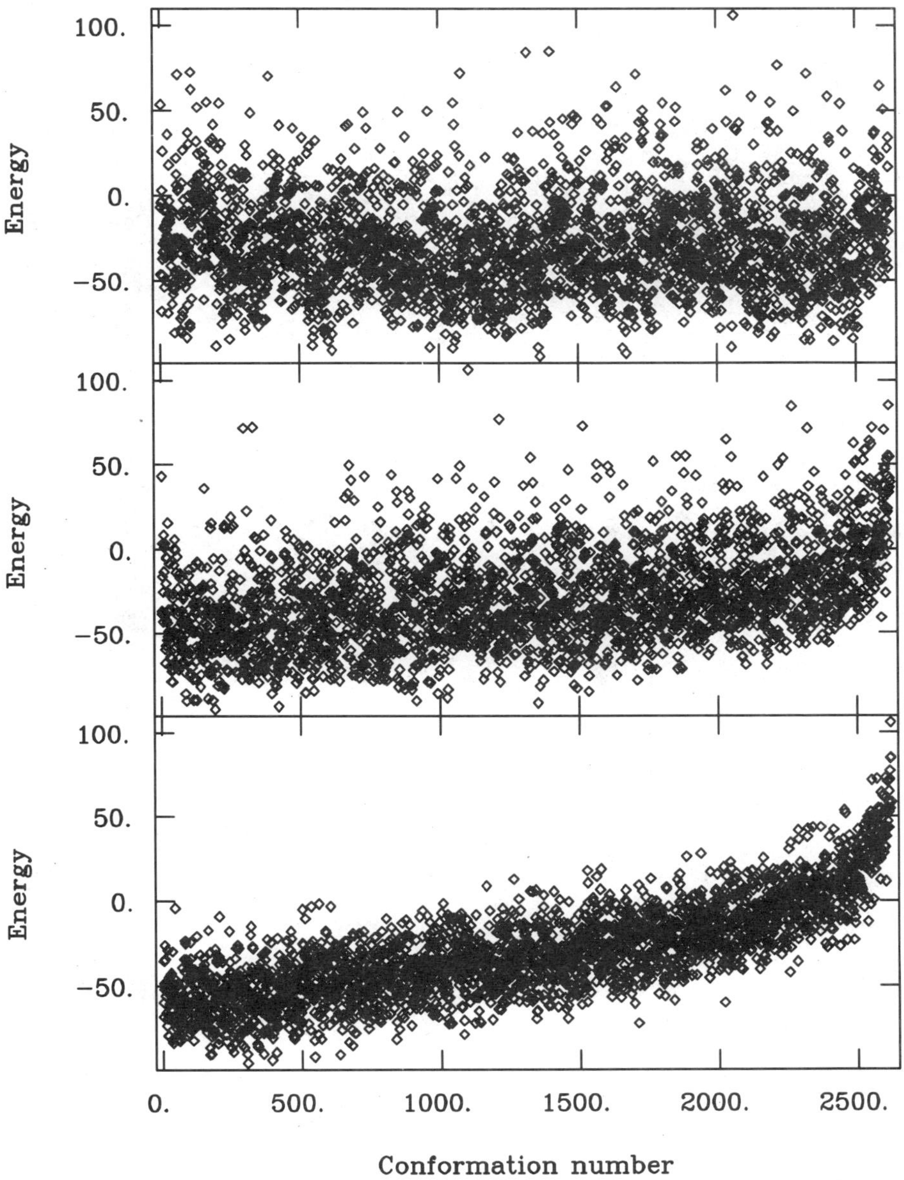


FIGURE 5. Antibody 17-9 H 100-107 Segment Conformation Energies. See the legend for figure 2 for details.

energy versus conformation number shows an overall trend in the energy and a highly variable spread in the energy values. It is highly probable that the lowest energy structure has been seen once the energy trend has moved up more than the energy spread. In the five test systems with appearance ratios below 0.015, the energy trend is fast enough to exceed the spread. Therefore, in the absence of an exhaustive search, it would have been possible to truncate these runs with reasonable assurance that the lowest conformation had been generated.

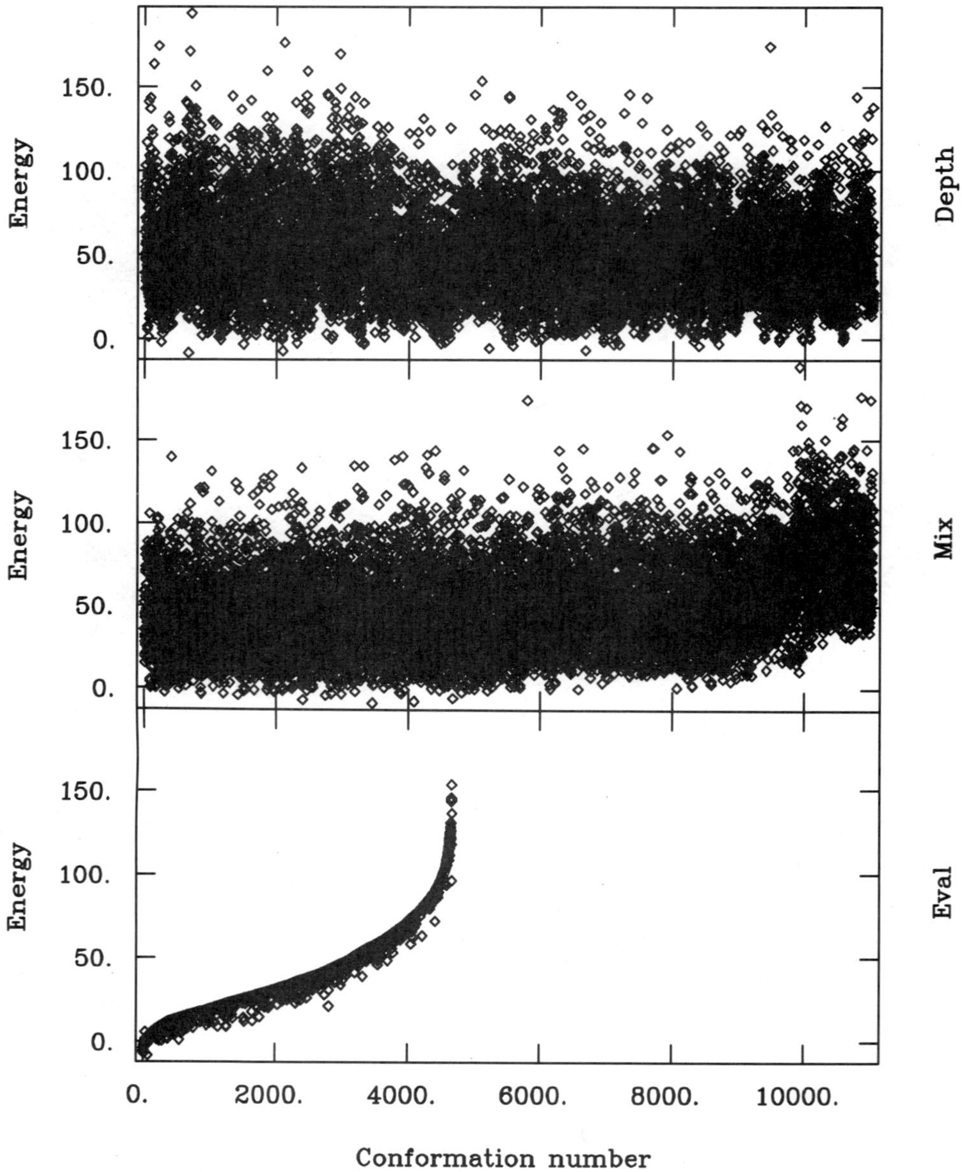


FIGURE 6. Antibody ANO2 H 49-57 Alternating Backbone Segment Conformation Energies. See the legend for figure 2 for details.

---

There is considerable variability in the energy spread in the conformations which result from the best first search. The ANO2 antibody loop runs have the most narrow and desirable spread, whereas the spread in the antibody 26-10 loop is the worst.

The energy trends and spreads in these runs reflect the interactions of the atoms in the degrees of freedom. In runs with flat energy trends, the initial degrees of freedom do not heavily influence the energies of the complete conformations. The

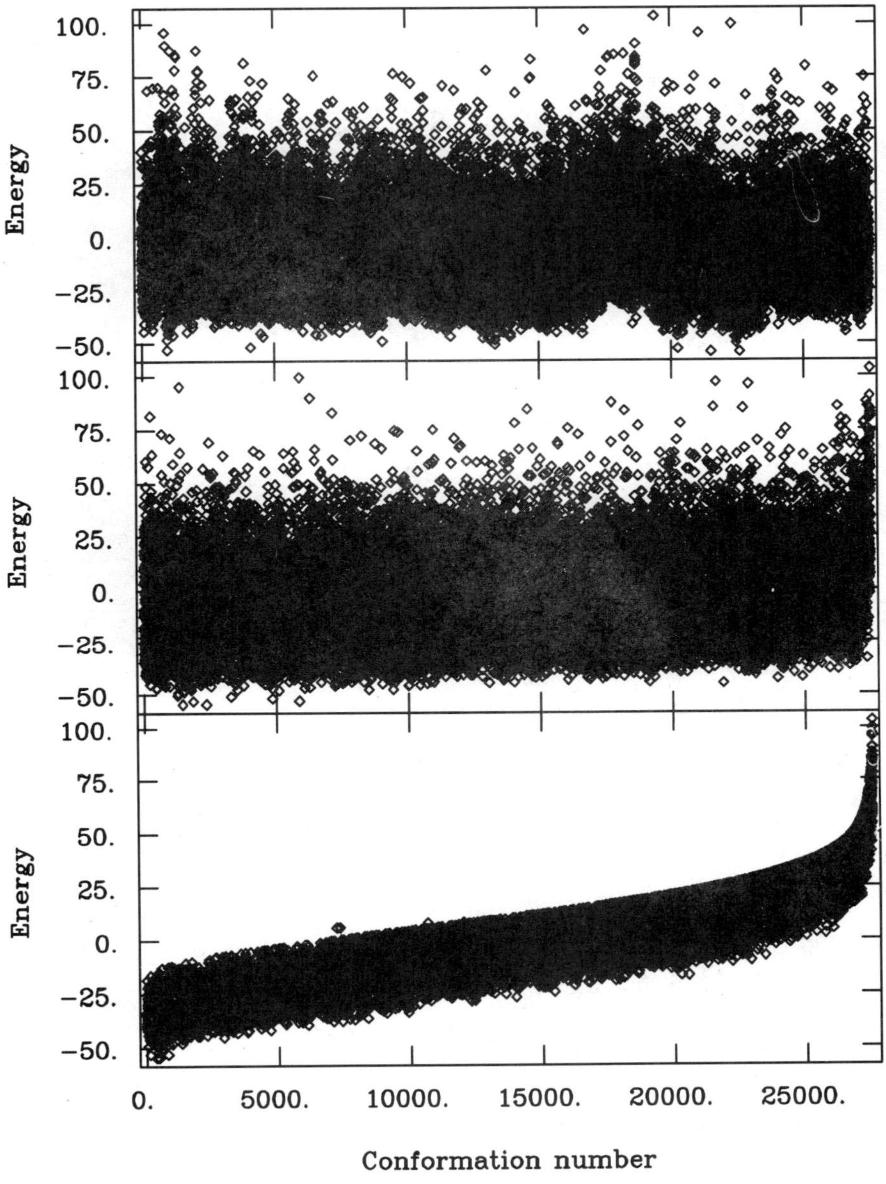


FIGURE 7. Carboxypeptidase A 53-59 Segment Conformation Energies. See the legend for figure 2 for details.

opposite is true for rising energy trends. The spread in the energies is largely due to the effects of adding the sidechains onto the backbone, which typically contributes the largest magnitude to the total loop energy. Future work on this methodology will explore how to improve the energy trends and decrease the spread to improve its applicability.

The major risk with using the directed search arises in cases where it does not serve to order the complete conformations. In such systems, it is possible that

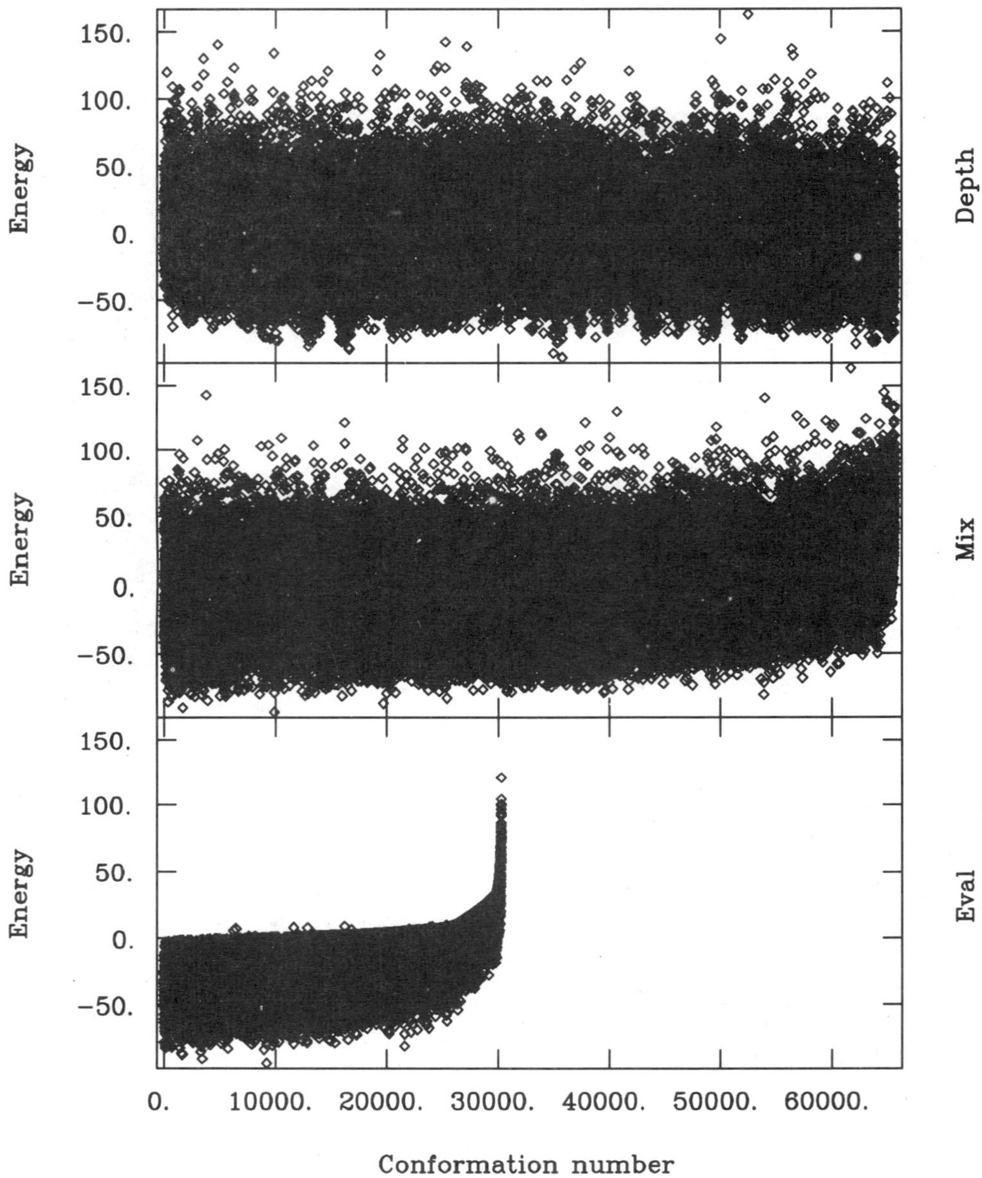


FIGURE 8. Antibody 26-10 H 26-34 Segment Conformation Energies. See the legend for figure 2 for details.

---

the tree pruning algorithms might discard an intermediate node which leads to the lowest energy structures. However, because the plot of output conformation energy versus conformation number is highly diagnostic of the ordering, this risk can be minimized either relaxing the pruning conditions or by not using directed searching when it does not appear to work.

In conclusion, the best first energy directed search methodology is a generally useful tool in applying systematic conformational search for global energy minimization of modest sized loops.

## References


1. Donna Bassolino, Robert E. Bruccoleri, and Shankar Subramaniam, *Modeling the antigen combining site of an anti-dinitrophenyl antibody, ANO2*, Protein Science **1** (1992), 1465–1476.
2. Donna Bassolino-Klimas and Robert E. Bruccoleri, *The application of a directed conformational search for generating 3-D coordinates for protein structures from α-carbon coordinates*, Proteins: Struc. Funct. Gen. **14** (1992), 465–474.
3. F. C. Bernstein, T. F. Koetzle, G. J. B. Williams, E. F. Meyer, M. D. Brice, J. R. Rodgers, O. Kennard, T. Shimanouchi, and M. Tasumi, *The Protein Data Bank: a computer-based archival file for macromolecular structures*, J. Mol. Biol. **112** (1977), 535–542.
4. B. R. Brooks, R. E. Bruccoleri, B. D. Olafson, D. J. States, S. Swaminathan, and M. Karplus, *CHARMM: A program for macromolecular energy minimization and dynamics calculations*, J. Comp. Chem. **4** (1983), 187–217.
5. Robert E. Bruccoleri, *CONGEN v2.0 manual*, Bristol-Myers Squibb Pharmaceutical Research Institute, 1990.
6. ———, *Application of systematic conformational search to protein modeling*, Mol. Sim. **10** (1993), 151–174.
7. Robert E. Bruccoleri and Martin Karplus, *Prediction of the folding of short polypeptide segments by uniform conformational sampling*, Biopolymers **26** (1987), 137–168.
8. Axel T. Brünger, D. J. Leahy, T. R. Hynes, and R. O. Fox, *2.9 Å resolution structure of an anti-dinitrophenyl-spin-label monoclonal antibody Fab fragment with bound hapten*, J. Mol. Biol. **221** (1991), 239–256.
9. S. H. Bryant and C. E. Lawrence, *An empirical energy function for threading protein sequence through the folding motif*, Proteins: Struct. Funct. Genet. **16** (1993), 92–112.
10. Joseph D. Bryngelson, Jose Nelson Onuchic, Nicholas D. Socci, and Peter G. Wolynes, *Funnels, pathways, and the energy landscape of protein folding: A synthesis*, Proteins: Struct. Funct. Genet. **21** (1995), 167–195.
11. Ken A. Dill, *Dominant forces in protein folding*, Biochemistry **29** (1990), 7133–7155.
12. Lila M. Gierasch and Jonathan King, *Protein folding: Deciphering the second half of the genetic code*, American Association for the Advancement of Science, Washington, D.C., 1990.
13. A. Godzik, A. Kolinski, and J. Skolnick, *De novo and inverse folding predictions of protein structure and dynamics*, J. Comput. Aided Mol. Design **7** (1993), no. 4, 397–438.
14. Phillip D. Jeffrey, Roland K. Strong, Larry C. Sieker, Chieh Ying Chang, Robert L. Campbell, Gregory A. Petsko, Edgar Haber, Michael N. Margolies, and Steven Sheriff, *26-10 Fab-digoxin complex: Affinity and specificity due to surface complementarity*, Proc. Natl. Acad. Sci. USA **90** (1993), 10310–10314.
15. D. Jones and J. Thornton, *Protein fold recognition*, J. Comput. Aided Mol. Design **7** (1993), no. 4, 439–456.
16. Cyrus Levinthal, *How to fold graciously*, Mossbauer Spectroscopy in Biological Systems (Urbana) (P. Debrunner, J. C. M. Tsibris, and E. Münck, eds.), University of Illinois Press, 1969, pp. 22–24.
17. M. Marquart, J. Deisenhofer, R. Huber, and W. Palm, *Crystallographic refinement and atomic models of the intact immunoglobulin molecule Kol and its antigen-binding fragment at 3.0 Å and 1.0 Å resolution*, J. Mol. Biol. **141** (1980), 369–391.
18. K. Morikawa, O. Matsumoto, T. Tsujimoto, K. Katayanagi, M. Ariyoshi, T. Doi, M. Ikehara T. Inaoka, and E. Ohtsuka, *X-ray structure of T4 Endonuclease V: An excision repair enzyme specific for a pyrimidine dimer*, Science **256** (1992), 523–526.
19. Jiri Novotny, Alexander A. Rashin, and Robert E. Bruccoleri, *Criteria that discriminate between native proteins and incorrectly folded models*, Proteins: Struct. Funct. Gen. **4** (1988), 19–30.
20. Judea Pearl and Richard E. Korf, *Search techniques*, Ann. Rev. Comput. Sci. **2** (1987), 451–467.

21. D. C. Rees, M. Lewis, and W. N. Lipscomb, *Refined crystal structure of Carboxypeptidase A at 1.54 Å resolution.*, J. Mol. Biol. **168** (1983), 367–387.
22. J. M. Rini, U. Schulze-Gahmen, and I. A. Wilson, *Structural evidence for induced fit as a mechanism for antigen-antibody recognition*, Science **255** (1992), 959–965.
23. Manfred J. Sippl, *Boltzmann's principle, knowledge-based mean fields and protein folding. an approach to the computational dtermination of protein structures*, J. Comput. Aided Mol. Design **7** (1993), no. 4, 473–501.
24. Robert Zwanzig, Attila Szabo, and Biman Bagchi, *Levinthal's paradox*, Proc. Nat. Acad. Sci. USA **89** (1992), 20–22.

Bristol-Myers Squibb Pharmaceutical Research Institute, P. O. Box 4000, Princeton, NJ 08543-4000

*E-mail address*: bruc@bms.com

DIMACS Series in Discrete Mathematics
and Theoretical Computer Science
Volume **23**, 1996

# Global Optimization Methods for Protein Folding Problems

Richard H. Byrd, Elizabeth Eskow, André van der Hoek, Robert B. Schnabel, Chung-Shang Shao, and Zhihong Zou

ABSTRACT. The problem of finding the naturally occurring structure of a protein is believed to correspond to minimizing the free, or potential, energy of the protein. This is generally a very difficult global optimization problem, with a large number of parameters and a huge number of local minimizers including many with function values near that of the global minimizer. This paper presents a new global optimization method for such problems. The method consists of an initial phase that locates some reasonably low local minimizers of the energy function, followed by the main phase that progresses from the best current local minimizers to even lower local minimizers. The method combines portions that work on small subsets of the parameters, including small-scale global optimizations using stochastic methods, with local minimizations involving all the parameters. In computational tests on the protein polyalanine with up to 58 amino acids (116 internal parameters), the method appears to be very successful in finding the lowest energy structures. The largest case is particularly significant because the lowest energy structures that are found include ones that exhibit interesting tertiary as opposed to just secondary structure.

## 1. Introduction

The naturally occurring three dimensional structure of a protein, called its "tertiary structure", is believed to be uniquely determined from its "primary structure", the sequence of amino acids of which the protein is composed. The problem of finding the tertiary structure given the primary structure is known as the protein folding problem. Scientists believe that the tertiary structure of a protein is the structure that minimizes its free, or potential energy. Thus the protein folding problem can be posed as an optimization problem. This optimization problem is an extremely challenging global optimization problem for even moderately sized proteins, because the number of optimization parameters for such problems is 100 or more, and the number of local minimizers is believed to be exponential in the number of parameters. Moreover, many of the local minimizers have function values near that of the global minimizer.

1991 *Mathematics Subject Classification.* Primary 65K10; Secondary 92C40.

Research supported by AFOSR Grants No. AFOSR-90-0109 and F49620-94-1-0101, ARO Grants No. DAAL03-91-G-0151 and DAAH04-94-G-0228, and NSF Grants No. CDA-8922510 and CCR-9101795.

This paper presents a new global optimization method for protein folding problems, and computational results for problems with up to 116 parameters. This problem of minimizing potential energy is an instance of the general global optimization problem of finding the lowest minimizer of a nonlinear function $f(x)$ that has multiple local minima, for $x$ in a domain $D$ defined by lower and upper bounds on each parameter $x_i$. The functions $f(x)$ used in this research are well known empirical potential energy functions which are described in more detail in later sections.

A major challenge in solving large-scale global optimization problems is that there is no clean mathematical basis for efficiently reaching a global minimizer, such as an analog to steepest descent or Newton's method for local minimization. Many methods have been developed for such global optimization problems (see the survey [**17**]), and these therefore tend to be heuristic and often require large amounts of computation time. This is true of the method presented here as well. We do aim in this research, however, to develop a global optimization methodology that can be applied to a wide range of large-scale problems, albeit with some application-dependent modifications. We also aim to assure that our methods can effectively utilize powerful parallel computers, although this aspect of our research is not addressed in this paper.

The new method described in this paper has the same top level structure as the global optimization methods for molecular cluster problems that we introduced in [**4, 2**]. These methods have two main phases. The first, initialization phase generates a set of configurations that are local minimizers of the energy function and have reasonably low energy values. The second phase, which accounts for the overwhelming majority of the computational effort, is an improvement phase that successively finds new and hopefully lower local minimizers from the current lowest local minimizers. The key feature of each step of this phase is the solution of a small-scale global optimization problem (typically with 3-9 parameters) by a stochastic global optimization method. That is, the relatively well developed methodology for small-scale global optimization problems is utilized within the large-scale global optimization algorithm. At the level described above, the methods for molecular cluster problems and the new methods for protein folding are similar; beneath this level they differ significantly.

The remainder of the paper motivates and describes our new global optimization method. Section 2 briefly summarizes our global optimization algorithm for molecular cluster problems, which is the starting point for our new method, and also very briefly summarizes our computational experience with this algorithm to justify that it appears to be a viable approach. Section 3 present our global optimization method for protein folding problems and computational results on the protein polyalanine. Section 4 gives some conclusions and directions for future research in this area.

## 2. Global Optimization Methods for Molecular Cluster Problems

The basic strategy of our algorithm was developed first to solve global optimization problems for finding the configurations of macro-molecular clusters. The approach was applied to potential energy functions for two different cluster problems. The first problem was that of identical atoms using the Lennard-Jones potential energy function, and the second problem involved clusters of water molecules using the Coker/Watts energy function [**6**]. These molecular cluster problems have some of the same challenges as the protein folding problem, including many parameters, a huge number of local minimizers, and many low local minimizers. In addition, the energy functions for both cluster and protein problems include $O(N^2)$ 2-body energy terms, composed of similar functional forms, for an $N$ atom problem. The characteristic that makes molecular cluster problems much different from protein folding problems is the lack of the chain structure that is inherent in proteins. This enables the movement of a single atom or molecule in the structure to a completely different location without compromising the integrity of the structure as would be the case in a chain of amino acids.

Lennard-Jones problems assume that the potential energy of the cluster is given by the sum of the pairwise interactions between atoms, with these interactions being Van der Waals forces given by the Lennard-Jones 6-12 potential. That is, if we define the position of the molecular cluster by

$$x = (x_1, x_2, ..., x_N)$$

where $x_i$ is a three-dimensional vector denoting the coordinates of the $i^{th}$ atom, then the potential energy function is

$$f(x) = \sum_{i=1}^{N} \sum_{j=1}^{i-1} \left[ \frac{1}{{d_{ij}}^{12}} - 2 * \frac{1}{{d_{ij}}^{6}} \right] \tag{2.1}$$

where $d_{ij}$ is the Euclidean distance between $x_i$ and $x_j$. In this formulation, the pairwise equilibrium distance is scaled to one and the pairwise minimum energy is scaled to -1. Lennard-Jones problems characterize the interaction of inert gasses such as argon; more importantly, Lennard-Jones terms are a key part of many empirical energy models including all commonly used energy functions for proteins.

The potential energy of water clusters that we use was developed by Coker and Watts [**6**], and has also been used in global optimization research by Long [**14**]. It has the form

$$f(x) = \sum_{i \neq j} u(x_i, x_j) + \sum_{i} v(x_i) \tag{2.2}$$

where each vector $x_i$ gives the coordinates of three atoms in the $i$-th molecule, and the functions $u$ and $v$ are composed of a number of terms that give the interaction energy between pairs of molecules and the internal energy of the molecule, respectively. Note that modeling water will be important in future research on protein folding, which will need to consider the behavior of proteins in a water solvent.

The basic framework of our global optimization methods for molecular cluster problems is outlined in Algorithm 2.1 below. As mentioned in Section 1, the method combines an initial phase that locates some low local minimizers with a second phase for progressing from low to even lower local minimizers that accounts for most of the computational effort, and the success, of the method. Both phases make use of techniques that vary only a small subset of the variables (an atom for Lennard-Jones problems, a molecule for water) at once. In the initial phase this approach is used to improve the sample points one atom or molecule at a time in step 1b. In the second phase, this approach is used to move one atom or molecule in an existing configuration to new positions via the one atom/molecule global optimization in step 2c. This global optimization finds the best possible position for the selected atom or molecule in the current configuration with the remainder of the configuration temporarily fixed. In both phases, the small subset of the variables that is selected is the one that contributes the most to the energy function (a quantity that is readily calculated due the partial separability of the energy function), an indicator that moving this atom or molecule may be a fruitful way to reduce the overall energy of the cluster. Both of the one atom/molecule steps are relatively inexpensive due to the small number of variables involved and the partial separability of the energy function, which allows the energy to be recalculated at $O(1/N)$ of the cost of a full energy evaluation if only one atom or molecule is moved. In both phases, the one atom/molecule steps are followed by local minimizations in all the variables; these steps allow the entire configuration to change. This overall approach has been found to lead to improvements in the sample points and local minimizers far more effectively and efficiently than strategies that always work on all the variables at once.

The overall progress of the improvement phase is governed by the heuristic in step 2a. This heuristic has been developed to explore the enormous search space of possible configurations in a manner that combines breadth and depth. Configurations that were passed to Phase 2 from Phase 1 are considered to be roots of trees that first are each grown to a specified depth, regardless of the energy function values produced at each consecutive level. After this initial balancing portion in which all of the trees have been explored to an equal (specified) depth, selections of configurations are based solely on energy criteria, which tends to encourage only a few trees (with the lowest energy values) to continue growing. At both stages, the energy values do not necessarily decrease monotonically as one descends a tree of configurations, since the best new configuration resulting from a given configuration may not necessarily have a lower energy value. The balancing of work over all of the initial configurations before working exclusively on the best few has proven to be an extremely beneficial strategy in the overall success of the algorithm.

A final important feature of Algorithm 2.1 is the expansion step 2b. It was found, particularly for water clusters, that by expanding the cluster prior to the one-molecule global optimization step, the small global optimization could locate better new positions for the molecule and the overall algorithm could make better progress. For Lennard-Jones problems this step is sometimes helpful as well. There is no analog to this step in our current protein methods.

**2.1. Results for Cluster Problems.** There has been extensive development of and experimentation with very special purpose methods for Lennard-Jones problems [**12, 16, 20**], and many researchers have applied more general approaches to these problems as well [**7, 15, 13, 18**]. As a result, the optimal configurations are believed to be known for most clusters with up to 100 or so atoms. Algorithm 2.1 has been applied to all Lennard-Jones clusters with 5 to 76 atoms (15 to 228 parameters), and finds the best known solutions in all cases. A new best minimizer was found for the 75 atom problem, with an energy value of -396.282, whereas the best previously reported value for this case was -396.239 [**20**]. Recently discovered lowest energy values for 66 and 72 cases [**7**] were also matched by Algorithm 2.1. These results appear to be the best currently reported by any global optimization method on Lennard-Jones problems.

For the water problem, we have mainly run our algorithm on clusters of 20, 21 and 32 water molecules (180, 189, and 288 parameters), because results of minimizing these same clusters and energy function, using a dynamic simulated annealing

**Algorithm 2.1 – Framework of the Large-Scale Global Optimization Algorithm for Molecular Cluster Problems**

1. **Initial Generation of Configurations Phase :**
   (a) **Sampling in Full Domain :** Randomly generate the coordinates of the sample points in the sampling domain, and evaluate $f(x)$ at each new sample point. Discard all sample points whose function value is below a global "cutoff level".
   (b) **One-atom/molecule Sampling Improvement :** For each remaining sample point : While the energy of the sample point is above the threshold value, Repeat:
      - Select the atom/molecule that contributes most to the energy function value
      - Randomly sample on the location of the selected atom/molecule
      - Replace this atom/molecule in the sample point with the new sample coordinates that give the lowest energy value.

   (c) **Start Point Selection :** Select a subset of the improved sample points from step 1b to be start points for local minimizations.
   (d) **Full-Dimensional Local Minimizations :** Perform a local minimization from each start point selected in step 1c. Collect some number of the best of these minimizers for improvement in Phase 2.
2. **Improvement of Local Minimizers Phase:** For some number of iterations:
   (a) **Select a Configuration :** From the list of full-dimensional local minimizers, select the local minimizer and atom/molecule to be optimized.
   (b) **Expansion :** Transform the configuration by multiplying the position of each atom relative to the center of mass of the configuration by a constant factor of between 1.0 (no expansion) and 1.75.
   (c) **One Atom/Molecule Global Optimization :** Apply a global optimization algorithm to the expanded configuration with only the atom/molecule chosen in step 2a as a variable.
   (d) **Full-Dimensional Local Minimization :** Apply a local minimization procedure, using all the atoms/molecules as variables, to the lowest configurations that resulted from the one-atom/molecule global optimization.
   (e) **Merge the New Local Minimizers :** Merge the new lowest configurations into the existing list of local minimizers.

procedure, have been obtained by Long [**14**]. We have obtained many configurations with significantly lower energies than those obtained in [**14**]. The best solutions obtained by running our algorithm have energies of -0.348183, -0.369011 and -0.585062 atomic units (a.u.) for 20, 21 and 32 molecules respectively. These energy values are approximately 0.005, 0.01 and 0.02 a.u. lower than the best structures found in [**14**], whereas at room temperature, only vibrational states with energies about 0.001 a.u. above the ground state are possible. The structures for 20 and 21 molecules have collapsed dodecahedral (for 20) and dodecahedral (for 21) shapes, which are in agreement with the shapes of the lowest energy clusters found in [**14**].

These results appear to indicate that the approach of Algorithm 2.1 is a useful one for the global optimization of energy functions for molecular clusters.

## 3. Methods for Protein Conformation Problems

The problem of finding the native three dimensional structure of a protein given its primary sequence of amino acids differs from the molecular cluster problems discussed in the previous section, due mainly to the inherent chain-like structure of proteins. Before expounding on the ramifications of this structural difference in our methods, we address some fundamental issues regarding the protein problem.

The basic building block of proteins is an amino acid. The amino acids are bonded together by peptide bonds to form a polypeptide chain, and the sequence of amino acids in the chain is referred to as the primary sequence of the protein. There are 20 amino acids that exist in nature, and all proteins are composed from these. The secondary structure of a protein refers to regular structure in portions of the polypeptide chain, such as alpha helices, beta sheets and the turns that link them, whereas tertiary structure describes the overall shape adopted by the polypeptide chain within a domain [**8**]. The tertiary structure is believed to be the structure that minimizes the free, or potential, energy of of the protein.

Chemists have developed a number of empirical potential energy functions to describe the interaction between atoms in proteins. In this study we use a version of the CHARMM energy function [**1**] to compute the potential energy. This function depends on the coordinates of the atoms in the protein and has the form

$$f(x) = \sum_{i,j} E_{bond} + \sum_{i,j,k} E_{angle} + \sum_{i,j,k,l} E_{dihedral} + \sum_{m,n} (E_{VdW} + E_{elec}) \,. \tag{3.1}$$

The first term, the bond length potential, is a sum over all pairs of of bonded atoms $i$ and $j$, and depends on the deviation of the distance $\|x_i - x_j\|$ from its equilibrium value. For each three atoms such that atoms $i$ and $k$ are both bonded to atom $j$, the second sum has a term depending on the bond angle formed by the three atoms, with vertex at $x_j$. The last bonded summation includes terms depending on certain dihedral angles formed by sets of four atoms connected by bonds. The dihedral angle is formed by two planes each containing a specified set of three of the four atoms, and is a measure of the torsion of the configuration. The final summation in (3.1) runs over all pairs of atoms not involved in bonds. It includes the Van der Waals (Lennard-Jones 6-12 as discussed above)) interactions and the electrostatic potential between pairs of atoms.

The CHARMM function computes potential energy as a function of the cartesian coordinates of the atoms, but it is convenient to express the protein configuration in terms of an equivalent set of internal parameters given in terms of bond lengths, bond angles and dihedral angles. This parameterization is useful for global optimization because the dihedral angles are the crucial parameters to be varied in the optimization. Since the energy to perturb bond lengths and bond angles from their equilibrium values is relatively large, it is conventional in the bio-chemistry community to consider these parameters as fixed, and to address the global optimization problem with just the dihedral angles as the variables, and this is how we have formulated the problem. The main advantage of this parameterization is that the number of variables is greatly reduced, generally by a factor of ten or more. A disadvantage is that the energy function is no longer partially separable in this parameterization, meaning that it is no longer much less expensive to re-evaluate

**Algorithm 3.1 – Framework of the Large-Scale Global Optimization Algorithm for Protein Problems**

1. **Initial Generation of Configurations Phase :**
    (a) **Protein Sample Point Buildup:** Build up sample configurations from one end of the protein to the other by sequentially generating each dihedral angle in the protein: randomly sample the current dihedral angle a fixed number of times and select the dihedral angle that gives the lowest energy function value for the partial protein generated so far.
    (b) **Start Point Selection :** Select a subset of the best sample points from step 1a to be start points for local minimizations.
    (c) **Full-Dimensional Local Minimizations :** Perform a local minimization from each start point selected in step 1b. Collect some number of the best of these minimizers for improvement in Phase 2.
2. **Improvement of Local Minimizers Phase :** For some number of iterations:
    (a) **Select a protein:** From the list of full-dimensional local minimizers, select a local minimizer and a small subset of dihedral angles from that minimizer to be optimized.
    (b) **Global Optimization on a small subset of variables :** Apply a fairly exhaustive small-scale global optimization algorithm to the energy of the selected configuration using the selected small subset of the dihedral angles as variables.
    (c) **Full-Dimensional Local Minimization :** Apply a local minimization procedure, with all dihedral angles as variables, to the lowest configurations that resulted from the global optimization of the step 2b.
    (d) **Merge the New Local Minimizers :** Merge the new lowest configurations into the existing list of local minimizers.

the energy if only a few parameters change than if they all change. This loss of partial separability has some ramifications for our algorithmic strategy that are mentioned below.

The framework of our global optimation method for protein problems is outlined in Algorithm 3.1 below. The highest level structure of the algorithm is the same as in the molecular cluster algorithm. It consists of an initialization phase that locates some low minimizers, followed by an improvement phase that finds better local minimizers from the current best local minimizers. As in the molecular cluster algorithm, the improvement phase utilizes a small-scale global optimization on a selected subset of the parameters followed by local minimizations on all the parameters. Beyond this level there are many differences between the two algorithms due to the chain structure of proteins.

One main difference of Algorithm 3.1 from the molecular cluster algorithm is in the sampling improvement strategy of the first phase, where the cluster methods move one particle at a time to random locations in order to improve the sample configurations. Because it is not possible to move an arbitrary atom of a chain without affecting the bond lengths and angles attached to neighboring atoms, the sample generation for proteins is done entirely differently. The chain is built up one dihedral angle at a time, by sampling dihedral angles in a sequential manner along the chain. Each dihedral angle is sampled a fixed number of times, and the value that gives the configuration under construction the best partial energy is selected. This initialization step seems to have worked well, and overcomes the lack of partial separability since the change in energy due to adding on to the end of a chain can be calculated efficiently. It is not the crucial part of the method, however, since it just supplies some starting configurations for the main, improvement phase. In fact, if a scientist knows some good candidate configurations, these could be used as input to the improvement phase, instead of or in addition to the ones supplied by the initialization phase.

The other key difference from the molecular cluster algorithm is that the small-scale global optimization in the improvement phase of the method can no longer select and move a single atom or molecule, for the reasons mentioned above, but instead optimizes over a small number of dihedral angles with the remaining dihedral angles temporarily fixed. The selection of the angles to optimize over is very important to the success of the method, and involves interesting new tradeoffs. Selecting a small number of dihedral angles that are dispersed throughout the protein chain offers the greatest possibility for change in the tertiary structure of the protein during the small-scale global optimization. However it means that the small-scale global optimization is relatively expensive because the energy evaluations are nearly as expensive as full energy evaluations. Conversely, selecting a small contiguous set of dihedral angles offers less possibility for change in the overall structure of the protein during the small-scale global optimization, but it allows the small-scale global optimization to be relatively inexpensive because the energy evaluations can performed efficiently, as in the small subproblems for molecular clusters. So far, our algorithms have been based on the first option, selecting a small, dispersed set of dihedral angles, but we intend to experiment with the second possibility as well.

Given this overall strategy of choosing a small set of possibly dispersed dihedral angles as the parameters for the small-scale global optimization, our algorithm must determine which angles to select. The method for doing this is different and more difficult than the analogous selection for cluster problems. Once again, we wish to use the energy function to indicate which parameters, if varied, offer the greatest potential for reduction in the energy function. But we can no longer calculate this directly by utilizing partial separability to compute the partial energy associated with each parameter. Instead, Algorithm 3.1 uses two different methods for selecting the parameters to optimize, both based upon the interaction energies between all the atoms to the left of a given dihedral angle and all those to its right. The methods differ in their normalization of the interaction energies. The first method computes the interaction energy just described for each possible dihedral angle, and normalizes it by the product of the number of atoms to the left times the number to the right. Some specified number (generally five) of dihedral angles with the highest normalized interaction energies are then selected. The second method computes the interaction energy in the same manner as the first, but normalizes it by the maximum of the number of atoms to the left and right. The first method exhibits a bias toward selecting dihedral angles in the middle part of the conformation, whereas the second method tends to pick dihedral angles closer to the ends. The first method appears to be more effective in the earlier stages of the computation when the conformation is still far from the global minimum, while the second method appears appropriate for polishing the ends of the conformation after the middle part has been optimized.

In our experiments with the 58 amino-acid polyalanine mentioned in Section 3.1, we have also used a third method for selecting the dihedral angles for the small-scale global optimization. This method, used only in the last stages of the calculcation, is to chose those dihedral angles that are close to where a hydrogen bond is expected to be but is not currently present in the conformation. So far, this choice has been made manually by visualizing the protein, although it should be possible to make it algorithmically. Of course, including this type of heuristic in the code makes it a less general-purpose global optimization method, but may be important in its ability to solve difficult protein problems. We discuss this issue futher in the section on future research.

Finally, the heuristics used in step 2a of the cluster method to balance the breadth and depth of the search have proven to be valuable in searching the space of possible protein conformations as well, and are used in essentially the same manner.

**3.1. Results for Protein Problems.** We have run Algorithm 3.1 on the protein polyalanine with 20, 30, 40 and 58 residues (amino acids). Polyalanine is a simple polypeptide in that each amino acid is the same and the side chain in the amino acid is simple and contributes no additional dihedral angles to the parameterization. It has been used commonly as a test problem in the bio-chemistry community (see e.g. [**19, 10**]). In internal parameters, there are two parameters per amino acid, meaning that the problems tested have up to 116 optimization parameters. Our calculations have mainly been performed on a KSR1 multiprocessor using up to 64 processors, using a parallelization strategy similar to that described in [**3**].

In the problems with 20, 30, and 40 amino acids, the best structure found by our algorithm in each case is a very regular alpha-helix. These results are expected, although we have no comparative energy values from other methods using the same input data and the CHARMM energy function to check against. Interestingly, in the 20 amino acid problem the initialization phase already finds the best configuration, although the improvement phase can also reach this configuration from poor starting configurations. In the 30 and 40 amino acid problems, the initialization phase does not produce the best configuration but it is found by the improvement phase.

The 58 residue problem is considerably more interesting. There are conflicting results in the literature regarding what shape to expect for this problem. Wilson and Cui [**19**] found alpha-helical structures for polyalanine with up to 80 residues using a simulated annealing method; however, Head-Gordon and Stillinger [**10**] show that for 58 residue polyalanine, a lower energy minimizer exists whose structure is two parallel alpha-helices connected by a turn. Our work on this problem is still in progress, but interestingly, so far our algorithm has found two very low energy minimizers with almost identical energies, and with the two different shapes just mentioned. These configurations were located using the normalized interaction energy strategies mentioned in the discussion of Algorithm 3.1. Following this, we have been able to further improve the energies of both of these conformations using the third technique for selecting parameters for the small-scale global optimization mentioned above, that is visually choosing dihedral angles near missing hydrogen bonds. At this point, the best straight conformation appears to be a totally regular alpha helix with all of the expected hydrogen bonds present. The best bent conformation has a somewhat lower energy value than the best straight conformation, and its shape has changed such that the two alpha helices are no longer parallel, but are almost crossing over each other. This conformation has a potential energy about 4 kcal/mole lower than that of the best straight configuration. Which of these two ultimately is "best" is not the key issue, and in fact it appears that this may vary if one uses different potentials or cartesian versus dihedral angle parameterizations. The most exciting aspect of these results is that they indicate that our algorithm has the potential to locate complex tertiary structures without knowing of them apriori.

## 4. Conclusions and Future Results

We have presented a new global optimization algorithm for locating the minimum energy configurations of proteins. The algorithm utilizes small-scale global optimization calculations on selected subsets of the parameters, performed by a stochastic global optimization method, as a key part of its approach. Its structure is related to our previous global optimization methods for molecular clusters but there are many important differences due to the chain structure of proteins. In computational tests on the protein polyalanine with up to 58 amino acids (116 internal parameters), the method appears to be very successful in finding the lowest energy structures. The largest case is particularly significant because the lowest energy structures that are found include ones that exhibit interesting tertiary as opposed to just secondary structure.

The research reported here is part of an ongoing project that is continuing in many directions. One of these is the extension of the methodology presented here to more complex proteins composed of different amino acids and more complex side chains. This raises a number of new algorithmic issues. A second is the exploration of new strategies for choosing the parameters to optimize over in the small-scale global optimizations. This includes the option discussed in Section 3 of selecting a set of consecutive dihedral angles, as well as automatic strategies that correspond to the use of visual information discussed at the end of Section 3. Another direction is exploring ways to incorporate additional partial information that scientists have about the structure of proteins. For example, scientists appear able to predict the secondary structure of portions of proteins with high but not perfect accurary [**11, 5**], and it would seem useful to be able to utilize these predictions in the global optimization algorithm in some manner.

## References

[1] B. R. Brooks, R. E. Bruccoleri, B. D. Olafson, D. J. States, S. Swaminathan and M. Karplus, *A program for Macromolecular Energy, Minimization, and Dynamics Calculations*, J. Comp. Chem. **4** (1983), pp. 187–217.

[2] R. H. Byrd, T. Derby, E. Eskow, K. Oldenkamp and R. B. Schnabel, *A New Stochastic/Perturbation Method for Large-Scale Global Optimization and its Application to Water Cluster Problems*, in Large-Scale Optimization: State of the Art, W. Hager, D. Hearn, and P. Pardalos, eds., Kluwer Academic Publishers, Dordrecht, The Netherlands, 1994, pp. 71–84.

[3] R. H. Byrd, E. Eskow, A. van der Hoek, R. B. Schnabel, K. P. B. Oldenkamp, *A Parallel Global Optimization Method for Solving Molecular Cluster and Polymer Conformation Problems*, Proceedings of the Seventh Siam Conference on Parallel Processing for Scientific Computing,

D. H. Bailey, P. E. Bjørstad, J. R. Gilbert, M. V. Mascagni, R. S. Shreiber, H. D. Simon, V. J. Torczon, L. T. Watson, eds., SIAM, Philadelphia, 1995, pp. 72–77.

[4] R. H. Byrd, E. Eskow and R. B. Schnabel, *A New Large-Scale Global Optimization Method and its Application to Lennard-Jones Problems*, University of Colorado Technical Report CU-CS-630-92.

[5] P. Prevelige Jr. and G. D. Fasman, *Chou-Fasman Prediction of the Secondary Structure of Proteins*, in Prediction of Protein Structure and the Principles of Protein Conformation, G. D. Fasman, ed., Plenum Press, New York, 1989, pp. 391–416.

[6] D. F. Coker and R. O. Watts, *Structure and Vibrational Spectroscopy of the Water Dimer Using Quantum Simulation*, J. Phys. Chem.,**91**, (1987), pp.2513–2518.

[7] T. Coleman, D. Shalloway, and Z. Wu, *A Parallel Build-up Algorithm for Global Energy Minimizations of Molecular Clusters Using Effective Energy Simulated Annealing* Technical Report, Center for Theory and Simulation in Science and Engineering, Cornell University, 1993.

[8] N. J. Darby, and T. E. Creighton, *Protein Structure*, Oxford University Press, New York, 1993.

[9] C. J. Epstein, R. F. Goldberger and C. B. Anfinsen, *The genetic control of tertiary protein structure: Studies with model systems.* Cold Spring Sympos. Quant. Biol.**28** (1963), pp. 439–449.

[10] T. Head-Gordon and F. H. Stillinger, *Enthalpy of Knotted Polypeptides*, J. Phys. Chem., **96**, (1992), pp. 7792–7796.

[11] T. Head-Gordon and F. H. Stillinger, *Optimal neural networks for protein-structure prediction*, Phys. Rev. E., **48**, (1993), pp. 1502-1516.

[12] M. R. Hoare and P. Pal, *Physical cluster mechanics: statics and energy surfaces for monatomic systems*, Adv. Phys., **20**, (1971) pp. 161–196.

[13] J. Kostrowicki, L. Piela, B. J. Cherayil and H. A. Scheraga, *Performance of the diffusion equation method in searches of optimum structures of clusters of Lennard-Jones atoms*, J. Phys. Chem., **95**, (1991) pp. 4113–4119.

[14] X.Long, *Molecular Dynamics Simulations of Clusters – Impure van der Waals and* $e^- - (H_2O)_n$ *Systems*, Ph.D. Thesis, Department of Chemistry, University of California, San Diego, 1992.

[15] C. D. Maranas and C. A. Floudas, *A global optimization approach for Lennard-Jones microclusters*, J. Chem. Phys., **97** (1992) p. 7667.

[16] J. A. Northby, *Structure and binding of Lennard-Jones clusters:*$13 \leq N \leq 147$ J. Chem. Phys., **87**, (1987) pp. 6166–6177.

[17] P.M. Pardalos, D. Shalloway and G.L. Xue, *Optimization Methods for Computing Global Minima of Nonconvex Potential Energy Functions*, Journal of Global Optimization, **4** (1994), pp. 117-136.

[18] D. Shalloway, *Packet Annealing: A Deterministic Method for Global Minimization: Application to Molecular Conformation*, in Recent Advances in Global Optimization, C. Floudas and P. Pardalos (eds.), Princeton University Press, Princeton, N.J., (1992) pp. 433-477.

[19] S. R. Wilson and W. Cui, *Applications of Simulated Annealing to Peptides*, Biopolymers, **29**, (1990), pp. 225–235.

[20] G. L. Xue, *Improvement of the Northby Algorithm for Molecular Conformation: Better Solutions*, J. Global Optimization, **4**, (1994), p. 425.

Department of Computer Science, University of Colorado at Boulder, Campus Box 430, Boulder, CO 80309-0430

DIMACS Series in Discrete Mathematics
and Theoretical Computer Science
Volume **23**, 1996

# Tracking Metastable States to Free–Energy Global Minima

Bruce W. Church, Matej Orešič, and David Shalloway

ABSTRACT. Proteins thermodynamically explore only limited regions of configuration space when they are annealed to a folded conformation, and it is believed that specific "folding pathways" exist. We discuss methods for identifying such pathways using a non–equilibrium thermodynamic approach. We computationally identify pathways by tracking the positions and sizes (in configuration space) of metastable macrostates (packet states) as temperature is reduced. "Packet conditions" derived from the Smoluchowski equation for diffusion in the presence of an external potential, provide a prescription for identifying metastable states and their temperature–dependent bifurcations. Since each "packet state" covers an extended region of configuration space, entropies and free–energies (as well as mean–energies) can be calculated. The resultant free–energy *vs.* temperature "trajectory diagrams" provide a new method for hierarchically characterizing energy landscapes and for designing, implementing and diagnosing global search strategies. In particular, energy landscape "scaling properties", which govern the feasibility of global minimization by potential–smoothing continuation methods, are described. The method is illustrated using Lennard–Jones microclusters. Analysis of the trajectory diagram for the pentapeptide Met–enkephalin illustrates the use of these diagrams as "road–maps" for global minimization algorithms and explains the performance of some current global minimization methods.

## 1. Introduction

Some, though not all, proteins can be physically annealed in solution from disordered to ordered conformations by reducing their temperature in the absence of any other molecular guidance. Equilibrium statistical mechanics describes this process in the limit of infinitely slow cooling. However, kinetic constraints on "folding pathways" probably play important roles as well. In particular, properties of molecular energy landscapes that "funnel" the ensemble down a limited set of pathways towards the folded state are probably required for sufficiently rapid folding of all but the smallest proteins [**1, 2**]. Funneling processes can reduce search times by many orders of magnitude, but do so at the expense of an incomplete search of the

---

1991 *Mathematics Subject Classification.* Primary 65K10, 90C30; Secondary 92C05, 92E10.

*Key words and phrases.* Global minimization, proteins, macromolecules, spatial scaling, energy landscape, folding pathways, simulated annealing, microclusters.

This work was supported by grants from the National Institutes of Health (GM48874) and the Air Force Office of Scientific Research (F49620).

energy landscape. The end–point of a funneling process will be a local free–energy minimum, but not necessarily the global free–energy minimum corresponding to the thermodynamic equilibrium state. Accordingly, some native proteins are believed to be in *meta*stable states which are kinetically stabilized by sufficiently large kinetic barriers in the energy landscape. Computational minimization methods for structure prediction must account for these restricting factors if they are to correctly predict kinetically–determined metastable conformations. Even when the folded state is the thermodynamic equilibrium state, computational methods can be accelerated by guiding them down the physical folding pathways.

Stochastic computational annealing procedures such as simulated annealing can, in the limit of infinite computational time, mimic the process of infinitely slow physical annealing. However, standard methods pay little attention to kinetic considerations. We are exploring ways in which this can be done. The first step is to develop methods for identifying the dominant search pathways used by physical ensembles of molecules. Central to our approach is the temperature–dependent dissection of the energy landscape in terms of *metastable macrostates*, spatially–localized regions that contain significant concentrations of ensemble probability density. By tracking the temperature–dependent variations of the parameters that describe the state locations and sizes, we obtain metastable state *trajectory diagrams* which model the potential folding pathways. These provide hierarchical, coarse–grained descriptions of energy landscapes that may be powerful tools for developing global minimization algorithms. In Sec. 2 we outline the conceptual basis for our method: the dissection of the energy landscape into packets which, as described in Sec. 3, correspond to metastable macrostates. In Sec. 4 we show how these packet states can be computationally identified using equations derived from the Smoluchowski equation for diffusion in an external potential. In Sec. 5 we show how trajectory diagrams are computed and illustrate their properties using a one–dimensional problem and Lennard–Jones microclusters as examples. In Sec. 6 we discuss how the packet annealing formalism can be adapted to geometrically complex protein torsion–angle coordinates. We calculate trajectory diagrams for the pentapeptide Met–enkephalin and show how they provide insight into the performance of different global minimization routines. We summarize in Sec. 7.

## 2. Packet annealing

Here we review the basis for the Packet Annealing Method from the viewpoint of non–equilibrium thermodynamics. More detailed discussion can be found in Refs. [**3**] and [**4**].

At equilibrium at temperature $T$, the ensemble probability density for occupancy of configuration space is given by the Gibbs/Boltzmann distribution

$$p_B(R) = \frac{e^{-\beta V(R)}}{\int e^{-\beta V(R)}\, dR} \equiv e^{-\beta V(R)}/Z\,, \tag{1}$$

where $\beta \equiv T^{-1}$, $V$ is the potential energy function, and $R$ represents all the internal degrees of freedom of the molecule.[1] For example, Fig. 1 displays $p_B$ at a sequence of decreasing temperatures for a model two–dimensional potential. The support of

---

[1] $p_B$ and most of the functions and variables that we will discuss are temperature–dependent. For notational simplicity, this is not explicitly indicated. In addition, we take the Boltzmann factor $k_B \equiv 1$ so temperature and energy have the same units.

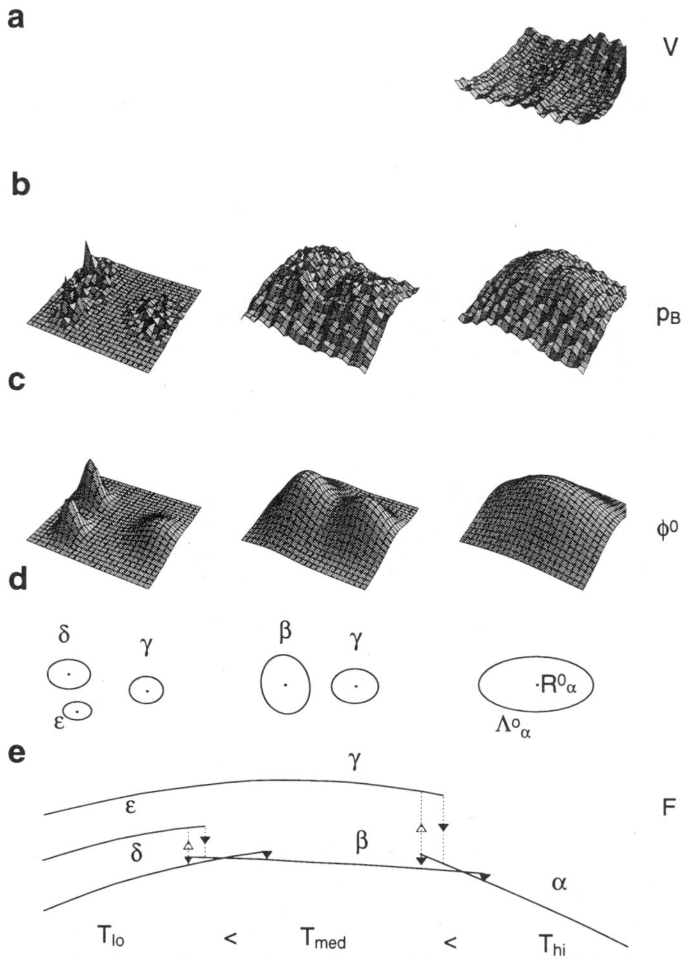


FIGURE 1. Annealing using metastable states. (a) A model two-dimensional potential, $V(r_1, r_2)$. (b) The corresponding Gibbs/Boltzmann probability distribution $p_B$ at three temperatures, $T_{hi} > T_{med} > T_{lo}$. (c) Superposition of the (squared) characteristic packets $(\phi_\alpha^0)^2$ that are solutions to the packet equations at the three temperatures. (A large number of characteristic packets, corresponding to the very small-scale fluctuations of $V$, will appear at lower temperatures.) (d) The characteristic packets are characterized by the positions of their center-of-masses ($R^0$) and by their mean-square fluctuation tensors ($\Lambda^0$), represented here by ellipses. (e) Free-energy vs temperature trajectory diagram for this temperature range. Solid lines represent metastable state trajectories and dotted lines represent transitions. The discontinuities in the trajectories correspond to branch points at which packets bifurcate. Solid arrowheads indicate "escape" or preferred "capture" transitions. Open arrowheads indicate unpreferred transitions that can be detected by the missing-mass procedure.

$p_B$ is dispersed at high temperature and converges at low temperature to a small region centered about the global minimizer of $V$. While the detailed form of $p_B$ at intermediate temperatures is complicated, the annealing process can be modeled by following the appearance, movement and disappearance of regions of concentrated probability density with temperature. As indicated in Fig. 1, at every temperature each such region $\alpha$ can be characterized by the position of its center-of-mass, $R^0_\alpha$, its spatial extent, $\Lambda^0_\alpha$, and a parameter governing its amplitude, $V^0_\alpha$.[2] Once these *characteristic parameters* have been identified, $p_B$ can be coarsely approximated as a superposition of terms confined by *characteristic packets*, $\phi^0_\alpha(V^0_\alpha, R^0_\alpha, \Lambda^0_\alpha)$:

$$(2)\quad p_B(R) \approx \sum_{\{\alpha\}} \phi^0_\alpha(V^0_\alpha, R^0_\alpha, \Lambda^0_\alpha; R)\, p_B^{\frac{1}{2}}(R) \qquad \text{(coarse approximation only)}\,.$$

Each term represents a concentration of probability density roughly confined to the region described by $R^0_\alpha$ and $\Lambda^0_\alpha$. While different characteristic packet shapes could be used, for systems where $V$ is the sum of a very large number of quasi-independent terms, the Central Limit Theorem suggests that Gaussians are appropriate. Thus, we take

$$(3)\qquad \phi^0_\alpha \equiv e^{-\frac{1}{2}\beta[V^0_\alpha + (R-R^0_\alpha)K^0_\alpha(R-R^0_\alpha)]}/Z^{\frac{1}{2}}\,,$$

where $K^0_\alpha \equiv (\Lambda^0_\alpha)^{-2}/\beta$. With this choice, each term in (2) is a product of a confining Gaussian characteristic packet with $p_B^{\frac{1}{2}}$.

As illustrated in Fig. 1, both the number and parameters of the characteristic packets vary with temperature. These variations can be summarized in trajectory diagrams (Fig. 1e) which provide a novel, alternative characterization of the energy landscape. As $T$ decreases, the packets continuously decrease in size and may split and thereby grow in number as small–scale variations in the potential become capable of localizing probability density. As $T \to 0$ each local minimum supports a packet having $\Lambda \to 0$. The probability contained within all but the lowest–energy packets will decrease as $T \to 0$ until, at $T = 0$ all the probability will be contained in the lowest–energy packet.

This visual procedure is codified in the Packet Annealing Method. Global minimization in that formulation consists of tracing a path from the extended high–temperature packet down to the point–like $T = 0$ packet of lowest energy. This process can also be terminated at non–zero temperatures to identify the metastable state having lowest free–energy at physiological temperature.

## 3. Packets and metastable states

The characteristic packet decomposition physically corresponds to a thermodynamic description of the system in terms of metastable macrostates (*states*, for short). Following classical statistical mechanics, each metastable state represents a weighted collection of individual configurations specified by $R$. That is, it specifies an extended region of configuration space. Unlike individual configurations, or microstates, which are characterized only by their energy $V(R)$, each metastable macrostate $\alpha$ is characterized both by its temperature–dependent mean energy, $E_\alpha \equiv \langle V \rangle$, and its entropy, $S_\alpha$, a measure of the volume the state occupies in

[2] $R^0_\alpha$ is a vector, $\Lambda^0_\alpha$ is a tensor, and $V^0_\alpha$ is a scalar. Vector algebraic operations are implied in all equations.

configuration space. $E_\alpha$ and $S_\alpha$ are related to the state's free–energy, $F_\alpha$, by the standard thermodynamic relationship [5]

$$F_\alpha = E_\alpha - \frac{1}{\beta} S_\alpha \,. \tag{4}$$

The free–energy determines the probability $\mathcal{P}_\alpha(\infty)$ that the system occupies metastable state $\alpha$ at equilibrium:

$$\mathcal{P}_\alpha(\infty) = \frac{e^{-\beta F_\alpha}}{Z} \,. \tag{5}$$

These thermodynamic concepts provide a coarse–grained characterization of the energy landscape that can be a valuable guide for designing, implementing and diagnosing global minimization algorithms. The connection is provided by Eq. (2). Once the characteristic parameters which determine the $\phi^0_\alpha$ have been calculated (as described in Sec. 4), configuration space can be dissected into metastable state regions using *window functions* $w_\alpha(R)$:

$$w_\alpha(R) \equiv \frac{\phi^0_\alpha(R)}{\sum_{\{\alpha\}} \phi^0_\alpha(R)} \,. \tag{6}$$

These are used for a temperature–dependent *packet state expansion* of the probability landscape [which, rather than (2), is used for precise computation]:

$$p_B(R) = \sum_{\{\alpha\}} p_\alpha(R) \tag{7}$$

where

$$p_\alpha(R) \equiv w_\alpha(R)\, p_B(R) \,. \tag{8}$$

Fig. 2 illustrates the relationship between $p_B$ and the $\phi^0_\alpha$, $w_\alpha$, and $p_\alpha$. Metastable state free–energies are calculated from

$$F_\alpha = -\frac{1}{\beta} \log \left( Z \int p_\alpha(R)\, dR \right) \,. \tag{9}$$

The mean energy is calculated from

$$E_\alpha = \frac{\int p_\alpha \, V \, dR}{\int p_\alpha \, dR} \,, \tag{10}$$

and the entropy can be calculated from (4). Eqs. (9) and (10) connect the packet formalism to thermodynamics.

## 4. Packet conditions and the reduced Smoluchowski equation

The efficiency of an annealing method for global minimization will critically depend on the way in which computational effort is assigned. The "cooling schedule" of simulated annealing provides an example of an assignment that varies with temperature but which is constant over all configuration space. This defect can be corrected by using the spatial decomposition provided by the packet expansion to adjust computational expenditure in a way that depends both on spatial location and temperature. Regardless of the specific algorithm used, the amount of computational effort expended in a configuration space region should correspond (in the physical analogy) to the amount of time needed for ensemble equilibration in that region. Physically, a metastable state corresponds to a region $\alpha$ that can be treated in relative isolation from the remainder of the system. In this case, the

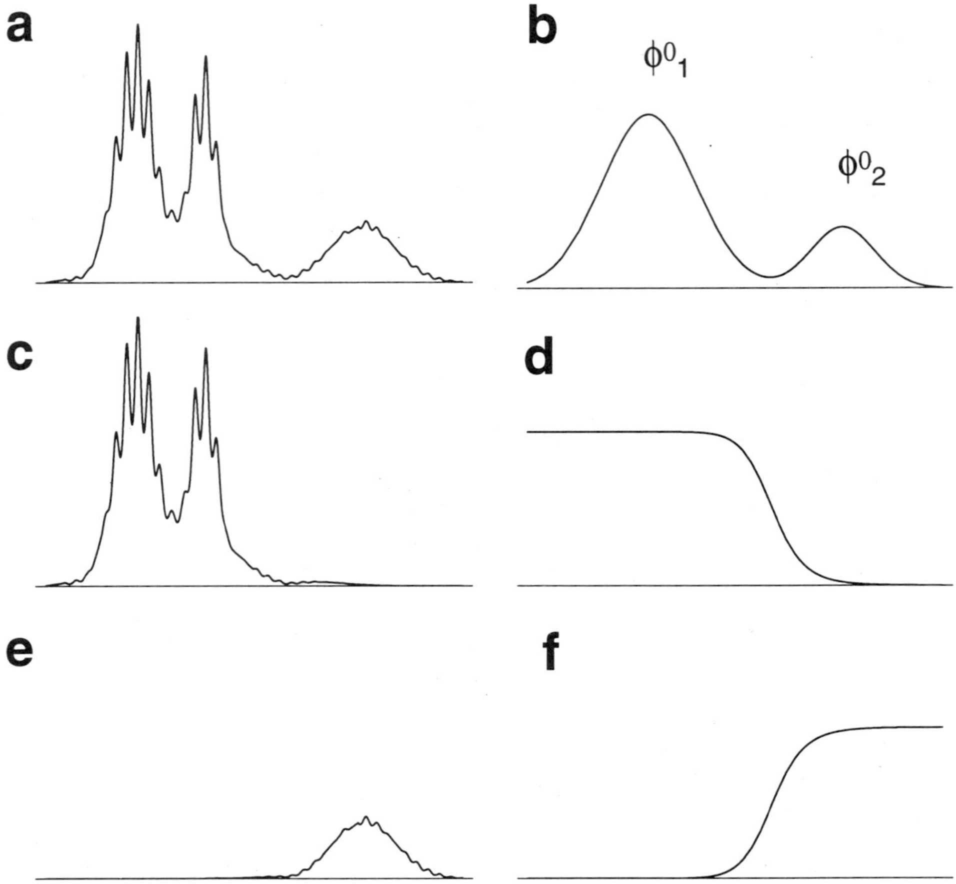


FIGURE 2. Characteristic packets, window functions and packet states. The Gibbs/Boltzmann distribution $p_B$ [shown in (a)] for a model one-dimensional potential is decomposed in a packet state expansion (7). Two characteristic packets are stable at the temperature under consideration. Their squares, $(\phi_1^0)^2$ and $(\phi_2^0)^2$, are plotted in (b). The corresponding window functions $w_1$ and $w_2$ are displayed in (d) and (f), respectively. The packet states $p_1$ and $p_2$ are displayed in (c) and (e).

time–scale for local equilibration, $\tau_\alpha^{local}$, determines the appropriate inverse cooling rate within region $\alpha$. The isolation requirement implies that transitions out of $\alpha$ into other regions must be much slower than internal equilibration. That is, they must occur on time–scale

$$\tau_\alpha^{global} \gg \tau_\alpha^{local} . \tag{11}$$

This is the fundamental condition which governs the existence of metastable states and, correspondingly, characteristic packets. To evaluate it we need a computational model for the system dynamics.

Proteins are complex, highly interactive, overdamped thermal systems, which suggests that their internal dynamics can be approximated as diffusive processes.[3] In this case the temporal evolution of the probability distribution function $p \equiv p(R,t)$ can be described by the Smoluchowski equation [**6**] with external potential $V \equiv V(R)$

$$\frac{1}{D}\frac{\partial p}{\partial t} = \nabla[e^{-\beta V}\,\nabla\,(e^{\beta V}p)]\,, \tag{12}$$

where $D$ is the diffusion constant. This can be reduced to an imaginary-time Schrödinger equation [**7, 8**], the *reduced Smoluchowski equation*:

$$\zeta\,\frac{\partial}{\partial t}\,\Psi(R,t)\,, = -\,\mathcal{H}(R)\,\Psi(R,t) \tag{13}$$

where

$$\Psi(R,t) \equiv p(R,t)/p_B^{\frac{1}{2}}(R) \tag{14}$$

is the *reduced distribution* and

$$\mathcal{H} \equiv D\zeta\,\left[e^{\frac{1}{2}\beta V}\left(\nabla^2 e^{-\frac{1}{2}\beta V}\right) - \nabla^2\right] \tag{15}$$

is the *reduced Hamiltonian.* $\zeta$, an arbitrary constant having the dimensions of action, has been introduced so that $\mathcal{H}$ has dimensions of energy, thereby enhancing the analogy to the Schrödinger equation.

The reduced Smoluchowski equation can be analyzed using an eigenfunction expansion. Its ground-state, with eigenvalue $E_0 = 0$, is

$$\psi_0(R) = p_B^{\frac{1}{2}}(R)\,. \tag{16}$$

All other normalizable eigenfunctions will have positive eigenvalues (corresponding to temporally decaying amplitudes) so that

$$\lim_{t\to\infty}\,\Psi(R,t) = \psi^0(R)\,, \tag{17}$$

or equivalently,

$$\lim_{t\to\infty}\,p(R,t) = p_B(R)\,. \tag{18}$$

As expected, the Gibbs/Boltzmann distribution is achieved after infinite time.

We use Eq. (13) to identify metastable states by finding Gaussian reduced distributions that are first-order stable in time. With $\phi_\alpha^0$ specified by (3), the conditions that the center-of-mass and the inertia tensor of the packet's probability distribution be first-order stationary are

$$\frac{\partial}{\partial t}\,\langle\psi_0|(R-R^0)_i|\phi_0\rangle_{t=0} = -\frac{1}{\zeta}\,\langle\psi_0|(R-R^0)_i\,\mathcal{H}|\phi_0\rangle = 0 \tag{19a}$$

$$\begin{aligned}\frac{\partial}{\partial t}\,&\langle\psi_0|(R-R^0)_i\,(R-R^0)_j\,|\phi_0\rangle_{t=0}\\ &= -\frac{1}{\zeta}\,\langle\psi_0|(R-R^0)_i\,(R-R^0)_j\,\mathcal{H}|\phi_0\rangle = 0\,.\end{aligned} \tag{19b}$$

[3] A more detailed discussion of the material in this section can be found in Ref. [**4**].

(Here we use quantum–mechanical bra–ket notation to indicate integration over configuration space.) We have shown that Eqs. (19a) imply

$$\text{(20a)} \qquad \langle\psi_0|R|\phi^0_\alpha\rangle/\langle\psi_0|\phi^0_\alpha\rangle = R^0_\alpha$$

$$\text{(20b)} \qquad \langle\psi_0|(R-R^0)(R-R^0)|\phi^0_\alpha\rangle/\langle\psi_0|\phi^0_\alpha\rangle = (2\beta K^0_\alpha)^{-1} .$$

Since the $\phi^0_\alpha$ are Gaussians, these are Gaussian–weighted integrals over $\psi_0$, the square–root of the Gibbs/Boltzmann distribution. Eqs. (20) can be solved by coupled iteration [**3**]. Once they have been solved, the third characteristic parameter $V^0_\alpha$ is determined by the requirement

$$\text{(21)} \qquad \langle\phi^0_\alpha|\phi^0_\alpha\rangle = \langle\psi_0|\phi^0_\alpha\rangle = \det{}^{\frac{1}{2}}(\pi/\beta K^0_\alpha)\, e^{-\beta V^0_\alpha}/Z .$$

Eqs. (20) and (21) are the *packet equations* that determine the characteristic parameters. Since they are non–linear self–consistent equations, their solutions can exhibit complicated temperature–dependent discontinuities which correspond to the temperature–dependent splitting of metastable states such as that displayed in Fig. 1. We note that Eqs. (20) are equivalent to the requirement that the characteristic parameters be a local minimizer of $\widetilde{F} \equiv \widetilde{F}(R^0_\alpha, K^0_\alpha)$, the free–energy of characteristic packet $\phi^0_\alpha$.[4] That is,

$$\text{(22a)} \qquad \frac{\partial}{\partial R^0}\widetilde{F}(R^0,K^0)\Big|_{\substack{R^0=R^0_\alpha\\ K^0=K^0_\alpha}} = 0$$

$$\text{(22b)} \qquad \frac{\partial}{\partial K^0}\widetilde{F}(R^0,K^0)\Big|_{\substack{R^0=R^0_\alpha\\ K^0=K^0_\alpha}} = 0 ,$$

where

$$\text{(23)} \quad \widetilde{F}(R^0,K^0) = -\frac{2}{\beta}\log\left\{\det{}^{\frac{1}{4}}\left(\frac{MK^0}{2\pi^2\zeta^2}\right)\int e^{-\frac{\beta}{2}V}\, e^{-\frac{\beta}{2}(R-R^0)K^0(R-R^0)}\,dR\right\} .$$

[M is the mass matrix and, while needed to define $\widetilde{F}$, does not enter Eqs. (22).] This recasts the packet equations as a temperature–dependent local minimization problem.

## 5. Packet trajectory diagrams

At high temperature a confined system like a protein will be in a highly fluctuating macrostate corresponding to a single packet solution with large $\Lambda^0$. Since this will be the only solution to the packet equations, it can be readily found and will provide a starting point for subsequent cooling to a folded state. Alternatively, in some cases experimental data may be available that provide zeroth–order starting points for finding low–temperature characteristic packets. These can be heated and then cooled to find the folded state.

In both cases, the variation of the characteristic packet parameters (and the thermodynamic properties of the corresponding metastable states) can be traced as temperature is varied by updating the solutions to (20) or (22). This procedure will generate *packet trajectories* which may terminate at *upper* and *lower transition temperatures*. As its upper transition temperature is approached from below, a *child* packet will become unstable (i.e., its characteristic packet will cease to be a minimum of $\widetilde{F}$) and its probability density will "escape" to the larger *parent* packet which contains it. This escape can be computationally traced using (22)

[4] Note that $\widetilde{F}_\alpha$ is distinct from $F_\alpha$, the free–energy of packet state $\alpha$.

by following the steepest-descent path in $\widetilde{F}(R^0, K^0)$ from the child's minimizer back to the parental minimizer. Conversely, as a lower transition temperature is approached from above, the parent will become unstable and will be "captured" by one of the smaller children that it contains. Children in turn may become parents and, as $T \rightarrow 0$ the total multiplicity of descendents will approach the number of local minima of $V(R)$. We will see that discontinuous transitions occur when the capture temperatures are below the escape temperatures; this is the phenomenon of *trajectory hysteresis.* Continuous transitions between parent and child states also occur, but are more difficult to detect computationally.

We suggest that this rich mathematical structure models the kinetically–constrained behavior of a physical ensemble during annealing with an optimized cooling schedule that allows just enough time for local, but not for global, equilibration. Each continuous trajectory segment represents a kinetically–constrained pathway; each transition point represents a physical branch–point where the ensemble must select a new pathway. However, the physical path–selection process will involve kinetic considerations that go beyond the discussion of Sec. 4. Whereas the steepest–descent solution of (22) identifies a unique *preferred* pathway, the physical process may distribute ensemble members between multiple pathways. The equilibrium allocation of probability, determined by the relative free–energies of the children, may provide a useful model of the physical process and a guide for allocating computational effort in multi–path search algorithms.

The connections between the parent and child trajectories/pathways constitute a hierarchical organization of the finite–temperature metastable states and zero–temperature local minima that reflects the underlying structure of the energy landscape. The connected trajectories constitute "road–maps" of potential folding pathways that can be used to guide computational annealing algorithms. In the remainder of this section we demonstrate these ideas using two examples: a model one–dimensional potential and Lennard–Jones microclusters. For pedagogical purposes, we compute complete trajectory diagrams including all metastable states. Of course, this will not be possible for large–dimensionality problems because of the expected exponential growth in the number of trajectories with dimensionality. But, it will not be necessary– in practical problems we anticipate exploring only those subsets of the trajectory diagrams that have a high likelihood of leading to the folded state.

**5.1. A one–dimensional example.** Fig. 3 illustrates the complete $R^0$ vs $T$ trajectory diagram for a model one–dimensional potential.[5] At high temperature, a single metastable state $\mu + \nu + \eta$ encompasses the entire accessible region of configuration space. As temperature decreases, $\mu+\nu+\eta$ is captured in a continuous manner by its child $\nu + \eta$.[6] This state is then discontinuously captured by its preferred child $\nu$ at the lower transition temperature, $T^-_{\nu+\eta}$. This occurs when

[5]The diagram was calculated by starting at $T = 0$ with the $\Lambda = 0$ local minima solutions, tracing trajectories during heating, and then tracing back down during cooling. Additional details can be found in Ref. [**3**].

[6]This transition is signaled by an inflection in the $\Lambda^0_{\mu+\nu+\eta}$ trajectory (see Fig. 3 in Ref. [**3**]). However, signals of this sort can be difficult to detect. The discontinuous $\mu \rightarrow \mu+\nu+\eta$ transition, detected during heating, provides an unambiguous signal.

$\widetilde{F}_{\nu+\eta} \approx \widetilde{F}_{\nu}$.[7] This first-order transition is characterized by discontinuous changes in $R^0_{\nu+\eta}$ and the other thermodynamic variables. Since the probability contained in $\nu+\eta$ is distributed between its children, once the preferred discontinuous transition to $\nu$ is detected, the unpreferred child $\eta$ can be found by looking for solution(s) in the region of the "missing-mass" (i.e., missing probability). Both children are stable down to $T = 0$.

To generate a complete trajectory diagram, we reverse the tracing process by heating the $T = 0$ $\mu$, $\nu$, and $\eta$ packets that are located at the three local minima. Each packet is stable up to its upper transition temperature, where it discontinuously escapes back to $\nu + \eta$ or $\mu + \nu + \eta$. The escapes reflect the increase of inter-state transition rates with temperature (i.e., the decrease of $\tau^{global}_{\mu}$, $\tau^{global}_{\nu}$ and $\tau^{global}_{\eta}$) and the resultant break-down of condition (11). The $\nu + \eta$ trajectory reversibly traces a continuous path back up to $\mu + \nu + \eta$.

**5.2. Global minimization and scaling properties.** Discontinuous transitions like $\nu + \eta \longrightarrow \nu$ only occur when the capture temperature is below the escape temperatures of both children; that is, when there is hysteresis. Otherwise the capture and (conversely) escape will be reversibly continuous as in the $\mu+\nu+\eta \leftrightarrow \nu+\eta$ transition. These phenomena may be useful in designing efficient global minimization algorithms. The discontinuities will provide clear signals of branching, and the hysteresis regions will provide finite temperature intervals in which an intelligent annealing algorithm can detect all the children using the missing-mass procedure. This will allow an algorithm to evaluate its branching options and assign computational effort appropriately. Algorithms that do not do this may be spontaneously captured by a single child in an unconscious, unregulated manner. In contrast, an intelligent algorithm may, like a good chess player, decide to track multiple pathways for some temperature intervals.

While it may be possible to detect even continuous transitions (and thereby instigate the missing-mass procedure to find unpreferred children) there will be many *disconnected* trajectory branches that will not be detected during cooling. This is the intrinsic price that must be paid by global minimization routines that focus computational effort on the kinetically-accessible regions of configuration space. This is appropriate for problems (e.g., as argued above, proteins) where the kinetically-inaccessible states are not of practical importance. In these cases, ignoring the disconnected branches improves computational efficiency without negative effect. While disconnected branches can, in principle, be detected by methods that jump over barriers with no concern for their effects on transition rates (e.g., global search or simulated annealing with random jump sizes), these methods will pay the price of reduced efficiency.

The topology of the trajectory diagrams reveals important hierarchical properties that would not otherwise be apparent. Importantly, the ability of a kinetically-constrained algorithm to find the global minimum will depend on the connectivity pattern. For example, the global minimum in Fig. 3 can be found by cooling along

[7] Consistent with general thermodynamic principles, it can be transitions between packet solutions only occur when free-energy is released. Thus, the $\nu + \eta \longrightarrow \nu$ transition can not occur until the free-energy of $\phi^0_{\nu+\eta}$ exceeds that of $\phi^0_{\nu}$.

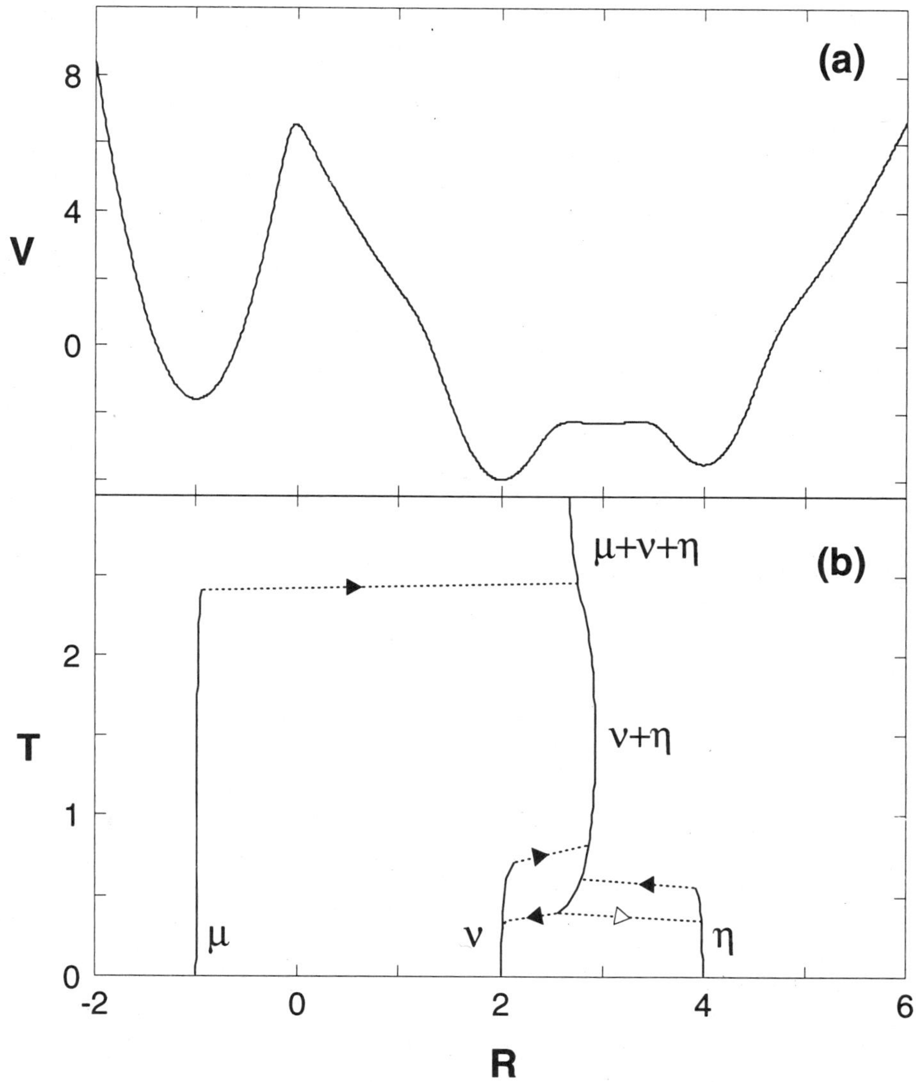


FIGURE 3. A one–dimensional model. (a) An example potential $V$ having three local minima. (b) A trajectory diagram showing the center positions ($R^0_\alpha$) of the characteristic packets as functions of temperature. Transitions are identified as described in Fig. 1. (Figure reprinted from Ref. [**3**].)

the preferred path down to state $\nu$. This is an example of *strong scaling*.[8] This property is not guaranteed. For example, the potential in Fig. 3 could be altered (by a narrow, deep extension within the $\eta$ catchment region) to make $\eta$ the global minimum without altering the trajectory diagram. In this case, global minimum $\eta$ could be found by tracing along a connected, but not preferred, path. We call this

[8] We call this a "scaling" property since it implies that the global minimum defined at large spatial–scale (i.e., by a large high–temperature packet) contains the global minima defined at smaller spatial–scales (i.e., by the smaller packets that are metastable at lower temperatures).

*weak scaling*. Energy landscapes having deep, narrow, isolated minima (i.e., "oil wells") will display *poor scaling*: their global minima are disconnected and can not be found by trajectory tracing.

**5.3. Lennard-Jones microclusters.** The packet equations and missing–mass procedure have been used to analyze the more complicated cases of Lennard–Jones microclusters with six, seven and eight atoms.[9] These have the potential energy function

$$V(\boldsymbol{r}_1, \ldots, \boldsymbol{r}_N) = \sum_{i<j} \left( \frac{1}{r_{ij}^{12}} - \frac{2}{r_{ij}^{6}} \right). \tag{24}$$

The energy and free–energy vs $T$ trajectory diagrams for six– and eight–atom microclusters are shown in Figs. 4 and 5. The $T = 0$ states are the local minimizers of $V(R)$ corresponding to the conformational isomers shown in Fig. 6. The six–atom trajectory has a discontinuous transition: the single $o1 + t1$ high–temperature metastable state bifurcates into $t1$ and $o1$ children at $T \approx 0.18$. Accordingly, a hysteresis interval exists for $0.18 < T < 0.24$. In contrast, the eight–atom diagram displays no discontinuous transitions.[10] In fact, all the states are kinetically disconnected and are not detected by annealing from the high–temperature metastable state $t1$. Both the qualitative and quantitative characteristics of these energy trajectory diagrams agree well with results obtained by molecular dynamics and Monte Carlo studies [**9, 10, 11**], supporting the assertion that characteristic packets correspond to metastable states and that state packets can be used to accurately calculate thermodynamic properties.

Both the six–, seven– (data not shown) and eight–atom trajectories display *strong scaling*, but this is not guaranteed for larger microcluster problems.

## 6. Trajectory diagrams for covalently–linked macromolecules

**6.1. Packets in distance space.** The gross motions of polypeptides are limited by fixed bond–angle and bond–length constraints, so their conformations are most efficiently described using torsion–angle rather than Cartesian space variables. However, the multidimensional formulation we have described is computationally impractical in torsion–angle space. The main problem is that there are large differences between the effects of different torsion–angles on conformation. Small changes in backbone angles near the center of the chain greatly affect conformation while changes in distal sidechain angles have little affect. Using a metaphor from numerical analysis, the torsion–angle representation suffers from large "condition numbers".

We bypass this difficulty by describing characteristic packets and metastable states using pair–wise interatomic distance variables, i.e., distance–geometry coordinates. This representation is valuable because most metastable states can be distinguished using only a few distance variables. For example, the two conformations of the pentapeptide Met–enkephalin (Fig. 7) shown in Fig. 8 have significantly different values of $d_{5CG,5HA}$, the distance between the $\gamma$-carbon and the $\alpha$-hydrogen

[9] For this example, the packet equations were evaluated using Monte Carlo integration. Further details are presented in Ref. [**3**].

[10] Because there are no covalent bonds in the system, atoms can vaporize from the microcluster. This is reflected in the abrupt termination of some trajectories (e.g., $t1$, $t3$ and $o3$ in Fig. 5).

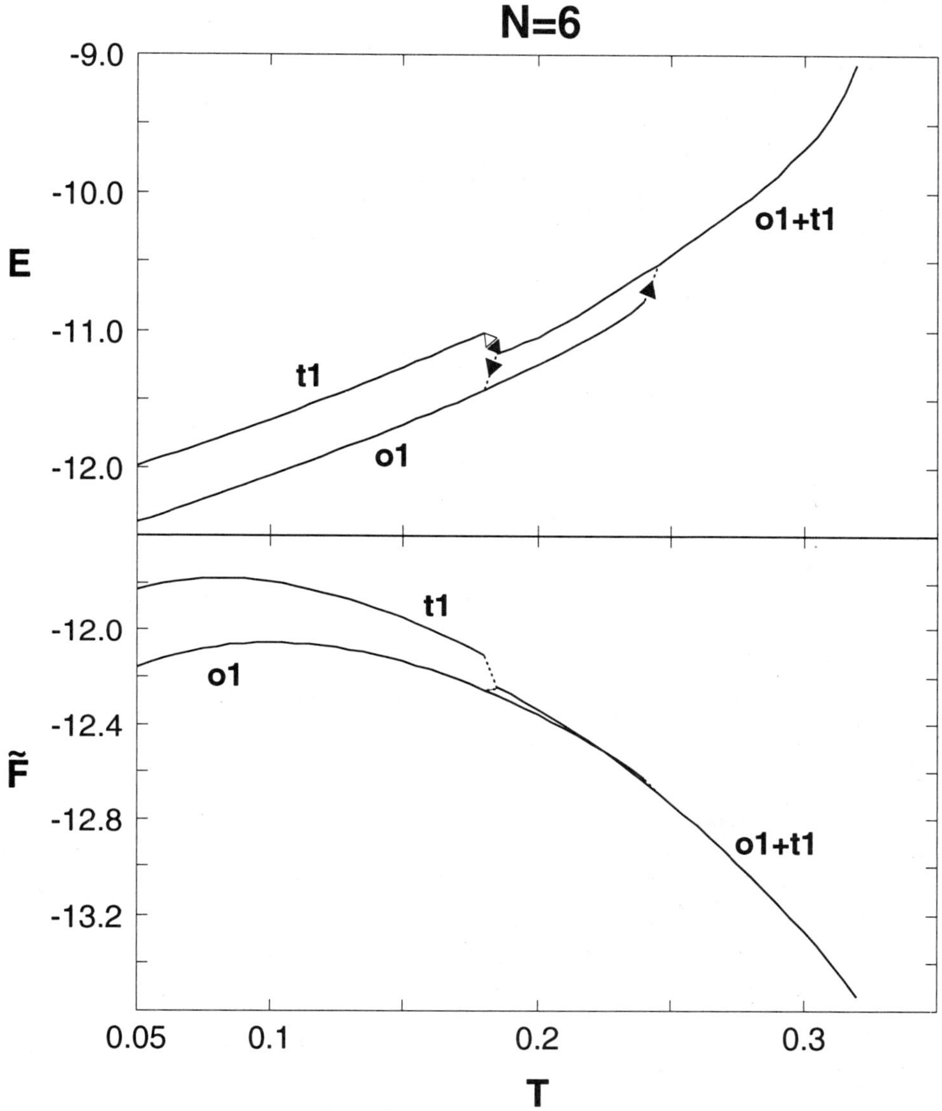


FIGURE 4. Six–atom microcluster trajectories. Approximate mean–energy and free–energy vs temperature trajectory diagrams for the six–atom Lennard–Jones microcluster. The open arrowhead indicting an unpreferred transition from $o1+t1 \rightarrow t1$ is partly obscured. Note that the horizontal scale begins at $T = 0.05$. (Figure reprinted from Ref. [**3**].)

in methionine 5. While $d_{5CG,5HA}$ alone is not sufficient to completely describe the peptide conformation, it and a relatively small number of other distances (in combination with the bond–angle and bond–length constraints) are sufficient to distinguish distinct configuration space regions.[11]

[11]These regions will have irregular shapes in both Cartesian and torsion–angle space.

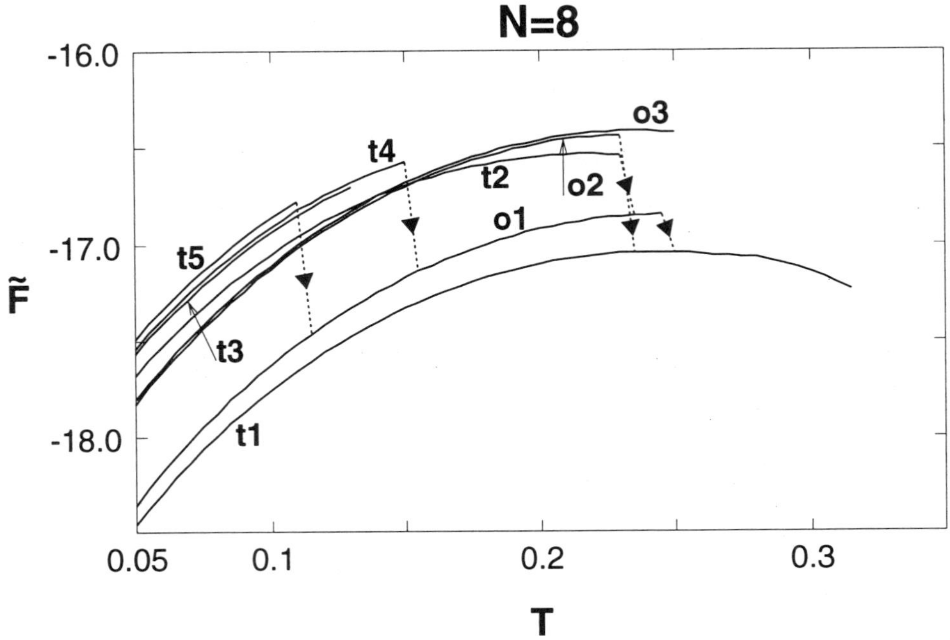


FIGURE 5. Eight-atom microcluster trajectories. Free-energy trajectory diagram for the eight-atom Lennard-Jones microcluster. (Figure reprinted from Ref. [**3**].)

A protein with $N$ atoms has $3N$ constrained Cartesian coordinates, $R \equiv \{\boldsymbol{r}_i\}$, roughly $N$ unconstrained torsion-angle variables, $\Theta \equiv \{\theta_n\}$, and $N(N-1)/2$ distance variables $D \equiv \{d_{ij}\}$, where $d_{ij}$ is the distance between atom $i$ and atom $j$. Clearly, the distance-geometry description is highly redundant and distance variables must vary in a highly correlated manner. Only distances between a small *representative* subset, $\mathcal{R}$, of atom-pairs need be used. For example, a rigid hinge-bending motion will alter the distances between all pairs of atoms selected from opposite arms of the hinge. However, a single inter-arm distance is sufficient to describe the bending angle. The distance between the distal atoms is the most appropriate representative since it is most sensitive to bending, but other representatives could be used. The many distance variables which are evaluated to compute the pairwise interactions in $V(R)$ provide a pool from which representatives needed to characterize packets can be selected. We denote by $\mathcal{R}_\alpha$ the representative subset of atom-pairs whose distances are used to bound metastable state $\alpha$.

Because the distance variables (even within $\mathcal{R}_\alpha$) are not independent, we can not express confining characteristic packets $\phi^0_\alpha$ as multidimensional Gaussians. However, the main role of the $\phi^0_\alpha$ is to define the window functions $w_\alpha$, which are used to calculate metastable state thermodynamic parameters via (9) and (10). The discussion above suggests that the $w_\alpha$ can be expressed as separable functions:

$$w_\alpha \approx \prod_{ij \in \mathcal{R}_\alpha} w^{ij}_\alpha(d_{ij}) \,. \tag{25}$$

Approximation (25) is equivalent to assuming that each fluctuation that splits a metastable state is approximately aligned with a system coordinate. This would

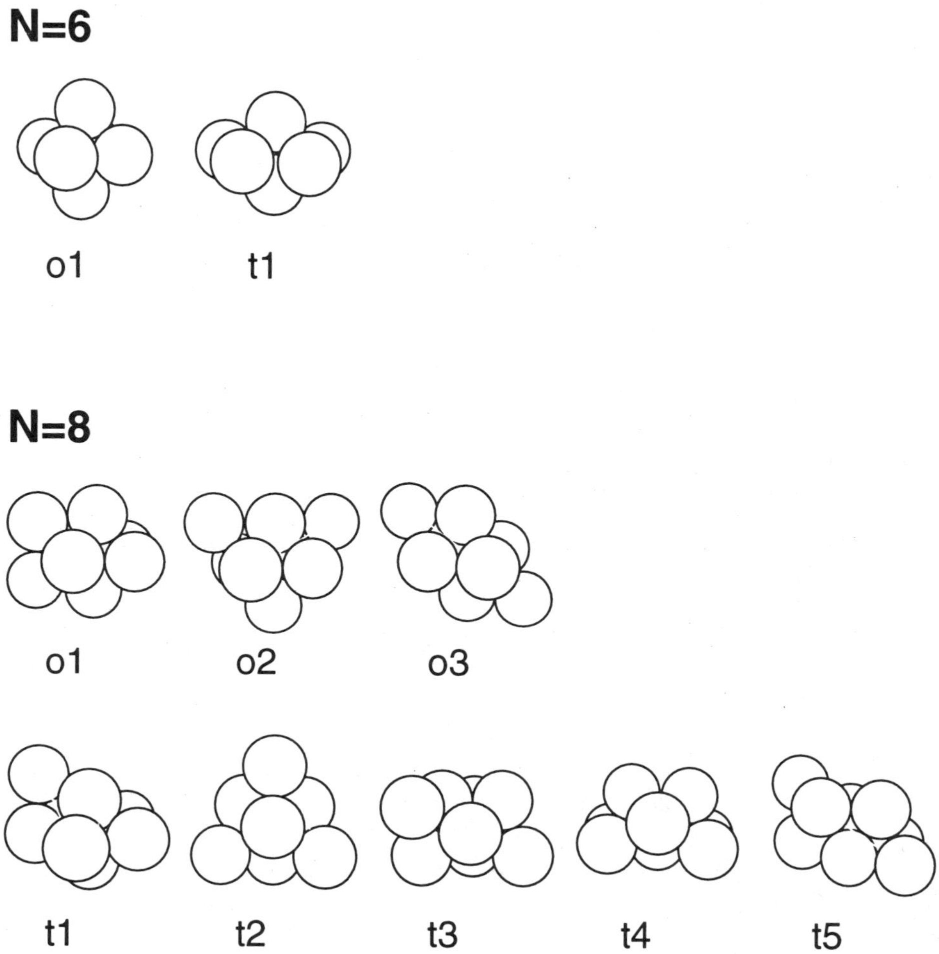


FIGURE 6. Isomers of six– and eight–atom Lennard–Jones micro-clusters at $T = 0$. (Figure adapted from Ref. [**3**].)

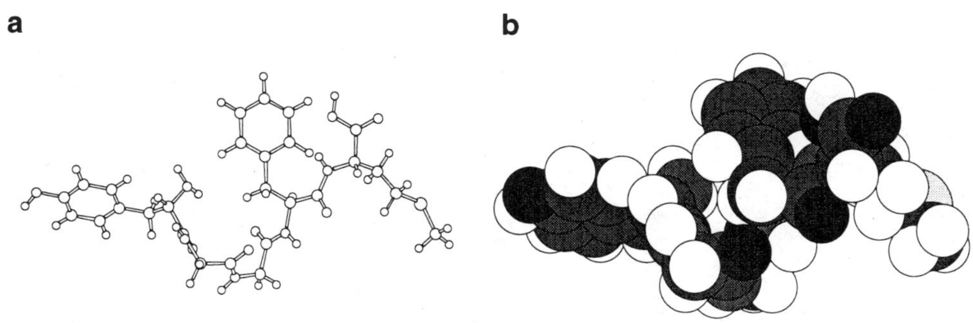


FIGURE 7. The pentapeptide Met–enkephalin: (a) ball and stick representation and (b) space filling representation.

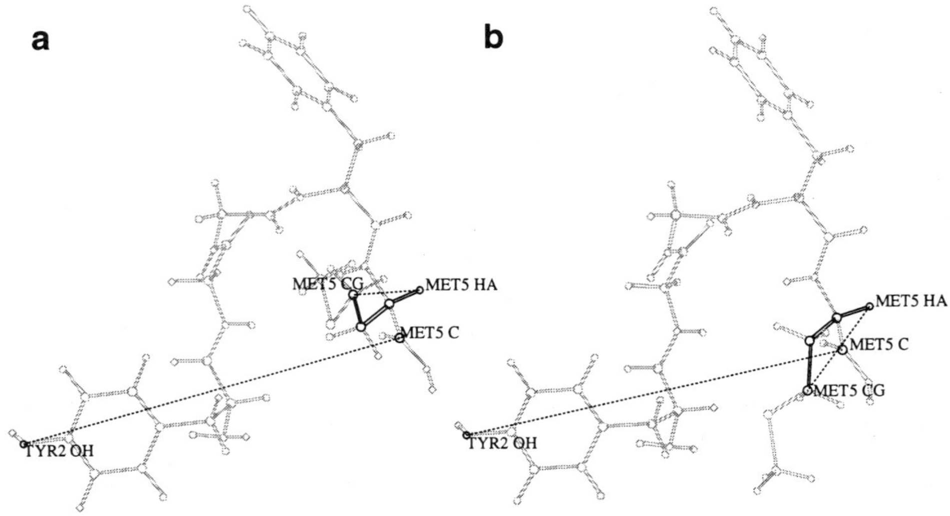


FIGURE 8. Pairwise distances differentiate two conformations of Met–enkephalin. a) Conformation of Met–enkephalin within $\alpha$ state of Fig. 10. b) Conformation of Met–enkephalin within $\beta$ state. The $\alpha$ and $\beta$ states are differentiated by a bifurcation on $d_{5HA,5CG}$.

be a very poor approximation if we were working in the Cartesian or torsion–angle representations, since metastable state boundaries are expected to be complicated functions in these variables. However, the approximation is acceptable in the distance representation where the existence of a very large number of $d_{ij}$ means that boundaries can be defined by superimposing constraints in selected representatives. As mentioned $d_{5CG,5HA}$ is a representative that discriminates between the two metastable states shown in Fig. 8. On the other hand, $d_{2OH,5C}$ is a good representative for discriminating between states separated by large fluctuations extending over the entire backbone. Both can be used together to specify substates. With this factorization, the problem of identifying metastable states reduces to a collection of one–dimensional subproblems in the $d_{ij}$.

Within metastable state $\alpha$, branching in the $d_{ij}$ coordinate is roughly governed by $p_\alpha^{ij}(d_{ij})$, the probability density projected onto $d_{ij}$:[12]

$$p_\alpha^{ij}(d) = \int p_\alpha[R(\Theta)]\,\delta(d - d_{ij}[R(\Theta)])\,\det{}^{\frac{1}{2}}[I(\Theta)]\,d\Theta\,, \tag{26}$$

where

$$I(\Theta)_{np} \equiv \sum_i m_i \frac{\partial R_i}{\partial\theta_n}\frac{\partial R_i}{\partial\theta_p} \tag{27}$$

is the generalized inertia tensor of the system. The $m_i$ are the atomic masses. They result from integrating over the momenta in statistical phase space in the presence of bond–length and bond–angle constraints [**12**]. The generalized inertia tensor and $m_i$ do not affect the $T = 0$ minima and so, from a purely mathematical point of

[12]In principle, a more sophisticated projection, which accounts for the non–linearity of the $\theta \to d$ projection should be used. However, since we only use the $p_\alpha^{ij}$ to identify bifurcations, this simple form is adequate.

view, are not part of the global minimization problem. However, from the broader physical perspective discussed in Sec. 1, they may be important in constructing algorithms that follow the physical folding pathways.

The $p_\alpha^{ij}$ for the representative distances can be tabulated (as histograms) during the course of a simulated annealing run through torsion–angle space. Since the $d_{ij}[R(\Theta)]$ must be calculated in any case to evaluate the potential $V$, the only additional costs are the calculation of the denominator factors and the $\det^{\frac{1}{2}}(I)$. The denominator factors can be calculated by by Monte–Carlo integration or by a recursive analytic procedure [**13**]. Since they are the same for all states $\alpha$ and do not depend on temperature they only need to be tabulated once for an entire annealing run which results in negligible additional cost. We expect that $\det^{\frac{1}{2}}(I)$ will be a relatively slowly varying function that need not be frequently evaluated, so its additional cost will also be small.

We could treat each $d_{ij}$ subproblem using the full characteristic packet formalism described in Sec. 4. However, using Met–enkephalin as a test case (see below), we have found that the $p_\alpha^{ij}$ can be well–approximated by single– or double–Gaussians (see Fig. 9). This is because short–wavelength fluctuations are smeared out by the $\Theta$ integration. In this approximation, packet branching can be determined by a simple algorithm on the parameters characterizing the $p_\alpha^{ij}$. At each temperature step, we apply the branching algorithm to all $p_\alpha^{ij} : ij \in \mathcal{R}$ and state packets are split as necessary. Distance variables involved in the bifurcation become part of the representative set of distances needed to describe the children. That is, if $\beta$ and $\gamma$ are children of $\alpha$ from a split on $d_{kl}$,

$$\mathcal{R}_\beta = \mathcal{R}_\gamma = \mathcal{R}_\alpha \cup (kl)\,, \tag{28}$$

and

$$w_\beta = w_\alpha\, w_\beta^{ij}(d_{ij}) \tag{29a}$$

$$w_\gamma = w_\alpha\, w_\gamma^{ij}(d_{ij})\,. \tag{29b}$$

Average thermodynamic properties of the children can be calculated from expressions like (10) and trajectory diagrams can be plotted as in the previous examples.

**6.2. Pilot studies with Met–enkephalin.** Met–enkephalin, $H_2N$–Tyr–Gly–Gly–Phe–Met–COOH, is a pentapeptide containing 75 atoms whose conformation is described by 24 torsion–angles. The peptide is shown in Fig. 7. Assuming peptide bonds to be rigid reduces the number of angles to 19. It is a brain peptide that is a natural ligand for opiate receptor sites [**14**]. The properties of real Met–enkephalin will be strongly influenced by the presence of solvent, but the simplified *in vacuo* model has been used extensively as a theoretical model by Scheraga's group who have identified a putative global minimum at -12.396 kcal/mol [**15**], and it has been used as a relatively small but non–trivial test case for a number of structure prediction methods [**16, 17, 18**].

The peptide energy can be modeled using the ECEPP/3 potential [**19, 15**] which is designed for use with torsion–angle variables with fixed bond–lengths and bond–angles. The energy $V$ is

$$V = \sum_{i<j}^{N_{atoms}} \frac{A_{ij}}{d_{ij}^{12}} + \frac{B_{ij}}{d_{ij}^{10}} + \frac{C_{ij}}{d_{ij}^{6}} + \frac{D_{ij}}{d_{ij}} + \sum_{n}^{N_\theta} V_i^{tor}(\theta_n)\,. \tag{30}$$

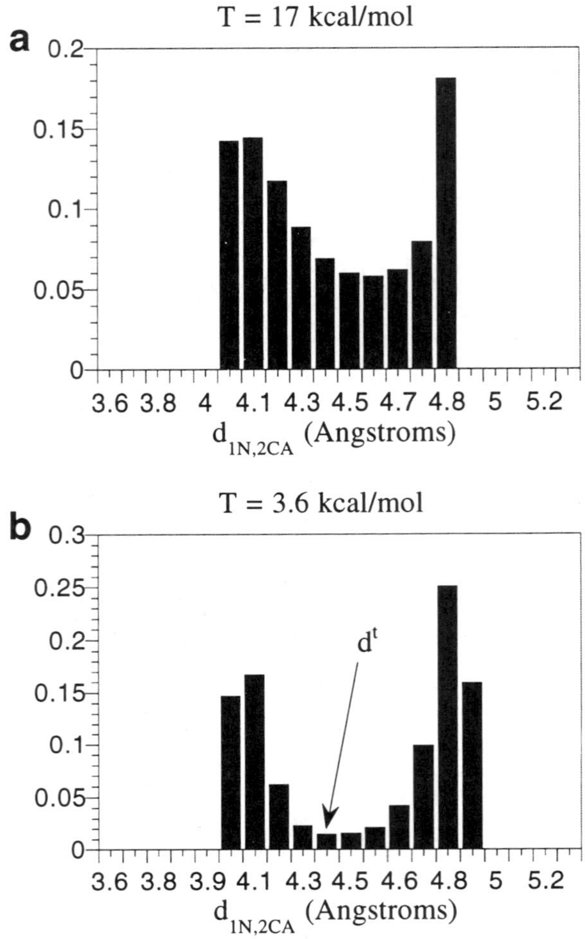


FIGURE 9. Projected probability distribution $p^{1N,2CA}$ at high temperature (a) and at low temperature (b). At $T = 3.6$ kcal/mol $d_{1N,2CA}$ satisfies the criteria for bifurcation into two metastable states: one with $d_{1N,2CA} < d^t$ and one with $d_{1N,2CA} > d^t$.

This includes Lennard–Jones $d_{ij}^{-12}$ and $d_{ij}^{-6}$ terms characterizing the van der Waals interactions, $d_{ij}^{-12}$ and $d_{ij}^{-10}$ terms characterizing hydrogen bond interactions, and $d_{ij}^{-1}$ terms characterizing electrostatic interactions. Additional $\theta$–dependent terms characterize torsion–angle dependent forces.

The putative global minimum of Met–enkephalin was first determined by the hybrid local minimization/Monte Carlo method of Li and Scheraga [**16**]. This uses the Metropolis algorithm [**20**] to sample the discrete lattice of local minima of the potential energy. The same minimum can be found directly using conventional continuous simulated annealing (our unpublished results). Met–enkephalin has also been used as a test case for the Diffusion Equation Method [**18, 21, 22**] which found a minimum of -8.06 kcal/mol.

We used the distance–geometry packet algorithm described in Sec. 6.1 to calculate the trajectory diagram of Met–enkephalin *in vacuo*. Since the potential probably has roughly $10^9$ local minima[13], a complete calculation is impossible. Instead, we began with the unique high–temperature metastable state ($T = 20$ kcal/mol) and and pruned the trajectory diagram by truncating the highest free–energy trajectories as the system was cooled. Our goal was to elucidate the general features of the trajectory diagram for a realistic (if small) peptide and, in particular, to determine its scaling properties. Thus, no attempt was made to reduce computational expense.

6.2.1. *Methods.* Since the peptide is small, all unconstrained interatomic distances were included in the representative set $\mathcal{R}$. The integrals in (26) were determined using the Vanderbilt and Louie simulated annealing method [**23**].[14] To determine when sampling was sufficient within a given state, the average and variance of each $d_{ij}$ was calculated for two separate sequences of $N_s$ samples. Convergence was assumed if the difference between the two averages was less than 20% of the combined standard deviation. Otherwise the two sequences were combined into a sequence of length $2N_s$ and another sequence of length $2N_s$ was also generated. The process was repeated, doubling the sequence length each time, until all the $d_{ij}$ had converged. This complete process typically required 64,000 to 1,024,000 evaluations of $V$ per temperature step per metastable state. Typical $p^{ij}_{\alpha}$ histograms are shown in Fig. 9.

Because the branching subproblems were only one–dimensional, we could replace the full packet equation machinery with a simplified algorithm based on our experience with one–dimensional packet branching (e.g., see Sec. 5 in Ref. [**24**]). Each state $\alpha$ was tested for branching on each $(ij)$ pair by checking for a binodal $p^{ij}_{\alpha}$ distribution. We used the criteria: 1) each node must contain at least 5% of the total probability, and 2) the ratio of $p^{ij}_{\alpha}(d^t_{ij})$, where $d^t_{ij}$ is the minimum point in the transition region, to the smaller of the two nodal peak probabilities must be less than 0.1. (A special treatment that allowed for the non–linear nature of the transformation between distance and angle variables was used for distances between atoms separated by only one variable torsion–angle.) This ratio condition identifies the lower of the two escape temperatures of the children. The other escape temperature and the parental capture temperature were not determined. Thus, hysteresis was ignored. Because of the inherent redundancy of the distance–geometry description, bifurcations appeared simultaneously in multiple interpair distances. The $d_{ij}$ having the largest fluctuation was added to the representative sets of the children according to Eq. (28).

Because the $p^{ij}_{\alpha}$ were smooth, it was possible to define simple hard–walled window functions using the $d^t_{ij}$ instead of using the characteristic packets described in Sec. 3. Parent states were separated into *near* (<) and *far* (>) children $\ell$ and $r$ according to (29) with

$$w^{ij}_{<}(d) = \theta(d - d^t_{ij}) \tag{31a}$$

$$w^{ij}_{>}(d) = \theta(d^t_{ij} - d) \; . \tag{31b}$$

Here, $\theta$ represents the step–function, not an angle. The child distributions $p^{ij}_{<}$ and $p^{ij}_{>}$ were then evaluated without changing temperature, and the branching test

[13] This rough estimate assumes three potential minima per torsion–angle.

[14] The $\det^{\frac{1}{2}}(I)$ term was assumed to be constant in these pilot studies.

criteria were again applied. This process was repeated until no more bifurcations were detected. Temperature was then lowered by steps of $\approx 5\%$.

6.2.2. *Results.* The resultant free–energy and energy vs $T$ trajectory diagrams for 3.5 kcal/mol $\leq T \leq$ 20 kcal/mol are shown in Fig. 10. At $T = 20$ kcal/mol a single metastable state was detected. Three bifurcations were detected at the lower transition temperature of 15 kcal/mol. The first occurred on the distance between the backbone oxygen of phenylalanine 4 (PHE4 O) and the $\alpha$–carbon hydrogen of methionine 5 (MET5 HA). Because peptide bonds were constrained to be rigid, these two atoms are separated by only one torsion angle. Similarly the second and third bifurcations also occurred on distances between "1–4" atom–pairs [on the (GLY3 CA, PHE4 HA) and (MET5 HA, MET5 CG) distances] that are separated by only one variable torsion angle. All of the high–temperature bifurcations (down to $T \approx 5$ kcal/mol) were of this type.

Child states of very high free–energy appeared in subsequent branches. For example, the free–energy of the child resulting from the (PHE4 HN, PHE 4HA) branch at $T \approx 13$ kcal/mol was about 30 kcal/mol higher than that of its sibling (see Fig. 10a). The diagrams were pruned by tracing trajectories of such high energy states only for small temperature intervals. The energy and free–energy ordering of states differed. For example, state $\gamma$ had higher free–energy, but lower energy, than state $\beta$. This implies that $\beta$ has greater entropy than $\gamma$, presumably because it occupies a larger region of configuration space. The narrowing free–energy gap between the two states as temperature decreases reflects the reduction in the entropic contribution to the free–energy as predicted by (4).

The number of trajectories rapidly increased as the system was cooled further. To survey the overall behavior, the trajectory tracing algorithm was continued down to $T = 1.9$ kcal/mol using a less strict criteria for convergence of the $d_{ij}$. The results of this low–resolution scan are shown in Fig. 11. The explosion of states is expected to continue down to $T = 0$, where each of the $\mathcal{O}(10^9)$ local minima of $V$ will confine a state.

6.2.3. *Using the Met–enkephalin trajectory diagram to analyze global minimization algorithms.* The high–temperature portion of the Met–enkephalin trajectory diagram can be used to analyze the behavior of global minimization algorithms. The diagram identifies variables which become "frozen–out" at high temperature and can not be easily altered during subsequent lower temperature search. For Met–enkephalin, the fate of an annealing algorithm may be determined at the relatively high transition temperature (15 kcal/mol) where branching occurs on $d_{5HA,5CG}$ to generate child states $\alpha$ and $\gamma$ (see Fig. 10). Metastable state $\gamma$ contains the global minimizer (i.e., the global minimizer lies within $w_\gamma$) but has higher free–energy at the transition point than state $\alpha$. Thus the $\gamma$ trajectory is an ancestor of the trajectory leading to the global minimizer and must, as temperature is lowered, cross below the descendents of the $\alpha$ trajectory. This implies that Met–enkephalin does *not* possess strong scaling. We do not yet know if it possesses weak scaling. The fact that the trajectory containing the global minimizer remains one of the low–lying trajectories as temperature is lowered (see Fig. 11) suggests, that Met–enkephalin displays weak–scaling.[15] This implies that it may not be too difficult to find its global minimum.

---

[15] Although we will have to trace the trajectory to $T = 0$ to be certain of this.

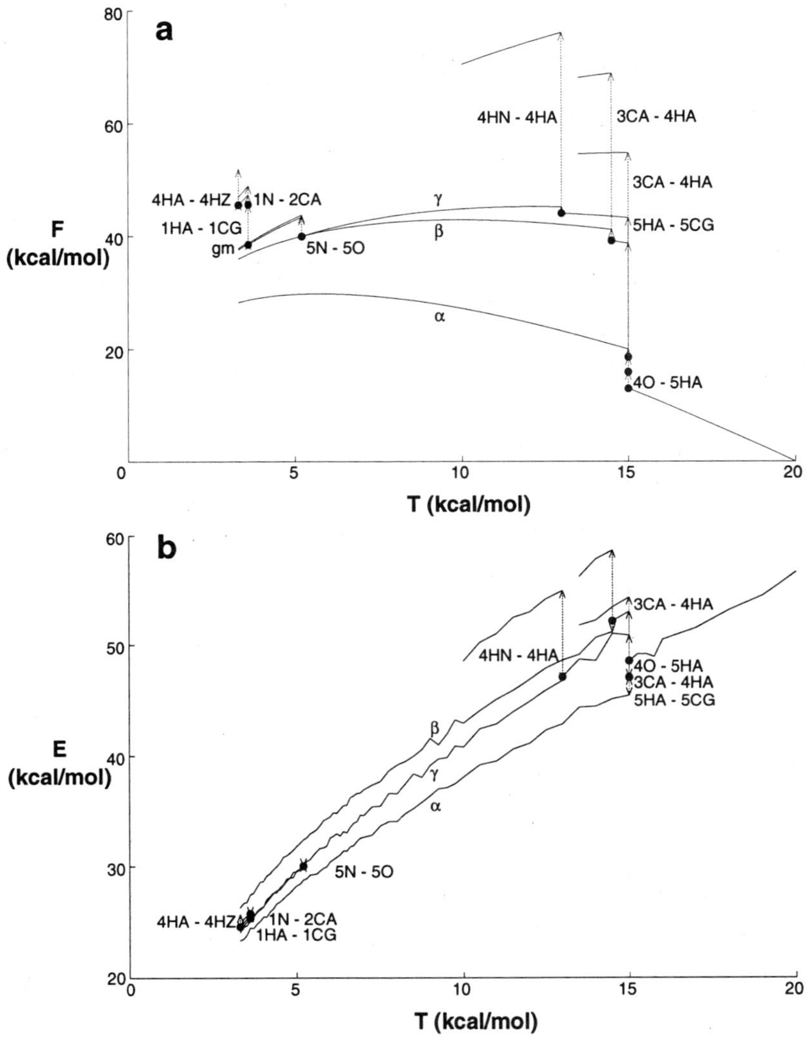


FIGURE 10. Free–energy (a) and energy (b) vs $T$ trajectory diagrams for Met–enkephalin. Bifurcations are indicated by large black dots and are labeled by the pair-wise distance that defines the split. The children of a bifurcated metastable state are connected to their parent by dashed lines. The free–energies of both children are higher than those of the parents at the transition points because conservation of probability implies that $e^{-\beta F_{child1}} + e^{-\beta F_{child2}} = e^{-\beta F_{parent}}$. The mean energy of the parent necessarily lies between the mean energies of the children. Several trajectories with high free–energy were abandoned to concentrate computational resources on the much more probable trajectories with low free energy. Metastable states on the $\gamma$ trajectory contain the global minimum (gm) conformation.

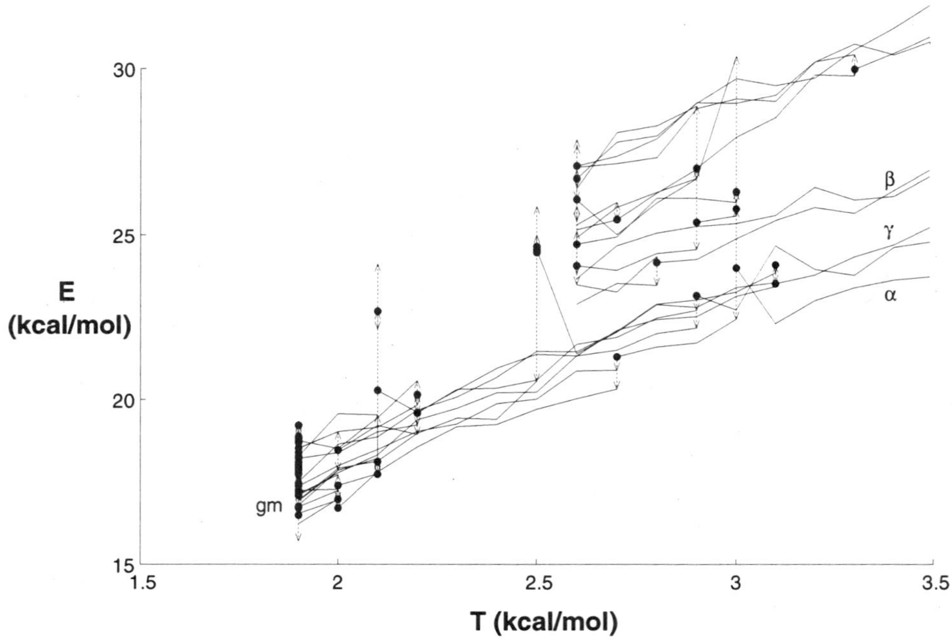


FIGURE 11. Met–enkephalin energy vs $T$ trajectory diagram at lower temperatures. As T is lowered the number of bifurcations, indicated by large black dots, increases rapidly. The trajectories labeled $\alpha$, $\beta$ and $\gamma$ correspond to trajectories with the same labels in Fig. 10. The jagged appearance of the trajectories is an artifact of the low resolution computation used in this temperature range.

As discussed in Sec. 5.2, we expect that annealing algorithms that focus attention on connected regions of configuration space will tend to follow the lowest–energy path at most branch points. We used the Vanderbilt and Louie method of simulated annealing to test this hypothesis. In this method [**23**], the transition function, which controls the sizes and directions of the jumps between sample points, is determined by the average local curvature of the energy landscape (calculated using a quasi–harmonic approximation). The method improves local efficiency but tends to remain confined in quasi–harmonic regions that roughly correspond to metastable states. Starting from $T = 20$ kcal/mol, we found that as predicted roughly 90% of the runs traveled down trajectory $\alpha$. This is somewhat higher than the ratio that would be predicted from thermodynamic partition,[16] implying that non–equilibrium factors play a part in the selection.

## 7. Summary

Trajectory analysis of metastable states provides a computational method for analyzing the potential folding pathways embedded in an energy landscape. States can be hierarchically characterized and the global scaling properties of the landscape can be determined. The analysis in terms of metastable states provides a finite–temperature dissection of conformation space that generalizes the zero–temperature

[16]The free–energy difference between $\alpha$ and $\beta + \gamma$ at $T = 15$ kcal/mol is about 16 kcal/mol, implying a thermodynamic partition of about $(1 + e^{-16/15})^{-1} = 0.74$.

dissection provided by the steepest–descent catchment regions of the local minima [**25**]. It effectively groups local minima that are separated by energy barriers that are small on the scale set by the current temperature. These linkages are embedded in the trajectory diagrams and provide a measure of the "distance" between local minima or between finite–temperature metastable states, even in complicated high–dimensionality problems. One can now relate local minima by determining the temperature intervals that they must traverse along trajectories to reach a common ancestor.

The fact that Met–enkephalin *in vacuo* does not possess strong scaling but probably possesses weak scaling emphasizes the need for methods that trace at least some non–preferred pathways. Additional studies are needed to understand the degree of multiple path tracing that is required. In particular, it will be important to determine the extent to which different algorithms are guided down the lowest or low–lying trajectories. Further investigation will be required to determine the relationship between the global minimum, the "kinetic minimum" found by selecting the lowest free–energy states at trajectory branch–points, and physical folded states. Solvent effects will have to be included. Such extensions will require the use of computational methods that are much more efficient than those used here. For example, we anticipate that pre–averaged "effective potentials" [**24**] will have to be employed.

Although we have analyzed the global minimization problem from the physically–motivated standpoint of diffusive kinetics and non–equilibrium thermodynamics, we believe it likely that the concepts described here will also be relevant to other potential–averaging continuation methods.

The multi-packet description is an important component of our approach. It permits different dominant spatial–scales to be recognized in different regions of configuration space. This is probably the most fundamental difference between the Packet Annealing Method and related methods such as the Diffusion Equation Method [**21, 22**] and the Gaussian Density Approximation [**26, 27, 28**] that focus attention on single coarse–grained states. The ability of the Packet Annealing Method to discriminate between different metastable state results from analyzing the Gibbs/Boltzmann probability density $p_B$ instead of the potential $V$. The price paid is that the Gaussian integrals over $p_B$ that govern annealing trajectories, for example in Eqs. (22) and (26), are more complicated than the Gaussian integrals over $V$ that appear in other methods.[17]

In our view, efficient pathway tracking annealing methods must solve two fundamental problems: 1) efficiently tracking along individual pathways, and 2) intelligently selecting pathways for exploration. We have previously [**24**] described potential averaging methods for solving the first problem. In this paper we have focused on the second problem. The next step will be to develop a minimization algorithm that combines all these elements.

[17]The integrals in the packet annealing method are also more complicated because we permit $\Lambda^0$ to be an arbitrary symmetric tensor. The use of a single parameter $t$ to describe averaging scale in the diffusion equation method is equivalent to requiring that $\Lambda^0$ be a multiple of the identity matrix. The averaging parameters in the Gaussian density approximation correspond to taking $\Lambda^0$ to be a diagonal matrix.

## Acknowledgments

We thank Harold Scheraga and his group for ECEPP/3 and Alex Ulitsky for many helpful conversations.

## References

[1] K. S. Dill, Curr. Opin. Struct. Biol. **3**, 99 (1993).
[2] J. D. Bryngelson, J. N. Onuchic, N. D. Socci, and P. G. Wolynes, (in press).
[3] M. Oresič and D. Shalloway, J. Chem. Phys. **101**, 9844 (1994).
[4] D. Shalloway, In preparation .
[5] L. D. Landau and E. M. Lifshitz, *Statistical Physics*, Addison–Wesley, 1969.
[6] S. Chandrasekhar, Rev. Mod. Phys. **15**, 1 (1943).
[7] H. Tomita, A. Ito, and H. Kidachi, Prog. Theor. Phys. **56**, 786 (1976).
[8] N. G. van Kampen, J. Stat. Phys. **17**, 71 (1977).
[9] S. Sawada and S. Sugano, Z. Phys. D **14**, 247 (1989).
[10] J. Kaelberer, J. D. Etters, and J. C. Raich, Chem. Phys. Lett **41**, 580 (1976).
[11] R. D. Etters, R. Danilowicz, and J. B. Kaelberer, J. Phys. Chem. **67**, 4125 (1977).
[12] N. Gō and H. A. Scheraga, J. Phys. Chem. **51**, 4751 (1969).
[13] J. Kostrowicki, Private communication.
[14] J. Hughes et al., Nature **258**, 577 (1975).
[15] G. Nemethy et al., J. Chem. Phys. **96**, 6472 (1992).
[16] Z. Li and H. A. Scheraga, Proc. Natl. Acad. Sci. USA **84** (1987).
[17] U. Hansmann and Y. Okamoto, J. Comp. Chem. **14**, 1333 (1993).
[18] J. Kostrowicki and H. Scheraga, J. Phys. Chem. **96**, 7442 (1992).
[19] M. J. Sippl, G. Nemethy, and H. A. Scheraga, J. Comp. Chem. **88**, 6231 (1984).
[20] N. Metropolis, A. W. Metropolis, M. Rosenbluth, A. H. Teller, and E. Teller, J. Chem. Phys. **21**, 1087 (1953).
[21] L. Piela, J. Kostrowicki, and H. Scheraga, J. Phys. Chem. **93**, 3339 (1989).
[22] J. Kostrowicki, L. Piela, B. Cherayil, and H. Scheraga, J. Phys. Chem. **96**, 4113 (1991).
[23] D. Vanderbilt and S. G. Louie, J. Comp. Phys. **56**, 259 (1984).
[24] D. Shalloway, J. Glob. Optim. **2** (1992).
[25] F. Stillinger and T. Weber, Phys. Rev. A **25**, 978 (1982).
[26] P. Amara, D. Hsu, and J. E. Straub, J. Phys. Chem. **97**, 6715 (1993).
[27] J. Ma and J. E. Straub, J. Chem. Phys. **101**, 533 (1994).
[28] J. E. Straub, J. Ma, and P. Amara, (submitted) .

*E-mail address*: bwc1@cornell.edu
*E-mail address*: matej@lancelot.bio.cornell.edu
*E-mail address*: dis2@cornell.edu

BIOPHYSICS PROGRAM, SECTION OF BIOCHEMISTRY, MOLECULAR AND CELL BIOLOGY, CORNELL UNIVERSITY, ITHACA, NY 14853.

DIMACS Series in Discrete Mathematics
and Theoretical Computer Science
Volume **23**, 1996

# A Multispace Search Algorithm for Molecular Energy Minimization

JUN GU AND BIN DU

March 12, 1995

ABSTRACT. Molecular energy minimization is one of the most challenging, unsolved problems in molecular biophysics. Due to numerous local minima in the search space, a traditional optimization method has a tendency to get stuck at some local minimum points. In this paper, for Lennard-Jones clusters, we give a multispace search algorithm for molecular energy minimization. Multispace search interplays structural operations in conjunction with the existing optimization methods. Structural operations dynamically construct a sequence of intermediate lattice structures by changing the original lattice structure. Each intermediate lattice structure is then optimized by the traditional optimization methods. Structural lattice operations disturb the environment of forming local minima, which makes multispace search a very natural approach to molecular energy minimization. We compare multispace search approach with the traditional optimization techniques for molecular energy minimization problem.

## 1. Introduction

Molecular energy minimization is one of the most challenging, unsolved problems in molecular biophysics [**1, 21**]. In recent years, this problem has attracted much attention from mathematicians and computer scientists because the problem is itself a benchmark for the existing optimization methods. One of the fundamental problems in molecular energy minimization is to minimize the Lennard-Jones (LJ) potential function [**3**], which can be stated as follows:

---

1991 *Mathematics Subject Classification.* Primary 03G10, 06B35, 90B40, 90C27, 92B05; Secondary 49M07, 51D20, 65K05, 70F10, 92C05.

*Key words and phrases.* Molecular energy minimization, Lennard-Jones cluster, combinatorial optimization, continuous optimization, multispace search, structural operations.

This work was supported in part by NSERC Strategic Grant MEF0045793 and NSERC Research Grant OGP0046423 and is presently supported in part by NSERC Strategic Grant STR0167029 and the Federal Micronet Research Grant.

Given a cluster of $n$ atoms interacting with two-body central forces, our goal is to find their configuration possessing the globally minimum total potential energy.

The total Lennard-Jones potential energy function of $n$ atoms $(a_1, a_2, ..., a_n)$ can be defined as follows:

$$E_n = \sum_{1 \leq i \leq j \leq n} V_{LJ}(\| a_i - a_j \|),$$

where $\| a_i - a_j \|$ is the Euclidean norm distance and $V_{LJ}(r)$ is the Lennard-Jones potential function of two atoms. It is defined as

$$V_{LJ} = \frac{1}{r^{12}} - \frac{2}{r^6}.$$

The Lennard-Jones function is widely used in molecular energy minimization. It provides a good model for the molecular conformational energy that governs the behavior of simple physical systems — molecular clusters of chemically inert atoms (e.g., argon) [**2**]. The difficulty of the LJ cluster problem arises from the fact that it is a global optimization problem with an exponential number of local minima. M.R. Hoare [**13**] claimed that the number of local minimizers of an $n$-atom cluster grows as fast as the function $O(e^{n^2})$. While L.T. Wille and J. Vennik [**25**] showed that the complexity of determining the global minimum energy of a cluster of particles interacting via two-body forces belongs to the class of NP-hard problems.

Much research work has been conducted to solve the LJ cluster energy minimization problem. In this research, for Lennard-Jones clusters, we give a multispace search algorithm for molecular energy minimization. The algorithm dynamically interplays structural operations in conjunction with traditional optimization algorithms for molecular energy minimization. Experimental results indicate that structural operations are effective to handle the pathological behavior of local minima.

The rest of paper is organized as follows: In the next section, we will briefly overview some existing molecular energy minimization methods for Lennard-Jones clusters. Section 3 describes the basic ideas of multispace search for optimization problem solving. A multispace search algorithm for molecular energy minimization is introduced in Section 4. The experimental results of the multispace search algorithm for Lennard-Jones energy minimization are shown in Section 5. We will compare our algorithm with other algorithms for Lennard-Jones energy minimization problem.

## 2. Previous Work

Many optimization methods have been developed for molecular energy minimization. These include lattice based search [**19, 27**], simulated annealing and genetic algorithms [**24**], diffusion equation method [**16**], effective energy methods [**22**], and packet annealing [**23**]. For a detailed survey of the optimization algorithms for molecular energy minimization, we refer readers to an excellent survey paper [**20**].

Lattice based search has achieved great success in the minimization of Lennard-Jones energy function. J.A. Northby developed an efficient lattice based search procedure [**19**]. In his approach, he proposed and used IC and FC lattices as the base lattice structures for constructing the initial Lennard-Jones clusters. The initial lattice configurations are allowed to relax freely on the Lennard-Jones pair potential to the adjacent energy minimum. Northby reported solution for Lennard-Jones clusters of size $13 \leq n \leq 150$. These were believed to be the most tightly bound configuration [**19**].

Later, Xue reduced the complexity of the Northby algorithm from $O(n^{\frac{5}{3}})$ to $O(n^{\frac{2}{3}})$ by introducing simple data structures [**26**]. He gave the detailed construction methods for the IC and FC lattice structures and improved the Northby's algorithm by relaxing every lattice local minimizer. Although his algorithm requires more computing time, Xue found lower energy configurations for n = 65, 66, 75, 76, 77, and 134.

The simulated annealing is a general optimization approach proposed by Kirkpatrick et al. [**15**]. Wille used simulated annealing and found a new minimum solution for a 24-atom Lennard-Jones cluster [**24**]. Further, Xue proposed a two-level simulated annealing algorithm that worked well for large Lennard-Jones clusters [**27**]. Coleman, Shalloway, and Wu used simulated annealing and found a better structure for the 72-atom Lennard-Jones cluster [**2**].

Genetic algorithms are a class of search algorithms based on the mechanics of natural selection and natural genetics. Judson et al. implemented both simulated annealing and genetic algorithm for Lennard-Jones energy minimization [**14**]. Experimental results showed that simulated annealing and genetic algorithms have complimentary strengths which implies that a hybrid GA-SA method would be more efficient than either one alone.

In addition to the above approaches, special techniques based on spatial averaging [**16, 22, 23**] and other optimization methods [**18**] have been proposed.

Recently *multispace search* approach has been developed for general optimization problem solving [**4, 8, 9**]. We have applied multispace search to the Lennard-Jones energy minimization problem. The algorithm interplays structural operations in conjunction with the existing optimization algorithms. During the optimization process, structural operations dynamically construct a sequence of intermediate lattice structures by changing the original lattice structure. Each intermediate lattice structure is optimized, producing an improved intermediate solution. Eventually the intermediate lattice structure approaches the given problem structure. Structural lattice operations disturb the environment of forming local minima, which makes multispace search a very natural approach to handle the molecular energy minimization problem.

In the next section, we will briefly describe the basic idea of multispace search for general optimization problem solving.

## 3. Optimization by Multispace Search

In this section, we briefly introduce the basic ideas of multispace search.

**3.1. Multispace Search.** Many search and optimization methods have been developed in combinatorial optimization, operations research, artificial intelligence, neural network, genetic algorithms, and evolution programming. An optimization algorithm seeks a value assignment to variables such that all the constraints are satisfied and the performance objective is optimized. The algorithm operates by changing values to the variables in the value space. Because value changing does not provide sufficient information from the problem structure and from the search space, it is difficult for a value search algorithm to handle the pathological phenomena in combinatorial optimization.

Multispace search is a new optimization approach developed in recent years [**4, 8, 9**]. The idea of multispace search was derived from principles of nonequilibrium thermodynamic evolution that structural changes are fundamental than quantitative changes and evolution depends on the growth of new structure in biological system rather than just information transmission. A search process resembles the evolution process and the structural operations are important to improve the performance of traditional value search methods [**4, 8, 9**].

In multispace search, any active component related to the given problem structure can be manipulated, and thus, be formulated as an independent search space. For the given search problem, for its variables, values, constraints, objectives, and key parameters, we define variable space, value space (i.e., the traditional search space), constraint space, objective space, parameter space, and other search spaces, respectively. The totality of all the search spaces constitutes a *multispace*.

The basic idea of multispace search is simple. Instead of being restricted in the value space, the multispace is taken as the search space. In the multispace, components other than value are manipulated and are optimized as well. During the search, a multispace search algorithm not only alters values in the value space; as shown in Figure 1, it also scrambles across the variable space and other active spaces, dynamically changes the problem structure in terms of variables, parameters, and other components, and incrementally *constructs* the final solution to the search problem. Only at the *last* moment of the search, the reconstructed problem structure approaches the original problem structure, and thus the final value assignment represents the solution to the given search problem. Because structural operations (related to variable, constraint, objective, and parameter changes) introduce new structural information, a multispace search method is more effective to handle difficult optimization problems.

Multispace search algorithm combines traditional optimization algorithms

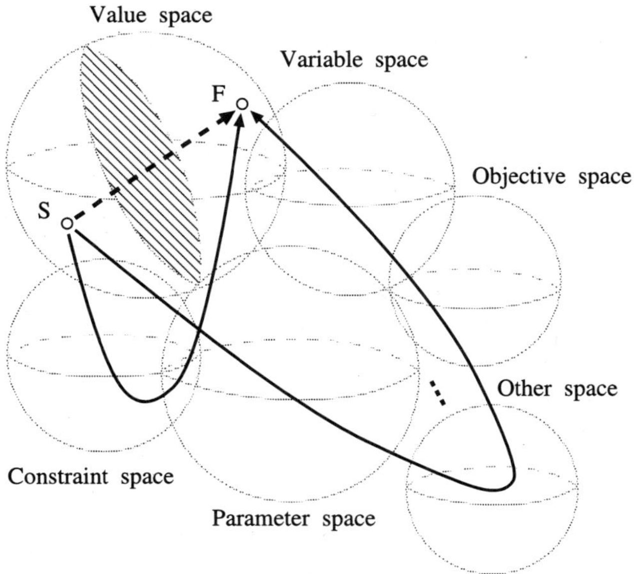


FIGURE 1. In the value space, a traditional search process (dashed line) cannot pass a "wall" of high cost search states (hatched region). It fails to reach the final solution state, $F$. A multispace search process (solid lines) scrambles across different search spaces. It could bypass this "wall" through the other search spaces.

```
procedure Multispace_Search (MS) ()
begin
    /* initialization */
    input the given problem instance;
    design initial problem instance and scrambling schedule;
    solve the initial problem instance;

    /* scrambling between the active search spaces */
    while (schedule has events) do
    begin
        enter an active search space according to the event;
        construct a new intermediate problem instance;
        /* traditional value search in the value space */
        begin
            enter the value search space;
            solve the intermediate problem instance with value search;
            transmit the intermediate solution to the next search phase;
        end;
        if the intermediate solution is not the final solution
            then optimize it in the next search phase;
        if the problem instance becomes the original problem instance
            then return the final solution;
    end;
end;
```

FIGURE 2. **MS:** A general multispace search algorithm.

with structural multispace operations. A typical multispace search algorithm is shown in Figure 2 [**4, 8, 9**]. It consists of the following three stages:

### Initialization

At the beginning of search, for the given problem instance, an initial problem instance — which is usually different from the given problem instance — is designed. This initial problem instance will approach the given problem instance through a number of the altered, intermediate problem instances in a finite number of search phases.

A *scrambling schedule*, or *scrambling scheme*, specifies a sequence of events as *when* to make *what* structural changes to the intermediate problem instances in *which* active search space. During each search phase, the intermediate problem instance will be solved by a value search algorithm and the solution obtained will be passed to the next search phase as determined by the scrambling schedule. The last event in the schedule returns the intermediate problem instance to the original, given problem instance.

### Search

This major search step involves two fundamental operations: a traditional value search and the structural reconfiguration of the intermediate problem instance during each individual search phase. According to the active event in the schedule, the search process enters a specified search space and performs structural operations to the intermediate problem structures, followed by a traditional value search that optimizes the constructed intermediate problem instance. The resulting intermediate solution is then used as the initial problem instance to the next phase of multispace search.

The major structural operations in multispace search include multispace scrambling [**4, 5, 8**], simulated evolution [**4, 5, 10, 11**], extradimension transition (e.g., air bridge, real dimension, and extra dimension) [**4, 6, 7, 12**], search space smoothing [**4, 5, 11**], preprocessors (e.g., compression, decomposition, reorganization, and semi-processing) [**4, 12**], and perturbations (e.g., jumping, tunneling, climbing, annealing, and indexing) [**4, 6, 7, 15**].

### Final state

At the *last* moment of multispace search, a dynamically constructed, intermediate problem structure must be replaced by the original problem structure. This results in the *final* search state. At this state, any value assignment to the variables gives the *final solution* to the given search problem.

**3.2. Initial State.** An important initial state for multispace search is the *uniform* state [**4, 5, 9**].

From nonequilibrium thermodynamic evolution, it is evident that symmetrical and asymmetrical interactions are two antagonistic processes operating simultaneously over the course of succession. At the beginning, homogeneity among species dominates the evolution. This suggests that a natural search process should initially start with a *relatively homogeneous state with almost symmetrical configurations.*

A typical homogeneous state is a *uniform state* where the initial structure of the search problem is completely symmetrical. In a uniform search state, there is no any local minima in the search space. Any solution point in the search space gives the same performance figure. For graph based optimization problems, two typical uniform initial states are a *null graph* and a *complete graph* [**5, 10**]. For path related optimization problems, three typical uniform initial states include the *average distances*, the *shortest distances*, and the *longest distances* [**5, 11**]. Multispace search starting from the uniform initial state may improve the performance of the traditional optimization methods.

In the next section, we will give a multispace search algorithm for the molecular energy minimization problem.

## 4. A Multispace Search Algorithm for Molecular Energy Minimization

In this section, we give a multispace search algorithm for an $n$-atom LJ cluster molecular energy minimization problem.

In the following discussion, let:

- $e_i$ be the energy of the $i$th atom. $e_i$ is the sum of LJ potential energy between the $i$th atom and the other atoms:

$$e_i = \sum_{1 \le j \le n, j \ne i} V_{LJ}(\| a_i - a_j \|).$$

- $E$ be the total energy of the $n$-atom LJ cluster:

$$E = \frac{1}{2} \sum_{i=1}^{n} e_i.$$

- $C$ be the *core* of the largest IC lattice which contains fewer than $n$ discrete lattice points. An IC lattice core may contain several spherical layers. If no confusion arises, we will use $C$ to denote the index set of the lattice points on the IC core.
- $S$ be the outer lattice surface covering the IC core. If no confusion arises, we will use $S$ to denote the index set of the lattice points on the surface.
- $I$ be a joint set of $C$ and $S$, i.e., $I = C \cup S$ and $|I| = |C| + |S|$. In this model, the number of the initial lattice points generated on the IC lattice is greater than the number of atoms in the LJ cluster, i.e., $|I| \ge n$.

```
procedure MEM()
begin
   /* initialization */
   complete_lattice := generate_a_complete_IC_lattice();
   problem_lattice := remove_extra_surface_atoms(complete_lattice);
   E := compute_LJ_energy(problem_lattice);

   /* local search */
   lattice = local_search(problem_lattice);

   /* scrambling search */
   optimal_lattice = scrambling_search(lattice);

   return optimal_lattice;
end;
```

FIGURE 3. A multispace search algorithm for molecular energy minimization (MEM).

- $S_f$ and $S_v$ be two subsets of $S$, representing the *filled* lattice positions and *vacant* lattice positions, on the surface $S$, respectively. In this model, $|S_f| = n - |C|$ [**27**] and $|S_v| = |I| - n$.

**4.1. The Algorithm.** Based on the reconstruction of molecular lattice structures, in Figure 3, we give a multispace search algorithm for molecular energy minimization (*MEM*). The *MEM* algorithm is divided into three major procedures, i.e., an initialization procedure, a local search procedure, and a scrambling search procedure.

### Initialization

Initially, procedure *generate_a_complete_IC_lattice*() generates the smallest IC lattice, $I$, that contains at least $n$ atoms ($|I| \geq n$). The surface layer of $I$ is $S$. Procedure *remove_extra_surface_atoms*() removes $|I| - n$ extra atoms from $S$, producing a lattice structure according to the given problem specification. Every time when an atom with the maximum energy is removed from the surface, the size of set $S_v$ is incremented by one and the size of set $S_f$ is reduced by one. Procedure *compute_LJ_energy*() calculates the energy of each atom and then evaluates the total amount of energy of all the atoms on the given LJ lattice structure.

### Local Search

Local search procedure contains mainly two parts, i.e., a discrete local search procedure that swaps atoms in the occupied set ($S_f$) and in the unoccupied set ($S_v$), reducing the LJ cluster energy. Following discrete local search, a continuous relaxation is performed by a conjugate gradient method which further minimizes the LJ cluster energy.

### Scrambling Search

Scrambling search procedure interplays a two-stage scrambling optimization procedure that removes/adds some atom(s) from/to the lattice structure, followed by a pattern search procedure that relaxes the lattice structure and produces the lower molecular energy.

In the following, we will discuss the major procedures in the multispace search algorithm in details.

**4.2. Local Search Procedure.** The algorithmic description of the local search procedure is shown in Figure 4.

### Discrete Local Search

The local search procedure is used to find a lattice structure with locally minimum LJ potential energy. For an atom in the $i$th lattice position, $a_i$, and an atom in the $j$th lattice position, $a_j$, if $e_i > e_j$, i.e., an atom in the $i$th lattice position possesses higher potential energy than in the $j$th lattice position, then the atom in the $i$th position will be moved to the $j$th position, followed by an update of memberships in sets $S_f$ and $S_v$. This is carried out by procedure *perform_swap*$(a_i, a_j)$. The local search process repeats to swap the atoms in two sets in a sequential order until no improvement occurs, i.e., a lattice structure with locally minimum energy is found.

From our experience, local search is very efficient in two aspects. First, at the beginning of the search, using a full assignment of lattice positions to all the atoms, it reduces down an exponential growth search space to a manageable one. Secondly, it searches for improvement within its local neighborhood as long as there is some improvement. A major weakness of local search is that it has a tendency of getting stuck at a locally optimum configuration (i.e., a local minimum point).

In the *MEM* algorithm, we have used structural operations to disturb the environment of forming local minima. So, in this *local_search*() procedure, we only use a simple local handler (Figure 5) to overcome the weakness of local search. Procedure *local_handler*() randomly swaps a pair of atoms in $S_f$ and $S_v$. The major iteration loop in the local search repeats some 40 times before the *MEM* algorithm enters into the scrambling search procedure. That is, constant

```
procedure local_search(problem_lattice)
begin
   lattice := problem_lattice;
   E_min := compute_LJ_energy(lattice);
   k_max := 40;

   k := 0;
   while k < k_max do
   begin
      /* discrete local search */
      for the ith atom in S_f do
         for the jth atom in S_v do
            if e_i > e_j then
               problem_lattice := perform_swap(a_i, a_j);

      /* continuous relaxation by conjugate gradient method */
      problem_lattice := conjugate_gradient(problem_lattice);
      E := compute_LJ_energy(problem_lattice);
      if E < E_min then
         lattice := problem_lattice; E_min := E;
      else
         problem_lattice := lattice;
      local_handler(problem_lattice);
      k := k + 1;
   end:

   /* return the improved lattice structure */
   return lattice;
end;
```

FIGURE 4. The local search procedure.

```
procedure local_handler(problem_lattice)
begin
   for a random pair of atoms in S_f and S_v do
   begin
      problem_lattice := perform_swap(a_i, a_j);
      E := compute_LJ_energy(problem_lattice);
   end;
end;
```

FIGURE 5. A simple local handler.

$k_{max}$ is initialized to be 40. The locally minimum lattice obtained from local search procedure is used as the initial lattice structure for scrambling search.

### Continuous Relaxation by Conjugate Gradient Method

Local search is much more efficient for discrete optimization. As the first processing step, it is effective to the minimization of molecular energy of the Lennard-Jones cluster. Local search finds a set of *discrete* lattice positions that may be close to the positions generating the minimum LJ energy. Local search, however, will not be able to locate the exact position with the minimum LJ energy since the problem is itself a global optimization problem in the real space. After each local search procedure, therefore, we use a continuous relaxation method to further search for an accurate *real* position that gives lower LJ potential energy.

In this work, with the function value and its gradient, for each lattice structure found by local search, we further use conjugate gradient method to minimize the molecular energy of the Lennard-Jones cluster. Instead of proceeding to the local gradient descent direction, we follow a direction that is somehow constructed to be *conjugate* to the old gradient, and, insofar as possible, to all previous directions traversed. Two most important conjugate gradient methods are the Fletcher-Reeves method and the Polak-Ribiere method (the later is an improved version of the former). We use the Polak-Ribiere method to perform unconstrained optimization to the LJ cluster since Polak-Ribiere approach accomplishes the transition to further iterations more gracefully. We found that, in most cases of our program running, the conjugate gradient method performed better than the quasi-Newton method, producing lower or the same energy as in the quasi-Newton method.

**4.3. Scrambling Search Procedure.** The scrambling search procedure, which is the third major part of *MEM*, is shown in Figure 6. It consists of a two-stage scrambling optimization procedure followed by a continuous relaxation using the pattern search method.

```
procedure scrambling_search(lattice)
begin
   E_min := compute_LJ_energy(lattice);
   optimal_lattice := lattice;

   /* scrambling on lattice structure */
   while there is improvement do
   begin
      /* Stage 1: scrambling with one atom */
      for each atom in S_v do
      begin
         lattice := lattice + atom;
         E := min_by_pattern_search(lattice);
         /* remove one atom with the maximum energy on the lattice */
         lattice := lattice − atom_(Max e_i);
         E := min_by_pattern_search(lattice);
         if E < E_min then optimal_lattice := lattice; E_min := E;
         else lattice := optimal_lattice;
      end;

      /* Stage 2: scrambling with two atoms */
      for two atoms in S_v do
      begin
         /* remove one atom with the maximum energy on the lattice */
         lattice := lattice − atom_(Max e_i);
         /* add two atoms in S_v to the lattice;
         lattice := lattice+ 2 atoms;
         E := min_by_pattern_search(lattice);
         /* remove one atom with the maximum energy on the lattice */
         lattice := lattice − atom_(Max e_i);
         E := min_by_pattern_search(lattice);
         if E < E_min then optimal_lattice := lattice; E_min := E;
         else lattice := optimal_lattice;
      end;
   end;
   return optimal_lattice;
end;
```

FIGURE 6. The scrambling search procedure.

## Two Stage Scrambling Optimization

Local search works better at earlier stage of optimization. When the molecular energy is reduced to a lower value, it is difficult for local search to proceed due to the pathological behavior of local minima. We use extra, structural operations to help the local search procedure to proceed.

During the initialization procedure (Figure 3), $|I|$ atoms on the initial IC lattice structure were generated. $|I| - n$ redundant atoms were removed from the initial lattice structure and their lattice positions were kept in the $S_v$ set. In the first stage of scrambling optimization, some minor structural operations are performed. Each time an atom is added to a lattice position in $S_v$, followed by a pattern search that reduces the molecular energy of this $(n+1)$-atom LJ cluster. The LJ system now has one extra atom. Then, an atom with the maximum energy is identified and is removed from the LJ cluster, followed by a pattern search that minimizes the molecular energy of this $n$-atom LJ cluster. From the above operations, if the molecular energy of the LJ cluster is reduced, the new lattice structure is kept and the minimum energy updated; otherwise, the previous lattice structure is recovered.

Structural operations in the second stage of scrambling search are similar to these in the first stage except slightly strong scrambling operations are performed, i.e., two atoms are scrambled with the LJ cluster to reduce the molecular energy.

## Relaxation by Pattern Search Method

Pattern search method, developed by Hooke and Jeeves in 1961, is similar to the coordinate ascent method. In pattern search, each coordinate is searched cyclically in order to identify the best local descent direction. Then the search process is accelerated along this favorable descent direction. Since pattern search is simple, the entire process can be executed fairly efficiently. Unlike optimization methods requiring partial derivatives, pattern search method does not require the calculation of the function derivatives. So it is simple to implement.

In our pattern search algorithm, we use the lattice structure from the scrambling search as the initial lattice structure. For a chosen step $d$, for each variable $x_i$ in the $k$th iteration, we cyclically test if $E(x_i^k \pm d)$ is less than $E(x_i^k)$. If it is, we have

$$\acute{x}_i^{k+1} := x_i^k \pm d,$$

otherwise,

$$\acute{x}_i^{k+1} := x_i^k.$$

This produces a vector $\acute{\mathbf{x}}^{k+1}$. From $\acute{\mathbf{x}}^{k+1}$, we can obtain an accelerated vector, $\ddot{\mathbf{x}}^{k+1}$, along the $\mathbf{d}$ direction:

$$\ddot{\mathbf{x}}^{k+1} := 2\,\acute{\mathbf{x}}^{k+1} - \mathbf{x}^k.$$

The next new solution vector, i.e., $\mathbf{x}^{k+1}$, is:

$$\mathbf{x}^{k+1} = \begin{cases} \acute{\mathbf{x}}^{k+1} & \text{if } E(\acute{\mathbf{x}}^{k+1}) < E(\ddot{\mathbf{x}}^{k+1}); \\ \ddot{\mathbf{x}}^{k+1} & \text{otherwise.} \end{cases}$$

The step $d$ will be further reduced if the pattern search process is retarded. The process is terminated if reducing $d$ does not have any effect, or the precision limit prohibits $d$ from being further reduced.

## 5. Experimental Results

The *MEM* algorithm for the minimization of the molecular energy of the Lennard-Jones cluster was written in C. The experiments were performed on a SUN SPARC 10 workstation.

For Lennard-Jones clusters with sizes varying from 50 to 400, Table 1 gives computational results of the multispace search algorithm for the minimization of the molecular energy of the LJ clusters. In the table, the first column indicates the number of atoms, $n$, in the LJ clusters. The second, third, and fourth columns, which are labeled with $E_0$, $E_{local}$, and $E_{scramble}$, denote the molecular energy of the LJ clusters after the initialization, local search, and scrambling search procedures, respectively. The last column gives the computing results of Xue's algorithm for the same size Lennard-Jones clusters [**26**]. Xue's algorithm was implemented on a 64-bit Cray-XMP computer at the Minnesota Supercomputer Center.

From this table we can see that the multispace search algorithm found the best known molecular energies for most LJ clusters [**27, 26, 17**]. For LJ clusters having 88, 98, and 115 atoms, the multispace search algorithm found lower molecular energy values than those reported earlier. The coordinates of the 98-atom LJ cluster are listed in Table 2.

The minimization of molecular energy for the LJ clusters is a difficult global optimization problem. We observed that the difficulty does not only depend on the size of the LJ cluster, $n$. It also depends on how close the cluster size $n$ is to the icosahedral "magic number" [**17**]. A larger distance between $n$ and the magic number makes this problem harder.

The experimental results for icosahedral structures $IC_2 - IC_8$ are shown in Table 3. In this case, the outer surface of IC lattice is fully occupied. So it is relatively easy to find an optimization solution. Our multispace search algorithm found all the best known molecular energy values for the LJ clusters in reasonable CPU time.

TABLE 1. Experimental results of the multispace search algorithm for molecular energy minimization of the Lennard-Jones clusters. $E_0$, $E_{local}$, and $E_{scramble}$ denote the LJ molecular energy after the initialization, local search, and scrambling search procedures, respectively.

| $n$ | $E_0$ | $E_{local}$ | $E_{scramble}$ | Xue [**26**] |
|---|---|---|---|---|
| 50 | -241.622 | -244.275 | -244.275 | -244.550 |
| 51 | -248.550 | -251.252 | -251.254 | -251.254 |
| 52 | -255.480 | -258.079 | -258.230 | -258.230 |
| 53 | -262.411 | -265.202 | -265.203 | -265.203 |
| 54 | -269.374 | -272.206 | -272.209 | -272.209 |
| 55 | -276.367 | -279.248 | -279.248 | -279.2485 |
| 56 | -280.589 | -283.643 | -283.643 | -283.643 |
| 57 | -284.918 | -288.139 | -288.343 | -288.343 |
| 58 | -289.894 | -293.223 | -294.378 | -294.378 |
| 59 | -294.912 | -299.302 | -299.738 | -299.738 |
| 60 | -300.290 | -304.490 | -305.875 | -305.876 |
| 61 | -305.430 | -310.042 | -312.009 | -312.009 |
| 62 | -310.827 | -316.234 | -317.354 | -317.354 |
| 63 | -316.118 | -321.550 | -323.490 | -323.490 |
| 64 | -322.408 | -327.849 | -329.620 | -329.620 |
| 65 | -328.955 | -334.446 | -334.447 | -334.972 |
| 66 | -335.112 | -340.046 | -340.047 | -341.111 |
| 67 | -341.300 | -346.350 | -346.351 | -347.252 |
| 68 | -346.518 | -352.683 | -352.684 | -353.395 |
| 69 | -352.736 | -359.017 | -359.018 | -359.726 |
| 70 | -359.184 | -365.350 | -365.351 | -366.892 |
| 71 | -365.379 | -371.713 | -371.714 | -373.350 |
| 72 | -370.731 | -377.294 | -377.294 | -378.524 |
| 73 | -376.963 | -383.638 | -383.639 | -384.789 |
| 74 | -383.434 | -390.008 | -390.009 | -390.908 |
| 75 | -388.833 | -395.733 | -395.734 | -396.037 |
| 76 | -395.232 | -402.083 | -402.084 | -402.385 |
| 77 | -401.807 | -408.463 | -408.463 | -408.518 |
| 78 | -408.007 | -414.680 | -414.681 | -414.681 |
| 79 | -413.365 | -421.480 | -421.482 | -421.811 |
| 80 | -419.600 | -427.829 | -427.829 | -428.084 |

| $n$ | $E_0$ | $E_{local}$ | $E_{scramble}$ | Xue [**26**] |
|---|---|---|---|---|
| 81 | -426.075 | -434.206 | -434.207 | -434.344 |
| 82 | -431.521 | -440.550 | -440.550 | -440.550 |
| 83 | -437.928 | -446.922 | -446.924 | -446.924 |
| 84 | -444.521 | -452.657 | -452.657 | -452.657 |
| 85 | -450.848 | -459.011 | -459.012 | -459.056 |
| 86 | -458.174 | -465.383 | -465.384 | -465.384 |
| 87 | -464.412 | -472.097 | -472.098 | -472.098 |
| 88 | -469.774 | -478.934 | -479.033 | -478.935 |
| 89 | -476.015 | -486.052 | -486.054 | -486.054 |
| 90 | -482.494 | -492.433 | -492.434 | -492.434 |
| 91 | -487.943 | -498.810 | -498.811 | -498.811 |
| 92 | -494.352 | -505.184 | -505.185 | -505.185 |
| 93 | -500.949 | -510.876 | -510.878 | -510.878 |
| 94 | -507.314 | -517.263 | -517.264 | -517.264 |
| 95 | -514.646 | -523.638 | -523.640 | -523.640 |
| 96 | -520.022 | -529.878 | -529.879 | -529.879 |
| 97 | -526.398 | -536.680 | -536.681 | -536.681 |
| 98 | -532.902 | -543.545 | -543.643 | -543.547 |
| 99 | -538.354 | -550.664 | -550.667 | -550.667 |
| 100 | -544.768 | -557.038 | -557.040 | -557.040 |
| 101 | -551.368 | -563.407 | -563.411 | -563.411 |
| 102 | -557.734 | -569.277 | -569.278 | -569.278 |
| 103 | -565.068 | -575.658 | -575.659 | -575.659 |
| 104 | -570.448 | -582.036 | -582.038 | -582.038 |
| 105 | -576.826 | -588.264 | -588.267 | -588.267 |
| 106 | -583.335 | -595.059 | -595.061 | -595.061 |
| 107 | -588.873 | -601.910 | -601.912 | -601.912 |
| 108 | -595.299 | -609.030 | -609.033 | -609.033 |
| 109 | -601.966 | -615.409 | -615.411 | -615.411 |
| 110 | -609.435 | -621.785 | -621.788 | -621.788 |
| 111 | -614.895 | -628.066 | -628.068 | -628.068 |
| 112 | -621.356 | -634.872 | -634.875 | -634.875 |
| 113 | -627.977 | -641.698 | -641.700 | -641.700 |
| 114 | -634.351 | -648.830 | -648.833 | -648.833 |
| 115 | -641.692 | -655.635 | -655.756 | -655.636 |
| 116 | -647.238 | -662.809 | -662.809 | -662.809 |

| $n$ | $E_0$ | $E_{local}$ | $E_{scramble}$ | Xue [26] |
|---|---|---|---|---|
| 117 | -653.709 | -668.280 | -668.283 | -668.283 |
| 118 | -660.396 | -674.765 | -674.770 | -674.770 |
| 119 | -667.869 | -679.556 | -679.559 | -681.419 |
| 120 | -674.250 | -686.686 | -686.689 | -687.022 |
| 121 | -681.630 | -691.111 | -691.114 | -693.820 |
| 122 | -687.180 | -698.298 | -698.301 | -700.939 |
| 123 | -693.657 | -705.488 | -705.490 | -707.802 |
| 124 | -700.358 | -712.676 | -712.678 | -714.921 |
| 125 | -707.838 | -717.493 | -717.497 | -721.303 |
| 126 | -713.407 | -724.679 | -724.682 | -727.350 |
| 127 | -719.976 | -731.864 | -731.869 | -734.480 |
| 128 | -726.844 | -739.055 | -739.059 | -741.332 |
| 129 | -734.409 | -744.194 | -744.197 | -748.461 |
| 130 | -742.005 | -751.743 | -751.753 | -755.271 |
| 131 | -748.547 | -758.931 | -758.934 | -762.442 |
| 132 | -756.088 | -765.976 | -765.979 | -768.042 |
| 133 | -763.738 | -773.674 | -773.678 | -775.023 |
| 134 | -771.496 | -781.471 | -781.476 | -782.206 |
| 135 | -780.254 | -790.274 | -790.278 | -790.278 |
| 136 | -787.326 | -797.448 | -797.453 | -797.453 |
| 137 | -794.401 | -804.626 | -804.631 | -804.631 |
| 138 | -801.479 | -811.809 | -811.813 | -811.813 |
| 139 | -808.557 | -818.989 | -818.994 | -818.994 |
| 140 | -815.635 | -826.172 | -826.175 | -826.175 |
| 141 | -822.716 | -833.355 | -833.359 | -833.359 |
| 142 | -829.794 | -840.537 | -840.539 | -840.539 |
| 143 | -836.875 | -847.718 | -847.722 | -847.722 |
| 144 | -843.957 | -854.901 | -854.904 | -854.904 |
| 145 | -851.038 | -862.085 | -862.087 | -862.087 |
| 146 | -858.122 | -869.269 | -869.273 | -869.273 |
| 147 | -865.209 | -876.450 | -876.461 | -876.461 |
| 148 | -869.676 | -881.069 | -881.073 | -881.073 |
| 149 | -875.142 | -886.688 | -886.693 | -886.693 |
| 150 | -880.426 | -893.307 | -893.310 | -893.310 |
| 200 | -1209.830 | -1229.179 | -1229.185 | -1229.185 |
| 250 | -1551.725 | -1579.790 | -1579.795 | - |
| 300 | -1914.944 | -1942.094 | -1942.107 | -1942.107 |
| 400 | -2602.038 | -2650.399 | -2650.430 | -2650.432 |

TABLE 2. The coordinates of an optimal configuration for 98-atom Lennard-Jones cluster. The molecular energy of this configuration is $E = -543.643$.

| i | $x_i$ | $y_i$ | $z_i$ | i | $x_i$ | $y_i$ | $z_i$ |
|---|---|---|---|---|---|---|---|
| 1 | 0.029 | 0.017 | 0.034 | 33 | -0.838 | -0.001 | -1.345 |
| 2 | 0.011 | 0.009 | 0.964 | 34 | -0.260 | -0.798 | -1.346 |
| 3 | 0.014 | 0.009 | -0.896 | 35 | 0.682 | -0.500 | -1.349 |
| 4 | 0.869 | 0.012 | 0.454 | 36 | 1.568 | 0.518 | 0.006 |
| 5 | 0.693 | 0.504 | -0.397 | 37 | 1.135 | 0.834 | 0.887 |
| 6 | 0.275 | 0.812 | 0.447 | 38 | 1.134 | -0.820 | 0.884 |
| 7 | -0.242 | 0.807 | -0.387 | 39 | 1.569 | -0.504 | 0.017 |
| 8 | -0.657 | 0.505 | 0.441 | 40 | 0.968 | 1.336 | 0.011 |
| 9 | -0.806 | 0.010 | -0.388 | 41 | 0.430 | 1.327 | -0.847 |
| 10 | -0.652 | -0.475 | 0.432 | 42 | 1.395 | 0.001 | -0.849 |
| 11 | -0.239 | -0.771 | -0.388 | 43 | 0.005 | 1.646 | 0.020 |
| 12 | 0.279 | -0.782 | 0.442 | 44 | -0.426 | 1.333 | 0.879 |
| 13 | 0.695 | -0.480 | -0.387 | 45 | -0.958 | 1.325 | 0.015 |
| 14 | -0.025 | -0.013 | 1.929 | 46 | -1.098 | 0.803 | -0.829 |
| 15 | -0.004 | -0.004 | -1.850 | 47 | -1.521 | 0.496 | 0.014 |
| 16 | -0.018 | -0.014 | -2.837 | 48 | -1.366 | 0.001 | 0.849 |
| 17 | 1.376 | 1.005 | -0.863 | 49 | -1.518 | -0.495 | 0.009 |
| 18 | 0.532 | 1.636 | 0.876 | 50 | -1.095 | -0.796 | -0.826 |
| 19 | -0.522 | 1.616 | -0.832 | 51 | -0.940 | -1.291 | 0.009 |
| 20 | -1.371 | 1.000 | 0.860 | 52 | -0.421 | -1.300 | 0.849 |
| 21 | -1.660 | -0.002 | -0.820 | 53 | 0.002 | -1.601 | 0.015 |
| 22 | -1.352 | -0.982 | 0.837 | 54 | 0.425 | -1.293 | -0.829 |
| 23 | -0.515 | -1.579 | -0.820 | 55 | 0.965 | -1.322 | 0.015 |
| 24 | 0.525 | -1.614 | 0.862 | 56 | -1.126 | -0.819 | -1.804 |
| 25 | 1.376 | -0.997 | -0.833 | 57 | -2.255 | -0.012 | 0.425 |
| 26 | 0.856 | 0.001 | 1.426 | 58 | -1.818 | -1.320 | -0.426 |
| 27 | 0.262 | 0.823 | 1.419 | 59 | 0.676 | -2.152 | -0.432 |
| 28 | -0.700 | 0.507 | 1.405 | 60 | -1.396 | 1.607 | -1.284 |
| 29 | -0.691 | -0.501 | 1.387 | 61 | -0.548 | 1.606 | -1.815 |
| 30 | 0.263 | -0.821 | 1.408 | 62 | 1.097 | -1.826 | -1.284 |
| 31 | 0.689 | 0.501 | -1.362 | 63 | 0.151 | -2.122 | -1.275 |
| 32 | -0.264 | 0.802 | -1.348 | 64 | -0.710 | -2.141 | 0.426 |

| i | $x_i$ | $y_i$ | $z_i$ | i | $x_i$ | $y_i$ | $z_i$ |
|---|---|---|---|---|---|---|---|
| 65 | -1.651 | -1.816 | 0.418 | 82 | -1.425 | -0.020 | 1.823 |
| 66 | -0.277 | -2.433 | -0.421 | 83 | -1.704 | -0.010 | -1.796 |
| 67 | 0.240 | -2.448 | 0.425 | 84 | -2.087 | -0.513 | 1.268 |
| 68 | -1.836 | 1.308 | -0.432 | 85 | -0.462 | -1.349 | 1.823 |
| 69 | -1.970 | 0.799 | -1.275 | 86 | -1.133 | 0.802 | -1.811 |
| 70 | -0.536 | -1.618 | -1.796 | 87 | -0.283 | 0.797 | -2.328 |
| 71 | 0.670 | -0.516 | -2.328 | 88 | -1.964 | -0.815 | -1.269 |
| 72 | 1.358 | -1.019 | -1.816 | 89 | -1.225 | -2.118 | -0.423 |
| 73 | -2.542 | -0.011 | -1.266 | 90 | -2.253 | 0.985 | 0.424 |
| 74 | -0.193 | -2.145 | 1.274 | 91 | -0.863 | -0.012 | -2.319 |
| 75 | -0.796 | -2.415 | -1.266 | 92 | -1.404 | -1.019 | 1.807 |
| 76 | 1.374 | -0.016 | -1.837 | 93 | -2.398 | 0.488 | -0.422 |
| 77 | -0.278 | -0.818 | -2.319 | 94 | -1.382 | -1.616 | -1.269 |
| 78 | 0.410 | 1.312 | -1.835 | 95 | -1.134 | -1.827 | 1.268 |
| 79 | 0.413 | -1.326 | -1.811 | 96 | -2.393 | -0.510 | -0.423 |
| 80 | 0.676 | 0.491 | -2.339 | 97 | -2.074 | -1.506 | 1.260 |
| 81 | -2.098 | 0.481 | 1.274 | 98 | -2.237 | -1.009 | 0.418 |

TABLE 3. Experimental results of the multispace search algorithm for molecular energy minimization of the Lennard-Jones clusters for IC lattice structures with 2 ($IC_2$) to 8 ($IC_8$) spherical layers. $E_0$, $E_{local}$, and $E_{scramble}$ denote the LJ molecular energy after the initialization, local search, and scrambling search procedures, respectively.

| $IC_i$ | $n$ | $E_0$ | $E_{local}$ | $E_{scramble}$ | Maier et al. [**17**] |
|---|---|---|---|---|---|
| $IC_2$ | 13 | -44.020 | -44.327 | -44.3268 | -44.3268 |
| $IC_3$ | 55 | -276.367 | -279.248 | -279.248 | -279.249 |
| $IC_4$ | 147 | -865.209 | -876.450 | -876.461 | -876.461 |
| $IC_5$ | 309 | -1978.958 | -2007.172 | -2007.219 | -2007.219 |
| $IC_6$ | 561 | -3785.915 | -3842.044 | -3842.393 | -3842.394 |
| $IC_7$ | 923 | -6454.332 | -6540.960 | -6552.722 | -6552.723 |
| $IC_8$ | 1415 | -10152.442 | -10307.042 | -10308.888 | -10308.89 |

## 6. Conclusions

For most molecular energy minimization problems, the energy function may be reduced rapidly at the early stage of optimization, showing an *easy phase* of progression. The *difficult phase* of the molecular energy minimization appears at the later stage of finding a globally optimum solution. During this stage, due to the existence of many local minima, it is difficult for most optimization algorithms to reduce the molecular energy function by even a very small percentage.

To cope with the exponential number of local minima, for Lennard-Jones clusters, we have developed a multispace search algorithm for molecular energy minimization. The algorithm interplays structural operations in conjunction with traditional optimization techniques. Structural lattice operations disturb the environment of forming local minima, which makes multispace search a natural approach to molecular energy minimization. Experimental results indicate that, in addition to achieving the best known results in most cases, for 88-atom, 98-atom, and 115-atom Lennard-Jones clusters, our algorithm found the new lattice structures whose molecular energies are lower than those found in the previous studies.

## Acknowledgements

Panos Pardalos, David Shalloway, and Guoliang Xue suggested that this article be written. They have informed us of recent research progress/publications in the area and have provided us many helpful discussions during the course of this research work.

## References


1. H.S. chan and K.A. Dill. The protein folding problem. *Physics Today*, pages 24–32, Feb. 1993.
2. T. Coleman, D. Shalloway, and Z. Wu. A parallel build-up algorithm for global energy minimizations of molecular clusters using effective energy simulated annealing. *Journal of Global Optimization*, 4(2):171–185, Mar. 1994.
3. I.Z. FisherPF64. *Statistical Theory of Liquids.* University of Chicago Press, 1964.
4. J. Gu. Optimization by multispace search. Technical Report UCECE-TR-90-001, Dept. of Electrical and Computer Engineering, Univ. of Calgary, Jan. 1990.
5. J. Gu. Optimization by simulated evolution. Technical Report UCECE-TR-90-004, Dept. of Electrical and Computer Engineering, Univ. of Calgary, Mar. 1990.
6. J. Gu. Local search for satisfiability (SAT) problem. *IEEE Trans. on Systems, Man, and Cybernetics*, 23(4):1108–1129, Jul. 1993.
7. J. Gu. Global optimization for satisfiability (SAT) problem. *IEEE Trans. on Knowledge and Data Engineering*, 6(3):361–381, Jun. 1994.
8. J. Gu. *Constraint-Based Search.* Cambridge University Press, New York, 1995.
9. J. Gu. *Optimization by Multispace Search.* Kluwer Academic Publishers, Massachusetts, 1995.
10. J. Gu and B. Du. Graph partitioning by simulated evolution. Technical Report UCECE-TR-92-001, Dept. of Electrical and Computer Engineering, Univ. of Calgary, Apr. 1992.
11. J. Gu and X. Huang. Efficient local search with search space smoothing. *IEEE Trans. on Systems, Man, and Cybernetics*, 24(5):728–735, May 1994.
12. J. Gu and R. Puri. Asynchronous circuit synthesis by Boolean satisfiability. *IEEE Transactions on CAD*, 1995, to appear.
13. M.R. Hoare. Structure and dynamics of simple microclusters. *Advances in Chemical Physics*, 40:49–135, 1979.
14. R.S. Judson, M.E. Colvin, J.C. Meza, A. Huffer, and D. Gutierrez. Do intelligent configuration search techniques outperform random search for large molecules? Technical Report SAND91-8740, Sandia National Laboratories, Center for Computational Engineering, Livermore, CA, Dec. 1991.
15. S. Kirkpatrick, C.D. Gelat, and M.P. Vecchi. Optimization by simulated annealing. *Science*, 220:671–680, 1983.
16. J. Kostrowicki, L. Piela, B.J.Cherayil, and H.A. Scheraga. Performance of the diffusion equation method in searches for optimum structures of clusters of Lennard-Jones atoms. *Journal of Physical Chemistry*, 95:4113–4119, 1991.
17. R.S. Maier, J.B. Rosen, and G.L. Xue. A discrete-continuous algorithm for molecular energy minimization. In *Proceedings of IEEE/ACM Supercomputing'91*, pages 778–786. 1992.
18. C.D. Maranas and C.A. Floudas. Global minimum potential energy conformations of small molecules. *Journal of Global Optimization*, 4(2):135–170, Mar. 1994.
19. J.A. Northby. Structure and binding of Lennard-Jones clusters : $13 \leq n \leq 147$. *Journal of Chemical Physics*, 87(10):6166–6177, Nov. 1987.
20. P.M. Pardalos and G.L. Xue. Optimization methods for computing global minima of non-convex potential energy functions. *Journal of Global Optimization*, 4(2):117–133, Mar. 1994.
21. F.M. Richards. The protein folding problem. *Scientific American*, pages 54–63, Jan. 1991.
22. D. Shalloway. Application of the renormalization group to deterministic global minimization of molecular conformation energy functions. *Journal of Global Optimization*, 2:283–311, 1992.
23. D. Shalloway. *Packet Annealing: A Deterministic Method for global Minimization. In* Recent Advances in Global Optimization, pages 433–477. Princeton University Press, 1992.
24. L.T. Wille. Minimum-energy configurations of atomic clusters: New results obtained by simulated annealing. *Chemical Physics Letters*, 133:405–410, 1987.
25. L.T. Wille and J. Vennik. Computational complexity of the ground-state determination of atomic clusters. *Journal of Physics A: Mathematic and General*, 18(8):L419–L422, June

1985.
26. G.L. Xue. Improvement on the Northby algorithm for molecular conformation: Better solutions. Accepted for publication in Journal of Global Optimization.
27. G.L. Xue. Molecular conformation on the CM-5 by parallel two-level simulated annealing. *Journal of Global Optimization*, 4(2):187–208, Mar. 1994.

DEPT. OF ELECTRICAL AND COMPUTER ENGINEERING, UNIVERSITY OF CALGARY, CALGARY, ALBERTA T2N 1N4, CANADA. PRESENTLY ON LEAVE AT: DEPT. OF COMPUTER SCIENCE, HONG KONG UNIVERSITY OF SCIENCE AND TECHNOLOGY, CLEAR WATER BAY, KOWLOON, HONG KONG.

*E-mail address*: gu@enel.ucalgary.ca, gu@cs.ust.hk

DIMACS Series in Discrete Mathematics
and Theoretical Computer Science
Volume **23**, 1996

# A Minimal Principle in the Phase Problem of X-Ray Crystallography

Herbert A. Hauptman

ABSTRACT. The intensities of a sufficient number of X-ray diffraction maxima determine the structure of a crystal. The available intensities usually exceed the number of parameters needed to describe the structure. From these intensities a set of numbers $|E_H|$ can be derived, one corresponding to each intensity. However, the elucidation of the crystal structure also requires a knowledge of the complex numbers $E_H = |E_H| exp(i\phi_H)$, the normalized structure factors, of which only the magnitudes $|E_H|$ can be determined from experiment. Thus, a "phase" $\phi_H$, unobtainable from the diffraction experiment, must be assigned to each $|E_H|$, and the problem of determining the phases when only the magnitudes $|E_H|$ are known is called the "phase problem". Owing to the known atomicity of crystal structures and the redundancy of observed magnitudes $|E_H|$, the phase problem is solvable in principle. The phase problem may be formulated as one in constrained global optimization. A method for avoiding the countless minima in order to arrive at the global minimum will be described.

## 1. The phase problem

The electron density function $\rho(r)$ in a crystal is a three-dimensional periodic function of the position vector **r** and may therefore be represented by the three-dimensional Fourier series:

$$\rho(\mathbf{r}) = \frac{1}{V}\sum_{\mathbf{H}} F_{\mathbf{H}} \exp(-2\pi i\mathbf{H}\cdot\mathbf{r}) \tag{1}$$

where $V$ is the volume of the fundamental parallelepiped, the so-called unit cell of the crystal, and the three components of the vector **H** range over all the integers. The Fourier coefficient $F_{\mathbf{H}}$, said to be the structure factor corresponding to the reciprocal lattice vector **H**, may then be calculated in the usual way:

$$F_{\mathbf{H}} = \int_V \rho(\mathbf{r}) \exp(2\pi i\mathbf{H}\cdot\mathbf{r}) dV \tag{2}$$

where the integration is carried out over the unit cell $V$. Clearly, each $F_{\mathbf{H}}$ is a complex number that may be written in polar form

---

1991 *Mathematics Subject Classification.* Primary 82D25; Secondary 42-XX.
Author was supported by a grant from NIH No. GM46733.

(3) $$F_{\mathbf{H}} = |F_{\mathbf{H}}|\exp(i\phi_{\mathbf{H}})$$

where $\phi_{\mathbf{H}}$ is the phase of the structure factor $F_{\mathbf{H}}$.

When a beam of monochromatic X-rays is incident on a crystal, the radiation is scattered in discrete directions determined by the crystal lattice and labeled by the reciprocal lattice vectors **H**. Both the amplitude and phase of each scattered ray, or reflection, depend on the crystal structure. The amplitude, or intensity, of a reflection leads in a straightforward way to the magnitude $|F_{\mathbf{H}}|$ of the complex structure factor $F_{\mathbf{H}}$. However, the phases $\phi_{\mathbf{H}}$, which are also needed if one is to determine $\rho(\mathbf{r})$ from Eq. 1, are lost in the diffraction experiment. If one uses the known values of the magnitudes $|F_{\mathbf{H}}|$ but arbitrary values for the phases $\phi_{\mathbf{H}}$ in Eq. 1, then density functions $\rho(\mathbf{r})$ consistent with the observed values of the diffraction intensities are obtained. Thus, diffraction intensities alone do not determine a unique density function $\rho(\mathbf{r})$. Even if the known nonnegativity of $\rho(\mathbf{r})$ is assumed, thus greatly restricting the values of the phases (Harker & Kasper, 1948; Karle & Hauptman, 1950) the observed diffraction intensities are, in general, still not sufficient to determine $\rho(\mathbf{r})$ uniquely. It follows that the phase problem, to determine the values of the phases $\phi_{\mathbf{H}}$ of the structure factors $F_{\mathbf{H}}$ when only the magnitudes $|F_{\mathbf{H}}|$ are given, is, in principle, unsolvable when formulated in these terms. It was this argument that led the crystallographic community, prior to 1950, to believe also that crystal structures could not, even in principle, be determined from the diffraction intensities alone. However, by invoking the prior structural knowledge that crystals consist of discrete atoms, one readily refutes this argument, as shown below.

If one replaces the real crystal, with continuous electron density $\rho(\mathbf{r})$, by an idealized one, the unit cell of which consists of $N$ discrete, nonvibrating point atoms, then the structure factor $F_{\mathbf{H}}$ is replaced by the normalized structure factor $E_{\mathbf{H}}$ and Eqs. 3, 2, and 1 are replaced by

(4) $$E_{\mathbf{H}} = |E_{\mathbf{H}}|\exp(i\phi_{\mathbf{H}})$$

(5) $$E_{\mathbf{H}} = \frac{1}{\sigma_2^{1/2}}\sum_{j=1}^{N} Z_j \exp(2\pi i\mathbf{H}\cdot\mathbf{r}_j)$$

$$\langle E_{\mathbf{H}}\exp(-2\pi i\mathbf{H}\cdot\mathbf{r})\rangle_{\mathbf{H}} = \frac{1}{\sigma_2^{1/2}}\left\langle\sum_{j=1}^{N} Z_j \exp[2\pi i\mathbf{H}\cdot(\mathbf{r}_j-\mathbf{r})]\right\rangle_{\mathbf{H}}$$

$$= \frac{Z_j}{\sigma_2^{1/2}} \text{ if } \mathbf{r} = \mathbf{r}_j$$

(6) $$= 0 \text{ if } \mathbf{r} \neq \mathbf{r}_j$$

respectively, where $Z_j$ is the atomic number and $\mathbf{r}_j$ is the position vector of the atom labeled $j$, and

(7) $$\sigma_n = \sum_{j=1}^{N} Z_j^n,\ n = 1,2,3,\ldots$$

In practice, the magnitudes $|E_{\mathbf{H}}|$ of the normalized structure factors $E_{\mathbf{H}}$ are obtainable (at least approximately) from the observed magnitudes $|F_{\mathbf{H}}|$, while the phases $\phi_{\mathbf{H}}$, as defined by Eqs. 4 and 5, cannot be determined experimentally. Since one now requires only the $3N$ components of the $N$ position vectors $\mathbf{r}_j$ rather than the much more complicated electron density function $\rho(\mathbf{r})$, it turns out that, in general, the known magnitudes are more than sufficient. This is most readily seen if we equate the magnitudes of both sides of Eq. 5, thus eliminating the unknown phases $\phi_{\mathbf{H}}$, in order to obtain

(8)
$$|E_{\mathbf{H}}| = \frac{1}{\sigma_2^{1/2}} \left| \sum_{j=l}^{N} Z_j \exp(2\pi i \mathbf{H} \cdot \mathbf{r}_j) \right|$$

a system of equations in which the only unknowns are the $3N$ components of the position vectors $\mathbf{r}_j$. Since the number of Eq. 8, equal to the number of reciprocal lattice vectors $\mathbf{H}$ for which the magnitudes $|E_{\mathbf{H}}|$ are observed, usually exceeds the number of unknowns, $3N$, by far, the system in Eq. 8 is redundant. Thus the phase problem is, in principle, solvable when reformulated in terms of fixed point atoms, as reference to Eq. 5 shows.

The system of Eq. 5 implies the existence of relationships among the normalized structure factors $E_{\mathbf{H}}$ since the (relatively few) unknown position vectors $\mathbf{r}_j$ may, at least in principle, be eliminated. By the term "direct methods" is meant that class of methods that exploits relationships among the normalized structure factors in order to go directly from the observed magnitudes $|E|$ to the needed phases $\phi$.

## 2. The structure invariants

Equation 6 implies that the normalized structure factors $E_{\mathbf{H}}$ determine the crystal structure. However, Eq. 5 does not imply that, conversely, the crystal structure determines the values of the normalized structure factors $E_{\mathbf{H}}$ since the position vectors $\mathbf{r}_j$ depend not only on the structure but on the choice of origin as well. It turns out, nevertheless, that the magnitudes $|E_{\mathbf{H}}|$ of the normalized structure factors are in fact uniquely determined by the crystal structure and are independent of the choice of origin, but that the values of the phases $\phi_{\mathbf{H}}$ depend also on the choice of origin. Although the values of the individual phases depend on the structure and the choice of origin, there exist certain linear combinations of the phases, the so-called structure invariants, whose values are determined by the structure alone and are independent of the choice of origin. The most important class of structure invariants, and the only one to be considered here, consists of the three-phase structure invariants (triplets),

(9)
$$\phi_{\mathbf{HK}} = \phi_{\mathbf{H}} + \phi_{\mathbf{K}} + \phi_{-\mathbf{H}-\mathbf{K}},$$

where H and K are arbitrary reciprocal lattice vectors.

## 3. The probabilistic background

It is assumed that (i) A crystal structure is specified, (ii) Three non-negative numbers $R_1, R_2, R_3$ are also specified; and (iii) The reciprocal lattice vectors H and K are the primitive random variables which are assumed to be uniformly and independently distributed in the subset of reciprocal space defined by

(10)
$$|E_{\mathbf{H}}| = R_1,\ |E_{\mathbf{K}}| = R_2,\ |E_{\mathbf{H+K}}| = R_3$$

where the magnitudes $|E|$ are defined by (5). Then the structure invariant $\phi_{HK}$ (Eq. (9)), as a function of the primitive random variables **H** and **K** (Eq. (5)), is itself a random variable.

## 4. The conditional probability distribution of $\phi_{HK}$, given $|E_H|$, $|E_K|$, $|E_{H+K}|$

For simplicity it will be assumed henceforth that all N atoms in the unit cell are identical. Under the three assumptions of §3. the conditional probability distribution of the triplet $\phi_{HK}$ (Eq. (9)), where $|E_H|$, $|E_K|$, and $|E_{H+K}|$ are given by (10), is known to be

$$P(\Phi / R_1, R_2, R_3) = \frac{1}{2\pi I_0(A_{HK})} \exp(A_{HK} \cos\Phi) \tag{11}$$

where $\Phi$ represents the triplet $\phi_{HK}$,

$$A_{HK} = \frac{2}{N^{1/2}} R_1 R_2 R_3 = \frac{2}{N^{1/2}} |E_H E_K E_{H+K}| \tag{12}$$

and $I_0$ is the Modified Bessel Function. Equation (11) implies that the mode of $\phi_{HK}$ is zero, and the conditional expectation value (or average) of cos $\phi_{HK}$, given $A_{HK}$, is

$$\varepsilon(\cos\phi_{HK} / A_{HK}) = \frac{I_1(A_{HK})}{I_0(A_{HK})} > 0, \tag{13}$$

where $I_1$ is the Modified Bessel Function. It is also readily confirmed that the larger the value of $A_{HK}$ the smaller is the conditional variance of $\cos\phi_{HK}$, given $A_{HK}$. It is to be stressed that the conditional expected value of the cosine, Eq. (13), is always positive since $A_{HK}>0$.

## 5. The minimal principle

**5.1. The heuristic background.** It is assumed that a crystal structure consisting of N identical atoms in the unit cell is fixed, but unknown, that the magnitudes $|E|$ of the normalized structure factors E are known, and that a sufficiently large base of phases, corresponding to the largest magnitudes $|E|$, is specified.

The mode of the triplet distribution (Eq. (11)) is zero and the variance of the cosine is small if $A_{HK}$ (Eq. (12)) is large. In this way one obtains the estimate for the triplet $\phi_{HK}$ (Eq. (9)):

$$\phi_{HK} = \phi_H + \phi_K + \phi_{-H-K} \approx 0 \tag{14}$$

which is particularly good in the favorable case that $A_{HK}$, (Eq. (12)), is large, i.e. that $|E_H|$, $|E_K|$, and $|E_{H+K}|$ are all large. The estimate given by Eq. (14) is one of the cornerstones of the traditional techniques of direct methods. It is surprising how useful Eq. (14) has proven to be in the applications especially since it yields only the zero estimate of the triplet, and only those estimates are reliable for which $|E_H|$, $|E_K|$, and $|E_{H+K}|$ are all large. Clearly the coefficient $2/N^{1/2}$ in Eq. (12), and therefore $A_{HK}$ as well, both decrease with increasing N, i.e. with increasing structural complexity. Hence the relationship (Eq. (14)) becomes increasingly unreliable for larger structures, and the traditional step-by-step sequential direct methods procedures based on Eq. (14) eventually fail. However, the phase-annealing approach of Sheldrick (1990) and the program SAYTAN, devised by Woolfson and Yao (1990), have to some extent overcome this limitation. Here a different approach is described.

In view of Eq. (13) and the previous discussion one now replaces the zero estimate (Eq. (14)) of the triplet $\phi_{\mathbf{HK}}$ by the estimate

$$\cos\phi_{\mathbf{HK}} \approx \frac{I_1(A_{\mathbf{HK}})}{I_0(A_{\mathbf{HK}})} \tag{15}$$

and expects that the smaller the variance, that is the larger $A_{\mathbf{HK}}$, the more reliable this estimate will be. Hence one is led to construct the function (the so-called minimal function), determined by the known magnitudes $|E|$,

$$R = R(\phi) = \frac{1}{\sum_{\mathbf{H},\mathbf{K}} A_{\mathbf{HK}}} \sum_{\mathbf{H},\mathbf{K}} A_{\mathbf{HK}} \left( \cos\phi_{\mathbf{HK}} - \frac{I_1(A_{\mathbf{HK}})}{I_0(A_{\mathbf{HK}})} \right)^2 \tag{16}$$

which is seen to be a function of all those triplets $\phi_{\mathrm{HK}}$ which are generated by a prescribed set of phases $\{\phi\}$. Recall that if the basis set of $\{\phi\}$ is sufficiently large then there are many more structure invariants $\phi_{\mathbf{HK}}$ than individual phases $\phi$, and a myriad of identities among these structure invariants must, of necessity, then be satisfied. In accordance with the principle of least-squares it is natural to suppose that that set of values for the structure invariants $\phi_{\mathbf{HK}}$ is best which minimizes the weighted sum of squares of residuals R, (Eq. (16)), subject to the constraint that all identities among the structure invariants are in fact satisfied.

Since the triplets $\phi_{\mathbf{HK}}$ are defined by Eq. (9) as functions of the individual phases $\phi$, Eq. (16) defines R implicitly as a function of the individual phases. One therefore naturally anticipates that that set of values for the individual phases is best which minimizes the residual R, Eq. (16), now regarded as a function of the individual phases $\phi$. The advantage of this formulation is that all identities among the structure invariants will then automatically be satisfied, and it is unnecessary to define in further detail what the nature of these identities must be.

**5.2. The minimal principle.** In order to derive the conditions under which the formulation of the minimal principle given in the previous paragraph is valid, one first defines $R_T$ as the value of R (Eq. (16)) obtained when the phases are equal to their true values for some choice of origin. One then defines $R_R$ as the value of R when the phases are assigned at random so that in this case

$$\langle \cos\phi_{\mathbf{HK}} \rangle_{\mathbf{H},\mathbf{K}} = \langle \cos 2\phi_{\mathbf{HK}} \rangle_{\mathbf{H},\mathbf{K}} = 0. \tag{17}$$

With these definitions for $R_T$ and $R_R$ it may then be shown that (DeTitta *et al.*, 1994)

$$R_T < \tfrac{1}{2} < R_R. \tag{18}$$

While the inequalities (18) tend to confirm the conjecture that the true values of the phases minimize R, the conjecture itself is in fact not true. As it turns out, there exist special values for the phases which yield values for R even less than $R_T$, so that $R_T$ is not the global minimum of R. It is for this reason that one must further constrain the phases to satisfy certain identities among them which are known to exist because their number generally exceeds by far the number of unknown parameters needed to specify the crystal structure (as described in §1.). In short it is the constrained global minimum of R which yields the desired phases (the minimal principle). The next section describes how this minimum is reached in practice.

## 6. The computer program "*Shake-and-Bake*" (Weeks *et al.*, 1994)

The six-part ***shake-and-bake*** phase determination procedure, shown by the flow diagram in Figure 1 (see next page), combines minimal-function phase refinement and real-space filtering. It is an iterative process that is repeated until a solution is achieved or a designated number of cycles have been performed. With reference to Figure 1, the major steps of the algorithm are described next.

**A. Generate invariants.** Normalized structure-factor magnitudes ($|E|$'s) are generated by standard scaling methods and the triplet invariants that involve the largest corresponding $|E|$'s are generated. Parameter choices that must be made at this stage include the numbers of phases and triplets to be used. The total number of invariants is ordinarily chosen to be at least 100 times the number of atoms whose positions are to be determined.

**B. Generate trial structure.** A trial structure or model is generated that is comprised of a number of randomly positioned atoms equal to the number of atoms in the unit cell. The starting coordinate sets are subject to the restrictions that no two atoms are closer than a specified distance (normally 1.2Å) and that no atom is within bonding distance of more than four other atoms.

**C. Structure-factor calculation.** A normalized structure-factor calculation based on the trial coordinates is used to compute initial values for all the desired phases simultaneously. In subsequent cycles, peaks selected from the most recent Fourier series are used as atoms to generate new phase values.

**D Phase refinement.** The values of the phases are perturbed by a *parameter-shift* method in which $R(\phi)$, which measures the mean-square difference between estimated and calculated structure invariants, is reduced in value. $R(\phi)$ is initially computed on the basis of the set of phase values obtained from the structure-factor calculation in step C. The phase set is ordered in decreasing magnitude of the associated $|E|$'s. The value of the first phase is incremented by a preset amount and $R(\phi)$ is recalculated. If the new calculated value of $R(\phi)$ is lower than the previous one, the value of the first phase is incremented again by the preset amount. This is continued until $R(\phi)$ no longer decreases or until a predetermined number of increments has been applied to the first phase. A completely analogous course is taken if, on the initial incrementation, $R(\phi)$ increases, except that the value of the first phase is decremented until $R(\phi)$ no longer decreases or until the predetermined number of decrements has been applied. The remaining phase values are varied in sequence as just described. Note that, when the $i$th phase value is varied, the new values determined for the previous $i$-1 phases are used immediately in the calculation of $R(\phi)$. This process, when convergent, yields the constrained global minimum of $R(\phi)$. The step size and number of steps are variables whose values must be chosen.

**E. Fourier summation.** Fourier summation is used to transform phase information into an electron-density map (Refer to Eq. (6)). Normalized structure-factor amplitudes, $|E|$'s, have been used at this stage (rather than $F$'s) because phases are

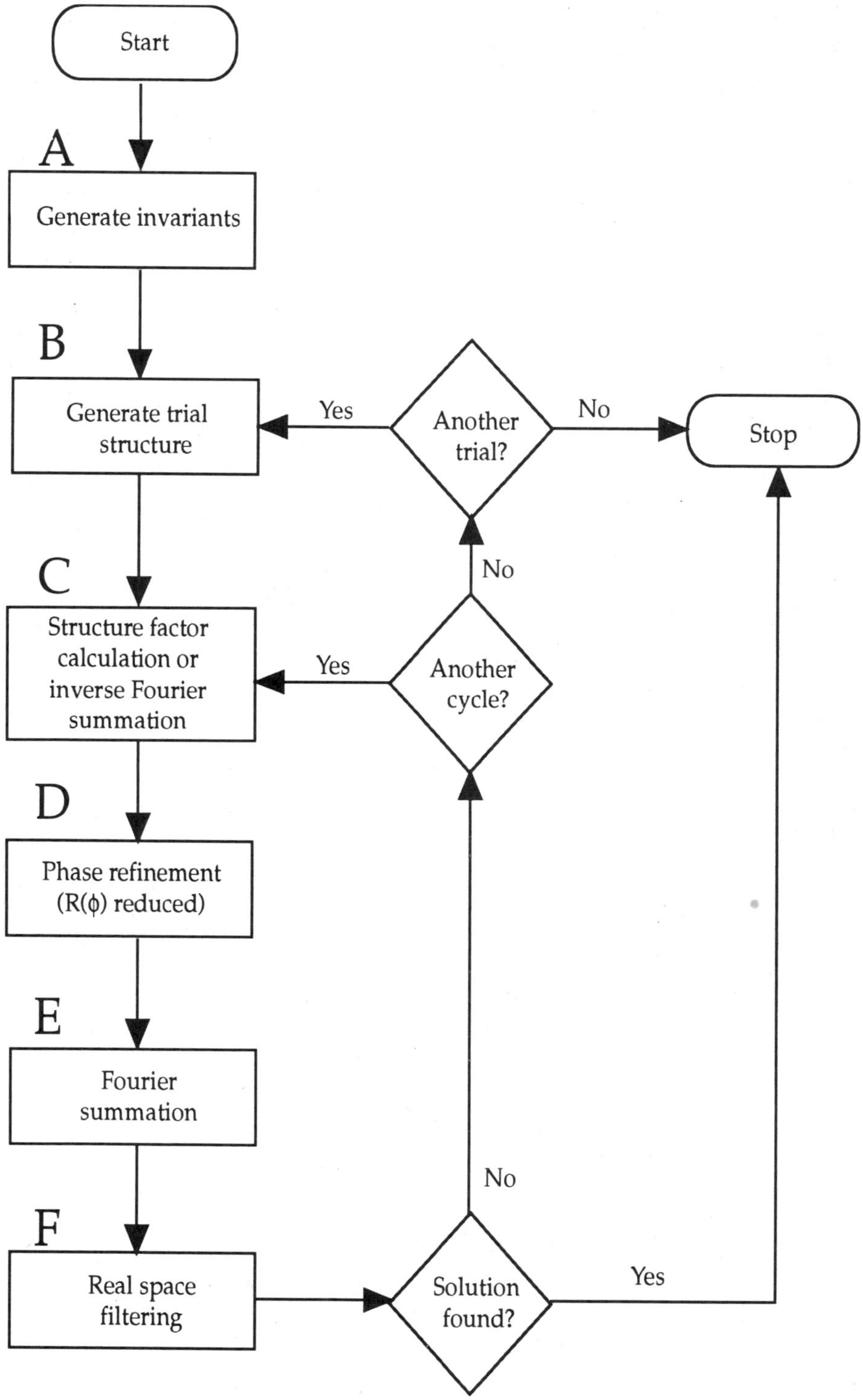


Figure 1. Flow chart for ***Shake-and-Bake***, the minimal-function phase refinement and real-space filtering procedure.

available for the largest $E$'s but not for all the largest $F$'s. The grid size must be specified.

**F. Real-space filtering (Identities among phases imposed).** Image enhancement has been accomplished by a discrete electron-density modification consisting of the selection of a specified number of the largest peaks on the Fourier map for use in the next structure-factor calculation. The simple choice, in each cycle, of a number of the largest peaks corresponding to the number of expected atoms has given satisfactory results. No minimum-interpeak-distance criterion is applied at this stage.

## 7. The applications

***Shake-and-Bake*** has been tested successfully, with no failure, using experimentally measured atomic resolution data for some 30 structures. These structures range in size from 25 to 500 independent non-hydrogen atoms, and four of these are in the 300 to 500 atom range. Although many of these structures had been previously known, this fact was not used in these applications. In all cases those trials which led to solutions were readily identified by the behavior of the minimal function $R(\phi)$.

## References

G.T. DeTitta, C.M. Weeks, P. Thuman, R. Miller, & H. Hauptman, *Structure Solution by Minimal-Function Phase Refinement and Fourier Filtering. I. Theoretical Basis*, Acta Cryst. **A50** (1994), 203-210.

D. Harker & J.S. Kasper, *Phases of Fourier Coefficients Directly from Crystal Diffraction Data*, Acta Cryst. **1** (1948), 70-75.

J. Karle & H. Hauptman, *The Phases and Magnitudes of the Structure Factors*, Acta Cryst. **3** (1950), 181-187.

G.M. Sheldrick, *Phase Annealing in SHELX-90: Direct Methods for Larger Structures*, Acta Cryst. **A46** (1990), 467-473.

C.M. Weeks, G.T. DeTitta, H. Hauptman, P. Thuman, & R. Miller, *Structure Solution by Minimal-Function Phase Refinement and Fourier Filtering. II. Implementation and Applications*, Acta Cryst. **A50** (1994), 210-220.

M.M. Woolfson & J.-X. Yao, *On the Application of Phase Relationships to Complex Structures. XXX. Ab Initio Solution of a Small Protein by SAYTAN*, Acta Cryst. **A46** (1990), 409-413.

*Current address*: Hauptman-Woodward Medical Research Institute, 73 High Street, Buffalo, NY 14203-1196; Adjunct Research Professor, State University of New York, University at Buffalo, Dept. of Biophysical Sciences and Computer Sciences, Buffalo, New York 14260.

*E-mail address*: hauptman@hwi.buffalo.edu

DIMACS Series in Discrete Mathematics
and Theoretical Computer Science
Volume **23**, 1996

# Knowledge Based Structure Prediction of the Light-Harvesting Complex II of *Rhodospirillum molischianum*

Xiche Hu, Dong Xu, Kenneth Hamer, Klaus Schulten, Juergen Koepke, and Hartmut Michel

ABSTRACT. We illustrate in this article how one proceeds to predict the structure of integral membrane proteins using a combined approach in which molecular dynamics simulations and energy minimization are performed based on structural information from conventional structure prediction methods and experimental constraints derived from biochemical and spectroscopical data. We focus here on the prediction of the structure of the light-harvesting complex II (LH–II) of *Rhodospirillum molischianum*, an integral membrane protein of 16 polypeptides aggregating and binding to 24 bacteriochlorophyll a's and 12 lycopenes. Hydropathy analysis was performed to identify the putative transmembrane segments. Multiple sequence alignment propensity analyses further pinpointed the exact sites of the 20 residue long transmembrane segment and the four residue long terminal sequence at both ends, which were independently verified and improved by homology modeling. A consensus assignment for secondary structure was derived from a combination of all the prediction methods used. The three-dimensional structures for the $\alpha$- and the $\beta$-apoprotein were built by comparative modeling. The resulting tertiary structures were combined into an $\alpha\beta$ dimer pair with bacteriochlorophyll a's attached under constraints provided by site directed mutagenesis and FT Resonance Raman spectra, as well as by conservation of residues. The $\alpha\beta$ dimer pairs were then aggregated into a quaternary structure through molecular dynamics simulations and energy minimization. The structure of LH–II, so determined, was an octomer of $\alpha\beta$ heterodimers forming a ring with a diameter of 70 Å. We discuss how the resulting structure may be used to solve the phase problem in X-ray crystallography in a procedure called molecular replacement.

## Introduction

One of the outstanding problems in structural biology is protein folding, i.e., given the linear sequence of amino acids that constitute a protein, what would be

1991 *Mathematics Subject Classification.* Primary 70F10,92-08; Secondary 92C05,92D20.

*Key words and phrases.* light-harvesting complex; purple bacteria; protein folding; protein structure; sequence analysis.

This work was supported by the Carver Charitable Trust and the National Institute of Health (NIH grant P41RR05969).

its native, or folded, state in three-dimensional space. The native conformation of a protein corresponds to a global minimum in free energy space. Mathematically, the problem can be formulated as global minimum search in multidimensional space. The solution to this problem has proven difficult because in theory it belongs to a class known as NP complete problems, and in practice the objective function contains too many local minima [**GJ79, CSW94, NVS91**]. As a result, current structure prediction methods are mostly empirical in nature [**Fas89a**]. We illustrate in this article how one proceeds to predict the structure of integral membrane proteins using a combined approach in which molecular dynamics simulations and energy minimization are performed based on structural information from conventional structure prediction methods and experimental constraints derived from biochemical and spectroscopical data.

With the advent of recombinant DNA technology, determination of primary sequences of proteins is proceeding at a much faster pace than determination of atomic resolution protein structures. While thousands of membrane protein sequences are available, there are few detailed three dimensional membrane protein structures known. Available structures, at present, include the photosynthetic reaction centers from *Rps. viridis* [**DEM$^+$85**] and *Rb. sphaeroides* [**AYKR87**], porins from *Rb. capsulatus* [**WKN$^+$91**] and *E. coli* [**CSR$^+$92**], bacteriorhodopsin from *H. halobium* [**HBC$^+$90**], and a plant light harvesting complex [**KWF94**]. Since structural information of membrane proteins is vital to understanding of their cellular functions, a great effort has been made to predict membrane protein structures from their primary sequences [**Whi94, PdVA94, vH94b, vH94a, vH92, CEHR92, Pop93, PdV90, PE90, Jäh89, TEPL94, AAM$^+$94, Fas89b, ARH82**]. Most structure prediction algorithms can be categorized into three main classes: statistical [**CF78, GOR78, Lev78, RA86, PA94, LSBW94, HK89, KL85**], physicochemical [**KD82, ARH82, ESG86, CCM$^+$87, RDE89, Whi94, Eis84, vH88**], and comparative [**ZBTS87, BSST87, CAKF86, PCC92, Bus86, JSSB94, SB93, RW88**]. Presently, prediction of tertiary structure is only of practical use when the structure of a homologous protein is already known. Protein homology modeling typically involves the prediction of side-chain conformations in the modeled protein while assuming a main-chain trace taken from a known tertiary structure of a homologous protein. However, the tertiary structures of proteins have been successfully predicted when experimentally derived constraints are used in conjunction with heuristic methods [**RC93**]. In such a knowledge based approach information, both from the three-dimensional structures of homologous proteins and from the general analysis of protein structure, is used to derive constraints for modeling a protein of known sequence, but unknown structure. The rest of this article will focus on our recent efforts in predicting the structure of the light-harvesting complex II (LH–II) of *Rs. molischianum*, an integral membrane protein of 16 polypeptides aggregating and binding to 24 bacteriochlorophyll-a (BChla) and 12 lycopenes.

## Light Harvesting Complex

Photosynthetic organisms radically increase the efficiency and decrease the complexity of their energy gathering apparatus by surrounding the complex photosynthetic reaction centers with simple, pigment-rich protein aggregates known as light harvesting complexes or antenna complexes. While the reaction center has a very rapid cycle time, on the order of $10^3$ photons per second [**vGS88**], it cannot independently collect enough photons to saturate itself. With the inclusion of the antenna complexes, the bacterium can collect and channel to each reaction center much more light energy.

In most purple bacteria there are two basic types of light harvesting complexes: the light harvesting complex I, or LH–I, is found directly surrounding the photosynthetic reaction centers, while the light harvesting complex II, or LH–II, surrounds the LH–I—reaction center aggregates. LH–I absorbs at longer wavelengths than LH–II, typically with a strong absorption band between 870 and 1015 nm, and is found in all types of purple photosynthetic bacteria [**Zub93**]. LH–II is found in some species (notable exceptions are *Rs. rubrum* and *Rps. viridis*) of bacteria, and typically has one absorption band between 820 and 860 nm, and another around 800 nm [**Zub85**]. For *Rs. molischianum*, the LH–II complex displays two peaks at 800 and 850 nm and is often referred to as the B800-850 complex. Since the photosynthetic reaction center absorbs in the deep infrared (960 nm for the *Rps. viridis* reaction center) there is a clear energetic hierarchy in the light-harvesting system, with the LH–II complex absorbing light at the highest energy, surrounding LH–I which absorbs at a lower energy, which in turn surrounds the reaction center which absorbs at the lowest energy. This arrangement naturally channels energy from the outer regions of the antenna complex to the reaction center.

Studies of the mechanism of the photosynthetic reaction center have been greatly enhanced by the determination of its three-dimensional structure [**DM89**]. However, structural information about light-harvesting complexes is still limited to spectroscopic and biochemical characterization [**HC91, SvG91, ZB91**]. The LH–II complex of *Rs. molischianum* has been crystallized and X-ray diffraction data have been collected up to 2.4 Å resolution [**KM, Mic91**]. To resolve a structure from measured diffraction intensities requires knowledge of phases which is unobtainable from a single diffraction experiment. Conventionally, the phase problem is solved by means of the multiple isomorphous replacement method. An alternative solution to the phase problem is to phase the structure by using a homologous structure in a procedure called molecular replacement [**Ros72, Lat85**]. In this method, a homologous probe structure is fit into the unit cell of the unknown structure and used to generate an initial phasing model for the unknown structure. At the time when this project was initiated, there existed no homologous structure to LH–II of *Rs. molischianum*. We attempted to predict the structure of *Rs. molischianum* and intended to use the predicted structure as a probe structure in the molecular replacement method to resolve the 2.4 Å X-ray diffraction data into an atomic structure. We report here (1) the predicted secondary structure for the $\alpha$- and $\beta$-polypeptides; and (2) the predicted quaternary structure for the aggregated LH-II complex of *Rs. molischianum* as well as the structure prediction methods. At the end, the current prediction will be compared with the recently published structure for LH–II from *Rps. acidophila* by McDermott et al., who have successfully solved the phase problem for their structure by conventional means [**MPF+95**].

All light-harvesting complexes display a remarkable similarity in the way they are constructed [**Zub85, ZB91**]. The basic structural unit is a heterodimer of two small polypeptides, commonly referred to as $\alpha$ and $\beta$, both shorter than 60 amino acids, which non-covalently bind BChla and carotenoid molecules. These heterodimers aggregate to a large complex, functioning as light harvesting antennae. The size of the aggregate depends on the type of light-harvesting complex and varies from species to species, ranging from a putative hexamer for LH–II of *Rb. sphaeroides* [**BVC+93**] to a hexadecamer for LH–I of *Rs. rubrum* [**KBG95**].

Various models have been proposed for the light-harvesting complex of purple bacteria [**OH94, Zub86, ZB91**]. The majority of these models are concerned with secondary structural features and the topology of the heterodimers. However, no atomic level modeling of the aggregated complex has been attempted before. Our goal is to build a model structure for LH–II of *Rs. molischianum* and to use it as a probe structure in the framework of the molecular replacement method. The ultimate correctness of the predicted structure can be tested by its ability to serve as a successful search model to resolve the X-ray diffraction data in terms of a consistent electron density profile to which an atomic structure can be configured.

## Method

In practice, the task of prediction is divided into three stages: (1) predict the secondary structure of $\alpha$- and $\beta$-polypeptides from their amino acid sequences; (2) build the tertiary structures for $\alpha$- and $\beta$-polypeptides by comparative modeling; (3) fold the tertiary structures into an aggregated complex (quaternary structure) by means of molecular dynamics simulations and energy minimization under the constraints of experimental data and the predicted secondary structure features. Finally, the molecular replacement test was performed using the predicted structure as a probe structure to resolve the unknown crystal structure. A flowchart of the entire procedure is provided in Figure 1.

The molecular dynamics simulations and energy minimizations described in this article were carried out using the program X-PLOR [**Brü92**]. All the simulation protocols were programmed with the versatile X-PLOR script language. An integration time step of 1 fs was chosen in the Verlet algorithm. The simulation of LH–II placed the protein in a vacuum. The parameters and charges used for the system were, respectively, the CHARMm all-atom parameter file `parallh22x.pro` and the CHARMm all-atom partial charge file `topallh22x.pro` [**BBO+83, A.** ] except for BChla. The partial charges and parameters for BChla were taken from those used in [**TSD+88**] for BChlb except for slight modifications to accommodate BChla. A cut-off distance of 12 Å for non-bonded interactions and a dielectric constant $\varepsilon = 1$ were employed.

## Prediction of Secondary Structure

The LH–II complex of *Rs. molischianum* consists of two BChla-binding polypeptides $\alpha$ and $\beta$ with the following sequences:

$\alpha$: `SNPKDDYKIWLVINPSTWLPVIWIVATVVAIAVHAAVLAAPGSNWIALGAAKSAAK`

$\beta$: `AERSLSGLTEEEAIAVHDQFKTTFSAFIILAAVAHVLVWVWKPWF`

The smallest compositional unit of LH–II contains a pair of $\alpha$ and $\beta$ polypeptides, three BChla and 1.5 lycopene molecules [**GLRM93**]. It has been determined by sedimentation equilibrium experiments that the native LH–II complex is an octomer of such $\alpha\beta$ units [**KGvdB+92**]. The space group for the crystal is $P42_12$ with cell dimensions of 92 × 92 × 209 Å.

Since the LH–II complex is an integral membrane protein, we performed hydropathy analysis to identify the putative transmembrane segments [**KD82, Whi94**]. The transmembrane segments of polypeptides are usually forced to adopt an $\alpha$-helical conformation due to constraints of the hydrophobic core of the membrane [**ESG86**].

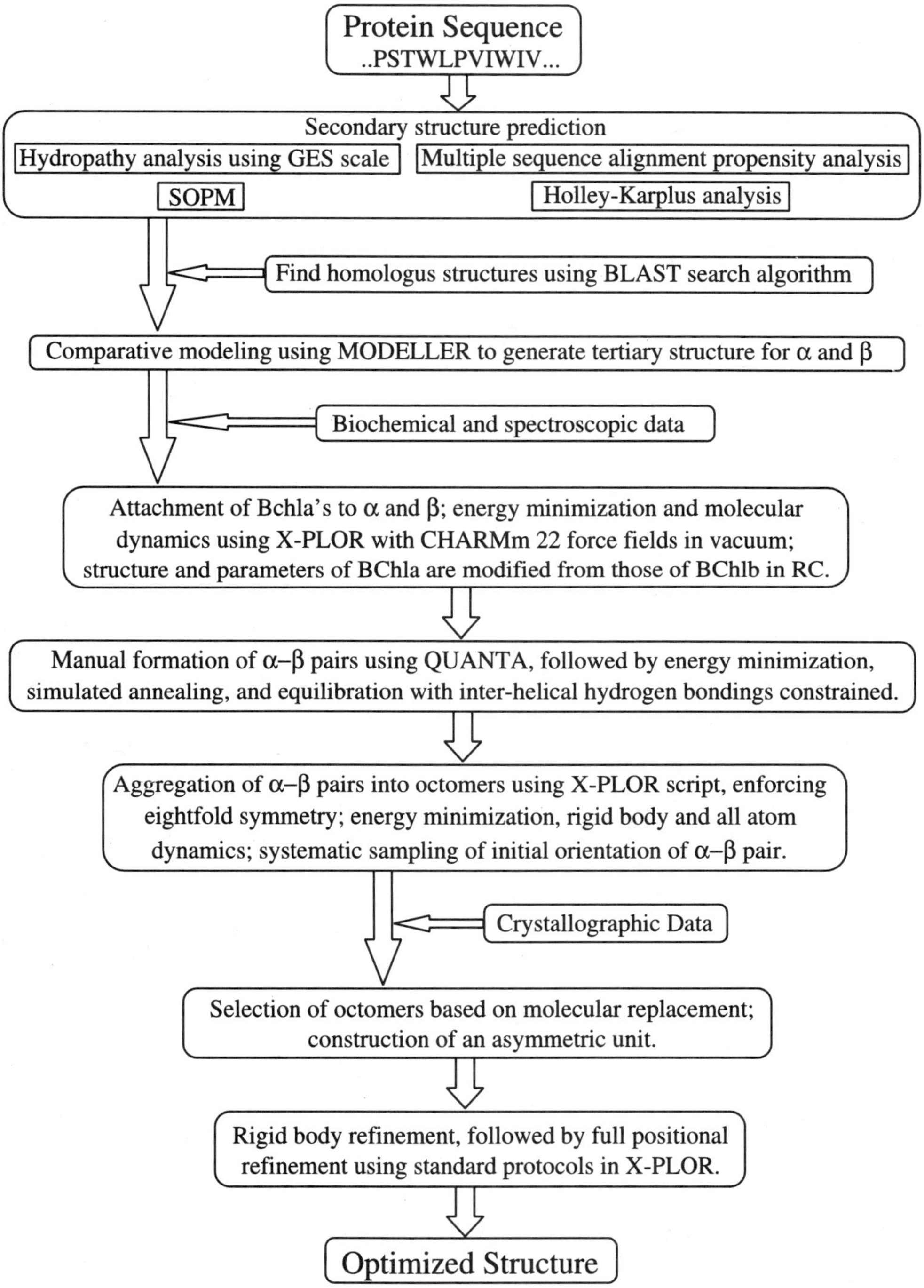


FIGURE 1. Flowchart outlining the entire structure prediction procedure.

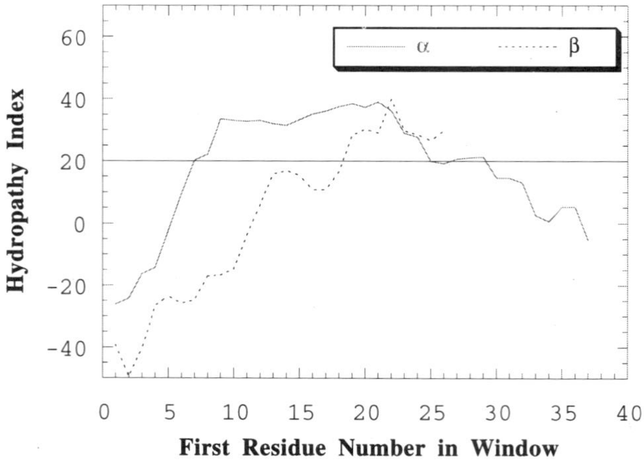


FIGURE 2. Hydropathy plot of $\alpha$- and $\beta$-polypeptides of the LH–II complex based on the GES hydrophobicity scale (see text) with a window size of 20 amino acids.

Hydropathy analysis assumes that transmembrane segments are comprised mainly of hydrophobic residues because of the low solubility of polar side chains in nonpolar lipid bilayers. It takes about 20 amino acids (in an $\alpha$ helix) to span the hydrocarbon regions of fluid bilayers that are typically 30 Å thick. Shown in Figure 2 is a hydropathy plot for the $\alpha$- and $\beta$-subunits based on the GES hydrophobicity scale [**ESG86**] with a window size of 20 amino acids. The GES scale is derived from the free energy cost for transferring amino acids from the interior of a membrane to its water surroundings. On such a scale, a peak of 20 kcal/mol or higher identifies a transmembrane segment. Figure 2 clearly shows that a transmembrane segment exists for both $\alpha$- and $\beta$-polypeptides. The highest peak occurs at a first residue number of 21 and 22 for the $\alpha$- and the $\beta$-polypeptide respectively. The transmembrane segments are thus identified as $\alpha$-Val-21:$\alpha$-Ala-40 and $\beta$-Thr-22:$\beta$-Trp-41. Hydropathy analyses with two other widely used hydrophobicity scales, i.e., Kyte & Doolittle [**KD82**] and Eisenberg consensus [**EWTW82**] scales, generate essentially the same hydrophobic core for the $\alpha$- and the $\beta$-polypeptide as the GES scale.

In addition to the hydropathy analysis, we have also carried out a multiple sequence alignment propensity analysis using the method of Persson and Argos [**PA94**] which combines two sets of propensity values (one for the middle, hydrophobic portion and one for the terminal region of the transmembrane span) to determine the transmembrane segments from multiply aligned amino acid sequences. A novel aspect of this method is the use of evolutionary information in the form of multiple sequence alignments as input in place of a single sequence. The method was shown to be more successful than predictions based on a single sequence alone. A total of 12 homologous sequences of LH–II and LH–I complexes have been aligned (see Figure 3 ) and analyzed. As shown in Figure 4, the transmembrane segment determined by this method spans from Trp-18 to Val-37 for the $\alpha$-polypeptide and from Thr-22 to Trp-41 for the $\beta$-polypeptide. The hydropathy analysis identifies the region of the transmembrane segment. Multiple sequence

```
                                    α-polypeptide
Rs. Molischianum B800-850              snpkddYKIWLVINPSTWLPVIWIVATVVAIAVHAAVLAAPGSNWIALGAAKsaak----
Rc. gelatinosus DSM 149 B800-850       -----mWRIWRLFDPMRAMVAQAVFLLGLAVLIHLMLLGTNKFNWLDGAKKApvasa---
Rs. rubrum B870                        -----mWRIWQLFDPRQALVGLATFLFVLALLIHFILLSTERFNWLEGASTQpvqts---
Rb. capsulatus B870                    --mskfYKIWLVFDPRRVFVAQGVFLFLLAVLIHLILLSTPAFNWLTVATAKhgyvaaaq
Rb. sphaeroides B870                   --mskfYKIWMIFDPRRVFVAQGVFLFLLAVMIHLILLSTPSYNWLEISAAKynrvavae
Rps. marina B880                       -----mWKVWLLFDPRRTLVALFTFLFVLALLIHFILLSTDRFNWMQGAPTApaqts---
Rp. viridis B1015                      eyrtasWKLWLILDPRRVLTALFVYLTVIALLIHFGLLSTDRLNWWEFQRGLpkaa----

Rs. Molischianum B800-850              snpkddYKIWLVINPSTWLPVIWIVATVVAIAVHAAVLAAPGSNWIALGAAKsaak----
Rp. viridis B1015                      eyrtasWKLWLILDPRRVLTALFVYLTVIALLIHFGLLSTDRLNWWEFQRGLpkaa----
Rc. gelatinosus DSM 149 B800-850       ---mnQGKVWRVVKPTVGVPVYLGAVAVTALILHGGLLAKTDWFGaywnggkkaaaa---
Rb. sphaeroides 2.4.1 B800-850         ---mtNGKIWLVVKPTVGVPLFLSAAVIASVVIHAAVLTTTTWLPayyqgsaavaae---
Rb. capsulatus B800-850                ---mnNAKIWTVVKPSTGIPLILGAVAVAALIVHAGLLTNTTWFAnywngnpmatvvava
Rps. acidophila Ac70.50 B800-850       ---mnQGKIWTVVNPSVGLPLLLGSVTVIAILVHAAVLSHTTWFPaywqgglkkaa----
Rps. palustris 2.6.1 B800-850          ---mnQGRIWTVVNPGVGLPLLLGSVTVIAILVHYAVLSNTTWFPkywngatvaapaaa-

                                    β-polypeptide
Rs. Molischianum B800-850              ----AERSLSGLTEEEAIAVHDQFKTTFSAFIILAAVAHVLVWVWKPWf--------
Rc. gelatinosus DSM 149 B800-850       --aeRKGSISGLTDDEAQEFHKFWVQGFVGFTAVAVVAHFLVWVWRPWl--------
Rs. rubrum B870                        --evKQESLSGITEGEAKEFHKIFTSSILVFFGVAAFAHLLVWIWRPWvpgpngys-
Rb. capsulatus B870                    -adkNDLSFTGLTDEQAQELHAVYMSGLSAFIAVAVLAHLAVMIWRPWf--------
Rb. sphaeroides B870                   -adkSDLGYTGLTDEQAQELHSVYMSGLWPFSAVAIVAHLAVYIWRPWf--------
Rps. marina B880                       aeidRPVSLSGLTEGEAREFHGVFMTSFMVFIAVAIVAHILAWMWRPWipgpegia-
Rp. viridis B1015                      --adLKPSLTGLTEEEAKEFHGIFVTSTVLYLATAVIVHYLVWTAKPWiapipkgwv
Rc. gelatinosus DSM 149 B800-850       addaNKVWPSGLTTAEAEELQKGLVDGTRIFGVIAVLAHILAYAYTPWlh-------
Rb. sphaeroides 2.4.1 B800-850         -ddlNKVWPSGLTVAEAEEVHKQLILGTRVFGGMALIAHFLAAAATPWlg-------
Rb. capsulatus B800-850                -mtdDKAGPSGLSLKEAEEIHSYLIDGTRVFGAMALVAHILSAIATPWlg-------
Rps. acidophila Ac70.50 B800-850       --adDVKGLTGLTAAESEELHKHVIDGTRVFFVIAIFAHVLAFAFSPWlh-------
Rps. palustris 2.6.1 B800-850          addpNKVWPTGLTIAESEELHKHVIDGSRIFVAIAIVAHFLAYVYSPWlh-------
```

FIGURE 3. Sequence alignment of the $\alpha$ and $\beta$-polypeptides of the LH–II complex of *Rs. molischianum* with $\alpha$- and $\beta$-polypeptides of LH–II and LH–I complexes from other photosynthetic bacteria. Color coding: Blue – Highly conserved; Magenta – Nearly conserved. Alignment done using program MACAW (Multiple Alignment Construction & Analysis Workbench) [**SAL91**].

alignment propensity analysis further pinpoints the most probable site of the 20 residue long transmembrane segment and the four residue long terminal sequence at both ends.

Table I. SOPM and Holley-Karplus Predictions

| | $\alpha$ – *polypeptide* |
|---|---|
| sequence | SNPKDDYKIWLVINPSTWLPVIWIVATVVAIAVHAAVLAAPGSNWIALGAAKSAAK |
| SOPM[a] | CCCCCCCCEEEEECCCCCEEEHHHHHHHHHHHHHHHHHHCCCCHEEHHHHHHHHHH |
| Holley-Karplus[b] | -------eEEEEe-------hhHHHHHHHHHHHHhhhhh-----hhHHHHHHHh-- |
| | $\beta$ – *polypeptide* |
| sequence | AERSLSGLTEEEAIAVHDQFKTTFSAFIILAAVAHVLVWVWKPWF |
| SOPM | HHHHHCCCCHHHHHHHHHHHHHHHHHHHHHHHHHHHHHHHHHHHH |
| Holley-Karplus | --------hHHHHHHHHhh-hhhh-hHHHHHHhhhhHHH------ |

a: Self optimized prediction method (SOPM) for protein secondary structure prediction [**GD94**]. b: Neural network based informational approach for protein secondary structure prediction[**HK91**].

To confirm the above secondary structure prediction, we performed more sequence analyses with various secondary structure prediction methods including SOPM (self optimized prediction method) and Holley-Karplus analysis [**GD94, HK89, HK91**]. SOPM takes structure classes into account and iteratively optimized prediction parameters to increase prediction quality. The Holley-Karplus

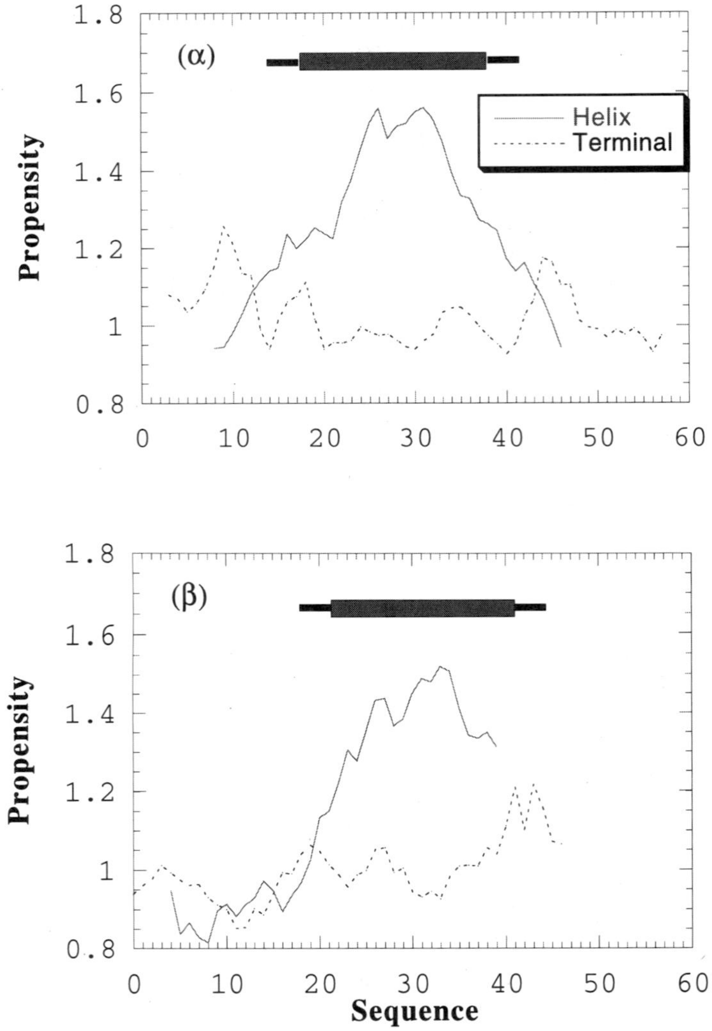


FIGURE 4. Propensity profile and assignment of 20 residue long transmembrane span (thick bar) and four residue long helical terminal (thin bar) for $\alpha$- and $\beta$-polypeptides of the LH–II complex based on the multiple sequence alignment propensity analysis method of Persson and Argos. Solid line - transmembrane helix; dashed line - terminal region.

prediction is an information-based neural network approach. SOPM and Holley-Karplus predictions are listed in Table I. The results are consistent with the assignment of transmembrane segments, for both the $\alpha$- and $\beta$-apoprotein, derived above. Both analyses demonstrate that the transmembrane segments have a high tendency to form an $\alpha$-helix.

The secondary structure assignments were further verified and improved by homology modeling. Although there exists no structure which is highly homologous to either $\alpha$- or $\beta$-apoproteins as a whole, we have found structures in the PDB (Protein Data Bank) which are homologous to multiple fragments of $\alpha$- and $\beta$-apoproteins. Table II lists some of the homologous fragments to $\alpha$- and $\beta$-apoproteins resulting from a PDB BLAST search [**AGM$^+$90**]. Detailed PDB BLAST search results are given in Table III. A homology with 26% identity and 50% positive exists between a segment of the $\alpha$-apoprotein [$\alpha$-Leu-11 to $\alpha$-Ala-40] and the transmembrane helix D of the M subunit of the photosynthetic reaction center of *Rb. sphaeroides* [M subunit, residues 196:225] [**MWG$^+$86**]. To establish the statistical significance of this alignment, we performed a statistical analysis with the BESTFIT program in the GCG package [**DHS84**]. Using a gap generating penalty of 3.0 and a gap extension penalty of 0.1, the BESTFIT program generated exactly the same alignment as shown in Table II with a quality of 19.2. The average quality for 100 randomized alignments in which the query sequence is randomly permuted (shuffled) is 12.7 with a standard deviation of 1.3. That gave rise to a Z-score of 5, which indicated a "possibly significant" alignment according to [**LP85, Pea90**]. Perhaps a more convincing support for this alignment is the fact that structurally, both proteins exist as $\alpha$-helical transmembrane segment, and functionally, both proteins contain bacteriochlorophyll-binding residues. Also, two short segments of $\beta$-apoprotein are highly homologous to two corresponding segments in the L subunit of the photosynthetic reaction center of *Rb. sphaeroides* [see Table II]. Reaction center L subunit sequence 122:133 `AFAILAYLTLVL` is located in the center of the transmembrane helix C [**MWG$^+$86**], which exhibits a close correspondence with our secondary structure assignment of the transmembrane segment for the $\beta$-apoprotein. Sequence `WVKLPWW` near the C-terminal end of the reaction center L subunit shows a close correspondence with sequence `WVWKPWF` of the $\beta$-apoprotein of LH–II.

Table II. Alignment of Homologous Sequences

| | | |
|---|---|---|
| alpha | | SNPKDDYKIWLVINPSTWLPVIWIVATVVAIAVHAAVLAAPGSNWIALGAAKSAAK |
| 1R1E\|E | 97:106 | ---KDDYGEWRVV------------------------------------------- |
| 2RCR\|M | 196:225 | ----------LFYNPFHGLSIAFLYGSALLFAMHGATILA--------------- |
| 1ACB\|E | 22:26 | --------------------------------------AVPGS------------- |
| 1CPC\|L | 68:72 | ---------------------------------------APGGN------------ |
| 1TYP\|A | 91:102 | -------------------------------------------NWKALIAAKNKA- |
| | | |
| beta | | AERSLSGLTEEEAIAVHDQFKTTFSAFIILAAVAHVLVWVWKPWF |
| 1AAM | 349:365 | ---SFSGLTKEQVLRLREEF------------------------- |
| 1GRA | 381:388 | ------GLTEDEAI------------------------------- |
| 256B\|A | 85:97 | ----------KEAQAAAEQLKTT---------------------- |
| 2RCR\|L | 122:133 | -------------------------AFAILAYLTLVL-------- |
| 1EPS | 22:36 | --------------------KTVSNRALLLAALAH---------- |
| 2BBQ\|A | 52:61 | ------------------------------LRSIIHELLW----- |
| 2RCR\|L | 266:272 | --------------------------------------WVKLPWW |

1R1E: Eco Ri endonuclease (E.C.3.1.21.4) complex with TCGCGAATTCGCG; 2RCR: Photosynthetic reaction center from *Rhodobacter sphaeroides*; 1ACB: Alpha-Chymotrypsin (E.C.3.4.21.1) complex with Eglin C; 1CPC: C-Phycocyanin; 1TYP: Trypanothione reductase (E.C.1.6.4.8); 1AAM: Aspartate aminotransferase (E.C.2.6.1.1) mutant; 1GRA: Glutathione reductase (E.C.1.6.4.2) (oxidized) complex; 256B: Cytochrome b562 (oxidized); 1EPS: 5-enol-pyruvyl-3-phosphate synthase (E.C.2.5.1.9); 2BBQ: Thymidylate synthase (E.C.2.1.1.45) complex.

A secondary structure analysis was performed for all the homologous proteins listed in Table II with the method of Kabsch and Sander which defines the secondary structure of a protein by means of pattern recognition of hydrogen-bonded and geometrical features [**KS83**]. The resulting secondary structure assignment for all the homologous fragments are listed in Table III along with PDB BLAST search results. As shown in Table III, all the homologous fragments located within the putative transmembrane region, except Eco Ri Endonuclease, exist in an $\alpha$-helical conformation.

Table III(A). PDB BLAST Search[a] Results

**$\alpha$-polypeptide**

(1) pdb|1R1E|E Identities = 6/10 (60%), Positives = 7/10 (70%)

```
Query:    4 KDDYKIWLVI 13
            KDDY W  V+
Sbjct:   97 KDDYGEWRVV 106
```

(2) pdb|2RCR|M Identities = 8/30 (26%), Positives = 15/30 (50%)

```
Query:   11 LVINPSTWLPVIWIVATVVAIAVHAAVLAA 40
            L  NP   L + ++  + +  A+H A + A
Sbjct:  196 LFYNPFHGLSIAFLYGSALLFAMHGATILA 225
2nd str^b:  TTS HHHHHHHHHHHHHHHHHHHHHHHHHT
```

(3) pdb|1ACB|E Identities = 4/5 (80%), Positives = 4/5 (80%)

```
Query:   39 AAPGS 44
            A PGS
Sbjct:   22 AVPGS 26
```

(4) pdb|1CPC|L Identities = 4/5 (80%), Positives = 4/5 (80%)

```
Query:   40 APGSN 45
            APG N
Sbjct:   68 APGGN 72
```

(5) pdb|1TYP|A Identities = 8/12 (66%), Positives = 9/12 (75%)

```
Query:   44 NWIALGAAKSAA 55
            NW AL AAK+ A
Sbjct:   91 NWKALIAAKNKA 102
2nd str^b:  -HHHHHHHHHHH
```

a. Basic local alignment search tool (BLAST) search [**AGM$^+$90**].

b. Secondary structure determined by DSSP [**KS83**]. H: 4-helix ($\alpha$-helix); E: extended strand, participates in $\beta$-ladder; T: H-bonded turn; S: bend.

Table III(B). PDB BLAST Search Results

**$\beta$-polypeptide**

(1) pdb|1AAM| Identities = 6/16 (37%), Positives = 12/16 (75%)

```
Query:     4 SLSGLTEEEAIAVHDQ 19
             S SGLT+E+ + + ++
Sbjct:   349 SFSGLTKEQVLRLREE 364
2nd str:     EEE---TTTTTTSSSS
```

(2) pdb|1GRA| Identities = 7/8 (87%), Positives = 8/8 (100%)

```
Query:     7 GLTEEEAI 14
             GLTE+EAI
Sbjct:   381 GLTEDEAI 388
2nd str:     E--HHHHH
```

(3) pdb|256B|A Identities = 7/13 (53%), Positives = 9/13 (69%)

```
Query:    11 EEAIAVHDQFKTT 23
             +EA A  +Q KTT
Sbjct:    85 KEAQAAAEQLKTT 97
2nd str:     HHHHHHHTHHHHH
```

(4) pdb|2RCR|L Identities = 7/12 (58%), Positives = 8/12 (66%)

```
Query:    26 AFIILAAVAHVL 37
             AF ILA + VL
Sbjct:   122 AFAILAYLTLVL 133
2nd str:     HHHHHHHHHHHT
```

(5) pdb|1EPS| Identities = 7/15 (46%), Positives = 11/15 (73%)

```
Query:    21 KTTFSAFIILAAVAH 35
             KT  +  ++LAA+AH
Sbjct:    22 KTVSNRALLLAALAH 36
2nd str*:    HHHHHHHHHHHHHHH
```

(6) pdb|2BBQ|A Identities = 4/10 (40%), Positives = 7/10 (70%)

```
Query:    30 LAAVAHVLVW 39
             L ++ H L+W
Sbjct:    52 LRSIIHELLW 61
2nd str:     HHHHHHHHHH
```

(7) pdb|2RCR|L Identities = 4/7 (57%), Positives = 5/7 (71%)

```
Query:    39 WVWKPWF 45
             WV  PW+
Sbjct:   266 WVKLPWW 272
2nd str:     HHT-TTS
```

Homology modeling was also used to improve secondary structure assignment. In case that no clear-cut secondary structure assignment can be made, the secondary structural features of the homologous structure were employed to establish the secondary structure identity of $\alpha$- and $\beta$-polypeptides. Specifically, the N and

C termini of the transmembrane helix for the $\alpha$-polypeptide were set to Ser-16 and Ala-41 in analogy to the homologous transmembrane helix D of the reaction center M subunit and in consideration of the known fact that all residues in the NPS (residue 14:16) and PGSN (residue 41:44) segments have a high tendency to form a reverse turn [**Lev78**]. Similarly, the C terminus of the transmembrane helix for the $\beta$-apoprotein was set to Lys-42 in analogy to the homologous reaction center L subunit sequence 266:272 `WVKLPWW` and in consideration of the proline residue. The N terminus of the transmembrane helix for the $\beta$-apoprotein was determined to be Tyr-10 in analogy to the highly homologous glutathione reductase sequence `GLTEDEAI`. This assignment of the secondary structure at the N terminus is consistent with both SOPM and Holley-Karplus predictions.

The final secondary structure assignment for both $\alpha$- and $\beta$-polypeptides is listed in Table IV. It is a consensus assignment derived from a combination of all the prediction methods used. It should be pointed out that in addition to a transmembrane helix, an interfacial helix of 10 residues [$\alpha$-Ile-46 to $\alpha$-Ala-55] has also been identified for the $\alpha$-apoprotein at the C-terminal. This assignment is supported by the following observations: (1) residues in the sequence `IALGAAKSAA` have a high propensity to form an $\alpha$-helix as evident from SOPM and Holley-Karplus analyses [see Tables I] and other propensity analyses we have performed; (2) the homologous fragment `KALIAAKNKA` from trypanothione reductase, as listed in Table II, is an $\alpha$-helix; and (3) as shown in Figure 5, the segment is highly amphiphilic and suitable for sitting at the interfacial region. The Trp and charged Lys residues face the lipid head group and all the hydrophobic residues face the interior of the membrane. This helical wheel representation of an amphiphilic helix can be quantified in terms of the hydrophobic moment [**EWTW82, Eis84, ESKW84**].

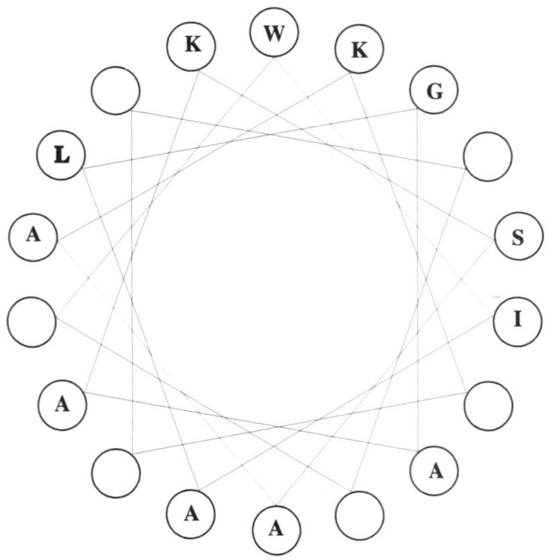


FIGURE 5. Helical wheel representation of the interfacial $\alpha$-helix from $\alpha$-Trp-45 to $\alpha$-Lys-56.

A reverse-turn segment for the $\alpha$-subunit, `PGSN`, is assigned based on a propensity analysis with Levitt's scale [**Lev78**]. Shown in Figure 6 is the propensity profile

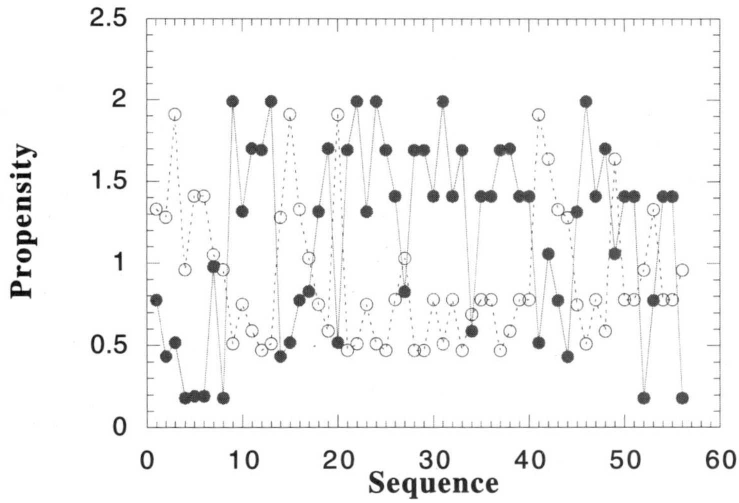


FIGURE 6. Propensity profile of transmembrane helix and reverse-turn for $\alpha$-polypeptide. Solid circles – transmembrane helix propensity base on Persson and Argos scale [**PA94**]; open circles – reverse-turn propensity by Levitt [**Lev78**].

of the transmembrane helix and reverse-turn for the $\alpha$-polypeptide. Solid circles indicate the transmembrane helix propensity (for a single sequence) based on the Persson and Argos scale [**PA94**]; open circles represent the reverse-turn propensity according to Levitt [**Lev78**]. All four residues show a reverse-turn propensity higher than one. In summary, we have identified a [transmembrane helix – reverse-turn – interfacial helix] motif for the $\alpha$-polypeptide. Such a motif has also been observed in other membrane proteins [**Whi94, KWF94**].

Table IV. Assignment of Secondary Structure and Topology

| | $\alpha - polypeptide$ |
|---|---|
| Topology | -------------TTTTMMMMMMMMMMMMMMMMMMMMTTTT----IIIIIIIIII- |
| Sequence | SNPKDDYKIWLVINPSTWLPVIWIVATVVAIAVHAAVLAAPGSNWIALGAAKSAAK |
| Secondary structure | CCCCCCCCEEEEEECCHHHHHHHHHHHHHHHHHHHHHHHHHCCCCHHHHHHHHHHC |
| | $\beta - polypeptide$ |
| Topology | --------------TTTTMMMMMMMMMMMMMMMMMMMMTTTT--- |
| Sequence | AERSLSGLTEEEAIAVHDQFKTTFSAFIILAAVAHVLVWVWKPWF |
| Secondary structure | CCCCCCCCCHHHHHHHHHHHHHHHHHHHHHHHHHHHHHHHHHCCC |

Note: M - Transmembrane region; T - Terminal (interfacial) region; I - Interfacial helix.

## Construction of $\alpha$ and $\beta$ Apoproteins

Based on the predicted secondary structure feature, tertiary structures for both the $\alpha$- and $\beta$-polypeptides were built by means of comparative modeling using the program MODELLER [**SB93**]. In comparative modeling, the homologous structure is used as a template for the unknown structure. In our case, homologous fragment structures, as aligned in Table II, were employed in such a procedure.

In MODELLER [**SB93**], the three dimensional model of the unknown protein is obtained by satisfying spatial restraints in the form of probability density functions (pdfs) derived from the alignment of the unknown structure with one or more homologous structures. The pdfs restrain $C_\alpha$-$C_\alpha$ distances, main-chain N-O distances, main-chain and side-chain dihedral angles. The optimization is carried out by the variable target method that applies the conjugated gradient algorithm to positions of all non-hydrogen atoms. Specifically, two transmembrane helices (one for the $\alpha$-apoprotein and another for the $\beta$-apoprotein) plus the interfacial helix of the $\alpha$-polypeptide were built with the program MODELLER. To build the [transmembrane helix – reverse-turn – interfacial helix] motif for the $\alpha$-polypeptide, the transmembrane helix and the interfacial helix were optimally linked with the reverse-turn fragment. In the normal practice of comparative modeling, the reverse-turn fragment is selected from a fragment library to generate the best fit between main secondary structure segments. Here, the reverse-turn fragment was built by superimposing two homologous fragments `AVPGA` and `APGGN` as listed in Table II. The rest of the terminal residues were also added using the corresponding homologous structures as templates. The $\alpha$- and the $\beta$-apoproteins so constructed were each optimized by energy minimization with fixed protein backbones.

### Placement of BChla's and Construction of the $\alpha\beta$ Heterodimer

BChla is an integral part of the pigment-protein complex. Placement of BChla in the complex is essential to model building. Resonance Raman spectra demonstrated that all $Mg^{2+}$ of all BChla molecules in the LH–II complex of *Rs. molischianum* are 5-coordinate [**GLRM93**]. Multiple sequence alignment of light-harvesting complexes (see Figure 3) reveals three highly conserved histidine residues. Most likely, the histidines form the binding site for the three BChla's, a hypothesis supported by site-directed mutagenesis experiments on related systems [**BRY88, CVF$^+$94, VCF$^+$94, Zub86**]. As can be seen from sequence alignment (Figure 3), the motif $\alpha$-Ala-30 ... $\alpha$-His-34 ... $\alpha$-Leu-38 is highly conserved (see Figure 7). It has been reported that the corresponding Ala (-4 residues of the conserved His) in LH–I of *Rb. capsulatus* can only be substituted with small residues, Gly, Ser and Cys [**BRY88**], indicating the structural importance of "smallness" of the residue in the -4 position to His. A similar binding pocket $\beta$-Phe-27... $\beta$-Ala-31...$\beta$-His-35 is also highly conserved (see Figure 7) in the $\beta$-polypeptide. The conserved Phe residue in the -8 position to His is also significant [**ZB91**]. However, there exists no such conserved binding pocket for $\beta$-His-17, making the assignment of $\beta$-His-17 as the binding site for B800 BChla less certain.

Residue conservation can not only provide clues about the helix-pigment interactions, but also provide clues about helix-lipid interactions. The lipid facing residues tend to be more evolutionarily variable than internal residues [**DOR$^+$93, vHM90**]. The highly conserved residues, for the $\alpha$-polypeptide ($\alpha$-Ala-30 ... $\alpha$-H is-34 ... $\alpha$-Leu-38) are four residues apart and thus aligned on the same side of the helical wheel, and so are the $\beta$-polypeptide ($\beta$-Phe-27... $\beta$-Ala-31...$\beta$-His-35). Taking into account the conserved residues in both the $\alpha$- and the $\beta$-polypeptide, the interhelical arrangement, as shown in Figure 7, is the most probable one. The side of highly conserved residues, on both the $\alpha$- and the $\beta$-polypeptide, face inward and the other side is in contact with lipid bilayers.

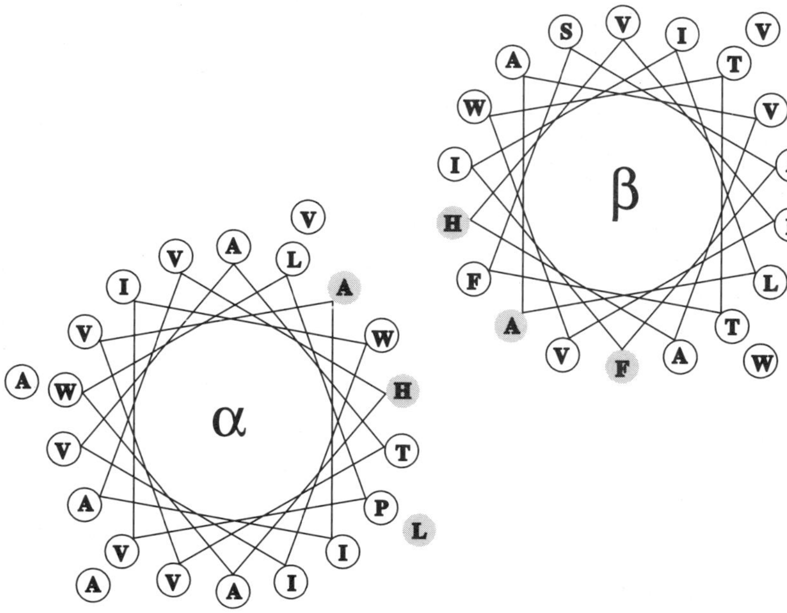


FIGURE 7. Proposed interhelical arrangement of $\alpha$- and $\beta$-polypeptides. The residues in shaded circles are highly conserved.

It has been observed that many transmembrane $\alpha$ helices are amphipathic, with opposing polar and nonpolar faces oriented along the long axis of the helix [**RDE89, vHM90, SLD$^+$90**]. Energetically, the apolar surfaces of transmembrane helices provide a good interface to the lipid bilayers, the polar side of inward-facing residues tend to be in the interhelical region to eliminate possible contacts between polar residue and lipid [**RDE89**]. In our case, transmembrane cores for both the $\alpha$- and $\beta$-subunit are highly hydrophobic as shown in the helical wheel representation of Figure 7. Except for the BChla binding histidine residues, the core of both transmembrane segments consists of nearly all nonpolar residues. On one hand, this potentially useful rule for arranging interhelical packing is not applicable to the current case. On the other hand, it may be a good indication of a lack of interhelical contacts in the transmembrane core for the LH–II complex.

As we build the heterodimer, the following well-known facts were also taken into consideration: (1) based on the observation of cleavage of part of the N-terminal domains of the $\alpha$- and $\beta$-polypeptides of LH–I of *Rs. rubrum* on the cytoplasmic side of the membrane by partial hydrolysis with proteolytic enzymes [**BWS$^+$84**], both $\alpha$- and $\beta$-polypeptides should be oriented with their N-terminals towards the cytoplasm; (2) strong circular dichroism (CD) signals suggest exciton interactions between pairs of BChla molecules [**Zub93**]; (3) linear dichroism data indicate that the B850 BChla's are approximately perpendicular to the membrane and the B800 BChla is parallel to the membrane [**KvGH$^+$84**]; (4) Fourier-transform Resonance Raman spectroscopy and site-directed mutagenesis [**FSRH94**] of a related light-harvesting complex indicate that another highly conserved residue, $\alpha$-Trp-45, is hydrogen-bonded to the 2-acetyl group of BChla. All these observations impose constraints on the structure. To put it simply: (1) the $\alpha\beta$ heterodimer should be arranged interhelically with two B850 BChla binding histidine residues facing each other as shown in Figure 7; (2) the two B850 BChla's are paired and oriented perpendicular to the membrane plane. Therefore, the basic unit of the LH–II complex is configured with the B850 BChla pair sandwiched between two helices of the $\alpha\beta$ heterodimer. We will come back to the consequence of hydrogen bond constraint in the following section.

## Construction of the Complete Octomer

The construction of the complete aggregated complex was based on the two stage model suggested in [**PE90**] which assumes that the individual helices are formed prior to the formation of the helix bundle. We developed a protocol to aggregate the transmembrane helices into an octomer of eight $\alpha\beta$ heterodimers by means of molecular dynamics simulations and energy minimization under the constraints of experimental data (see Figure 1).

Our procedure consists of three essential steps: In a first step, the $\alpha$- and $\beta$-apoproteins were constructed by comparative modeling based on information obtained through homology and secondary structure analyses as described above. The optimized tertiary structures for the $\alpha$- and the $\beta$-apoprotein were preserved in the subsequent simulations by applying for the helical backbone harmonic restraints to the distances between the $i$-th carbonyl and the $(i+4)$-th amide nitrogen and to the distances between the $i$-th and the $(i+4)$-th $C_\alpha$; by applying in the turn region the harmonic restraints to the dihedral angles of the main chain. No restraint was applied to coiled terminal residues. Two BChla's binding to $\beta$-His-17 and $\beta$-His-35 were manually attached to the $\beta$-apoprotein using the program QUANTA, and a third BChla was attached similarly to $\alpha$-His-34. We employed the heavy atom coordinate of BChla from the crystal structure of BChlb in the photosynthetic reaction center of *Rps. viridis* [**DEM$^+$85**] and added explicit hydrogens using X-PLOR's function *hbuild.* The binding conformation between BChlb and His in the crystal structure of *Rps. viridis* was employed for placement of the BChla's. The X-PLOR utility *patch* was used to build the ligand bond between magnesium and the nitrogen of His, and was followed by rigid body minimization between the polypeptide and the BChla's. Subsequently, energy minimization runs were performed with harmonic restraints, followed by three 1 ps molecular dynamics runs at consecutively increasing temperatures of 100, 200 and 300 $^o$K to equilibrate the system. The lycopenes were not included in the model structure.

In the second step, the complete octomer was constructed enforcing an eightfold symmetry. A self rotational search [**Brü92, McR93, Lat85**] of the X-ray diffraction data indicated that, most likely, the LH–II complex possesses a pseudo-eightfold symmetry. Thus, the task of constructing the octomer was reduced to building a protomer for the eightfold octomer. There were two possible choices for the protomer: the $\alpha\beta$ heterodimer or the neighboring $\alpha\beta$ pair as shown in Figure 8. Since we intended to perform an intermediate step optimization for the $\alpha\beta$ dimer, the neighboring $\alpha\beta$ pair was chosen as the protomer. The rationale behind our choice was that in the $\alpha\beta$ heterodimer ($\alpha_2\beta_1$ in Figure 8), the two helices separated by the B850 BChla pair are too far apart to generate any significant interaction to be optimized. In contrast, in the neighboring $\alpha\beta$ pair ($\alpha_1\beta_1$ in Figure 8), inter-unit hydrogen bonds exist. As mentioned above, the $\alpha$-Trp-45 is hydrogen-bonded to the 2-acetyl group of BChla. However, there is no direct information about which of the two B850 BChla's is involved in such a hydrogen bond [**FSRH94**]. Based

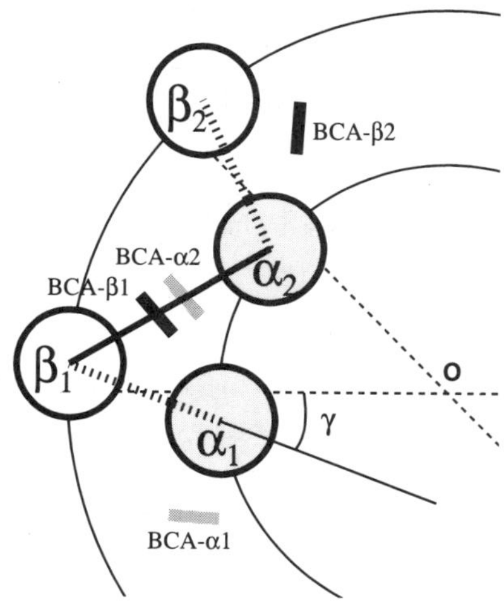


FIGURE 8. Definition of $\alpha\beta$ pair geometries. Heterodimer: $\alpha\beta$ dimer formed between $\beta_1$ and $\alpha_2$, linked by a solid line; Neighboring pair: $\alpha\beta$ dimer formed between $\beta_1$ and $\alpha_1$, linked by a dashed line; $\gamma$ is defined as the angle between the vector linking two helices and the radial vector from the center of the octomer to the center of the neighboring pair

on the optimized tertiary structure for both $\alpha$ and $\beta$ subunits,we found that it is more feasible spatially to form an inter-unit hydrogen bond, i.e., a hydrogen bond between $\alpha$-Trp-45 and the 2-acetyl group of BChla binding to $\beta$-His-35. Since open hydrogen bonds are extremely unstable in a lipid environment, we also arranged to have $\beta$-Trp-41 hydrogen bonded to the 2-acetyl group of BChla binding to $\alpha$-His-34. We used QUANTA to place the neighboring pair ($\beta_1$ and $\alpha_1$ as shown in Figure 8) so that: (1) the $\alpha\beta$ heterodimers are arranged in a configuration as shown in Figure 7; (2) hydrogen bonds were established between $\alpha$-Trp-45 and the 2-acetyl group of BChla attached to $\beta$-35 His, and between $\beta$-Trp-41 and the 2-acetyl group of BChla binding to $\alpha$-His-34. We applied the "soft van der Waals" option in X-PLOR to minimize the energy content of the neighboring $\alpha\beta$ pair, followed by rigid body minimization with two helices and three BChla's as five rigid bodies. Then, a simulated annealing was applied by heating the system to 2000 $^o$K and slowly cooling it down. This procedure was followed by a 100 ps molecular dynamics run at 300 $^o$K, obtaining an equilibrated structure of the neighboring $\alpha\beta$ pair.

In a third step, the equilibrated neighboring $\alpha\beta$ pair with all three bound BChla's were combined into an octomer by means of long time molecular dynamics simulations and energy minimization with eightfold symmetry and all the restraints as described above enforced throughout the entire simulation. An iterative protocol consisting of multiple cycles was employed to optimize the octomer structure. Each cycle started with a 200 step rigid body minimization (with the entire protomer as a rigid body), followed by a 2.5 ps rigid body dynamics run at 600 $^o$K, then a 5 ps molecular dynamics run at 300 $^o$K, ending with a 200 step Powell minimization. The radius of the octomer was monitored to detect convergence. Initially, the octomer was constructed with an outer diameter as large as 100 Å to avoid any close inter-unit contact. During the first cycle, a large cutoff distance for non-bonded interactions was required (we chose 15 Å), since the initial inter-unit distances were large. The iterative process was terminated when the final radii at the end of two consecutive runs differed by less than 0.25 Å. At the end, the octomer was minimized again with the Powell algorithm until it converged to a minimum energy configuration or until a limit of 700 cycles was reached. It normally took about 50-100 ps of equilibration (10 to 20 iterations) before the radii converged. Test runs with longer times indicated that the radius of the octomer begins to fluctuate around its average value after 50 ps. The underlying physical principle behind this simulation protocol is energy minimization perturbed by successive dynamic equilibration, which can be viewed as a dynamical analogy to simulated annealing.

Protein folding is related to the problem of global minimization, the practical solution of which still relies heavily on proper initial configurations. Starting from a heterodimer built within the constraints of all the biochemical and spectroscopic data dramatically reduces the phase space to be sampled, enhancing the chance of placing the initial structure in the basin of attraction of the global minimum. A systematic sampling of initial configurations has also been attempted by (1) arranging two kinds of helical assembly; (2) varying the relative orientations (angle $\gamma$ in Figure 8) of the vector linking two helices and the radial vector from the center of octomer to the center of neighboring pair. There are two options to assemble the eightfold octomer ring in terms of spatial positions of the $\alpha$- and the $\beta$-apoproteins. One is $\alpha$-polypeptide inside and $\beta$-polypeptide outside, and another is $\beta$-polypeptide inside and $\alpha$-polypeptide outside. Since there exist no direct biochemical data in favor of one of the two, both arrangements were considered in our modeling. Angle $\gamma$ was sampled from -50$^o$ to 50$^o$ with a step size of 10$^o$. Thus, a total of 22 octomers were constructed and optimized. Twelve of the 22 optimized structures were rejected either because the size of the octomer was too large to fit the unit cell dimension or because of severe distortion of pairing of B850 BChla's. The rest of the optimized structures were fed to the molecular replacement procedure. The best predicted structure was determined by the molecular replacement procedure as described in the following session. The structure that gave rise to the lowest R value was deemed the most optimal one.

Figure 9 shows one of the octomers we optimized. The octomer forms a ring with an outer diameter of about 70 Å with $\alpha$-polypeptide inside and $\beta$-polypeptide outside. This 70 Å outer diameter covers the side chains of the $\beta$-apoprotein. Typically, the radius of the eight $\beta$-apoproteins (measured from the center of the helices) varied between 27.8 Å and 32.8 Å; the radius of the eight $\alpha$-apoproteins varied between 15.1 Å and 21.0 Å. The $Mg^{2+}$ to $Mg^{2+}$ separation in the $\alpha\beta$ heterodimer was about 8.8 Å. The $Mg^{2+}$ to $Mg^{2+}$ separation between adjacent heterodimers was 16.0 Å. The inner helices were tilted against the membrane plane normal by about 6$^o$ and the outer helices were tilted by about 9$^o$. The resulting BChla's are optimally oriented to capture light coming from all directions and are close enough for efficient energy transfer. As depicted in Figure 10, the B850 BChla pair are sandwiched between two helices of the $\alpha\beta$ heterodimer. Such an arrangement has recently been suggested for the LH–I complex of *Rs. rubrum* to interpret the ring–shaped 8.5 Å resolution projection map [**KBG95**]. It should be pointed out that our model shows the expected placement of the two B850 Bchla's perpendicular to the membrane, near the periplasmic side. However, the B800 BChla is not oriented parallel to the cytoplasmic side of the membrane as expected

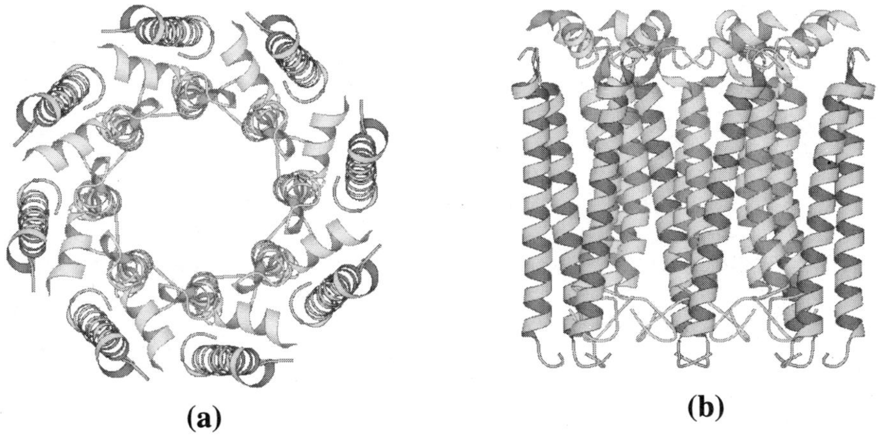


FIGURE 9. Schematic structure of the LH–II complex: (a) top view with C terminal pointing upward, showing a ring structure with 70 Å diameter, with $\alpha$-polypeptide inside and $\beta$-polypeptide outside; (b) side view with C terminal on top. [produced with MOLSCRIPT [**Kra91**]].

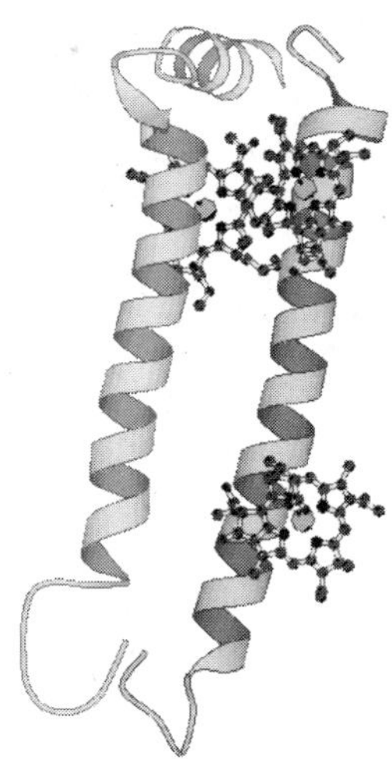

FIGURE 10. Schematic drawing of the $\alpha\beta$ heterodimer, i.e., the monomer "unit" of LH–II. Shown are the predicted secondary structure and locations of three Bchla's binding to $\alpha$-34 His, $\beta$-17 His and $\beta$-35 His. The C terminal is on top.

from LD data due to the lack of helix turn residues near $\beta$-His-17. As stated above, the placement of BChla at $\beta$-His-17 is itself problematic. The reason we placed it there is that we did not have any better clues at the time when the work reported here was carried out.

## Future Work

It is worth mentioning that the optimized octomers were not selected based on the criterion of energy minimization. Although the vacuum environment is similar to a lipid bilayer in terms of hydrophobicity, the peptide-lipid interactions are not explicitly included in the conformational energy. Instead, the molecular replacement test was employed as the ultimate test of the correctness of the model. The predicted structures are presently employed as a search model in the framework of the molecular replacement method as implemented in the program X-PLOR [**Brü92**]. In the molecular replacement method, a six-dimensional search is required to find the best match between observed and calculated diffraction data. In practice, the method is implemented as a three-dimensional rotational search followed by a three-dimensional translational search [**Lat85, Ros72**]. We also performed a Patterson correlation refinement preceding the translational search to filter the peaks of the rotational search [**Brü90**]. The preliminary results have been encouraging as demonstrated by the model's ability to probe the position of molecules in the crystal. The rotational search oriented the $\alpha$ helices of the octomers in a direction parallel to the c-axis of the crystal unit cell. The translational search further placed the octomers in the site of the four-fold axis of the crystal. Currently, for our best model with 6-12 Å resolution data, the R value after rigid body refinement is 49.8% and decreases to 29.8% after positional refinement. Work is in progress on two fronts to resolve the X-ray diffraction data for the LH–II of *Rs. molischianum*: (1) a more systematic sampling of initial octomer configurations will be undertaken; (2) a rigid body simulated annealing protocol is being developed and will be implemented in both the Patterson correlation refinement and the rigid body refinement procedures [**Brü90, Brü92**] in place of the rigid body minimizer to improve the convergence radius of the minimizer. It is our hope that the structure will be further refined to resolve the 2.4 Å diffraction data. If successful, this work will be the first demonstration of use of an *ab initio* predicted structure as a probe structure in the framework of molecular replacement to solve the phase problem in X-ray crystallography structure determination. The methodology developed may be useful for structure prediction of other integral membrane proteins.

## Comparison with the Structure of LH–II of *Rps. acidophila*

This article is based on an article we recently submitted for publication in Protein Science [**HXH$^+$**].[1]

As this article was written, the crystal structure of LH–II from *Rps. acidophila* determined by conventional multiple isomorphous replacement method was published by McDermott et al. [**MPF$^+$95**]. There exists significant sequence homology

[1] Modified versions of this article have been submitted to *Protein Science* and will also appear as a chapter in "Membrane Structure and Dynamics", K. M. Merz and B. Roux, Eds., (Birkhäuser, Cambridge, to be published).

between the $\alpha$-subunit from LH–II of *Rps. acidophila* strain 10050 and *Rs. molischianum.* For the $\beta$-subunit, the two sequences are homologous, but to a lesser extent. In Table V, a comparison between the predicted secondary structure assignment for *Rs. molischianum* and the X-ray resolved secondary structure for *Rps. acidophila* [**MPF$^+$95**] is given.

Table V. Comparison of Secondary Structures

| | $\alpha$ – *polypeptide* |
|---|---|
| *Rs. molischianum* | SNPKDDYKIWLVINPSTWLPVIWIVATVVAIAVHAAVLAAPGSNWIALGAAKSAAK |
| Prediction | CCCCCCCCEEEEECCHHHHHHHHHHHHHHHHHHHHHHHHHCCCCCHHHHHHHHHHC |
| *Rps. Acidophila* | ---MNQGKIWTVVNPAIGIPALLGSVTVIAILVHLAILSHTTWFPAYWQGGVKKAA |
| X-ray | ---C333333333HHHHHHHHHHHHHHHHHHHHHHHHHHHCCHHHHHHH??????? |
| | $\beta$ – *polypeptide* |
| *Rs. molischianum* | AERSLSGLTEEEAIAVHDQFKTTFSAFIILAAVAHVLVWVWKPWF |
| Prediction | CCCCCCCCCHHHHHHHHHHHHHHHHHHHHHHHHHHHHHHHHHCCC |
| *Rps. Acidophila* | -----ATLTAEQSEELHKYVIDGTRVFLGLALVAHFLAFSATPWLH |
| X-ray | -----CCCCCChhhhHHHHHHHHHHHHHHHHHHHHHHHHHHHCCCC |

Note: H - Helix; E - $\beta$-Sheet; C - Coil; 3 - 3 Turn Helix; ? - Unresolved.

Given the fact that substantial sequence homology exists between LH–II of *Rs. molischianum* and that of *Rps. acidophila*, the predicted secondary structure of LH–II of *Rs. molischianum* compares well with that of the *Rps. acidophila* crystal structure. Interestingly, the predicted [transmembrane helix – reverse-turn – interfacial helix] motif for the $\alpha$-polypeptide was observed in the crystal structure.

For comparison, the overall structure of the aggregated LH–II complex of *Rps. acidophila* is briefly outlined below. For a complete description, interested readers are referred to the original publication [**MPF$^+$95**]. The LH–II complex of *Rps. acidophila* is a ring-shaped aggregate of nine $\alpha\beta$ heterodimers with nine-fold symmetry. "The transmembrane helices of nine $\alpha$-apoprotein are packed side by side to form a hollow cylinder of radius 18 Å. The nine helical $\beta$-apoproteins are arranged radially with the $\alpha$-apoproteins to form an outer cylinder of radius 34 Å. The $\alpha$-apoprotein helices are parallel to the ninefold axis to within $2^o$, and the $\beta$-apoprotein helices are inclined by $15^o$ to this axis" [**MPF$^+$95**]. The two B850 BChla's are bonded to the two conserved histidine residues at the periplasmic side, sandwiched between the $\alpha\beta$ heterodimer and oriented perpendicular to the membrane plane. The B800 BChla is oriented parallel to the membrane plane, and bonded to the formyl group of the f-Met residue at the N terminus of the $\alpha$-apoprotein. Overall, the B850 BChla's form almost a symmetrical ring, with an $Mg^{2+}$ to $Mg^{2+}$ distance of 8.7 Å within the $\alpha\beta$ heterodimer and 9.7 Å between the adjacent heterodimers. It should be pointed out that the aggregated LH–II complex of *Rs. molischianum* consists of eight $\alpha\beta$ heterodimers while that of *Rps. acidophila* consists of nine heterodimers. The overall helical assembly of the LH–II complex of *Rs. molischianum* described in the previous section corresponds well with that of *Rps. acidophila.* The biggest discrepancy lies in the placement of BChla's. While our predicted structure places the B850 BChla pair unevenly, the X-ray structure shows that all B850 BChla's are arranged nearly symmetrically in a ring conformation. As expected, the B800 BChla is oriented parallel to the membrane plane in the crystal structure. Perhaps the biggest surprise is the binding of the B800 BChla to the formyl group of the fMet residue at the N terminus of the $\alpha$-apoprotein, which for the purpose has to dive into the interior of the membrane by as much as 9 Å. As stated before, our difficulties in placing the B800 BChla are attributed to a lack of pertinent biochemical information. It should also be born in mind that although in the majority of LH–II, and even LH–I complexes, there exists a corresponding Met residue at the N terminus of the $\alpha$-apoprotein, the N terminus of the $\alpha$-apoprotein of *Rs. molischianum* is a Ser residue. As a result, a slightly different conformation for *Rs. molischianum* is expected. The difference in the number of the constituting $\alpha\beta$ dimers ( eight vs. nine) is a clear indication of a difference as well. The extent of this difference will be clear once the 2.4 Å resolution X-ray diffraction data for LH–II of *Rs. molischianum* are resolved.

## References


[A. ] A. MacKerell *et al.*, To be published.

[AAM+94] I.T. Arkin, P.D. Adams, K.R. MacKenzie, M.A. Lemmon, A.T. Brünger, and D.M. Engelman, *Structural organization of the pentameric transmembrane α-helices of phospholamban, a cardiac ion channel*, Embo Journal **13** (1994), 4757.

[AGM+90] S. F. Altschul, W. Gish, W. Miller, E. W. Myers, and D. J. Lipman, *Basic local alignment search tool*, J. Mol. Biol. **215** (1990), 403–410.

[ARH82] P. Argos, J.K. Rao, and P.A. Hargrave, *Structural prediction of membrane-bound proteins.*, European Journal of Biochemistry **128** (1982), 565.

[AYKR87] J.P. Allen, T.O. Yeates, H. Komiya, and D.C. Rees, *Structure of the reaction center from Rhodobacter sphaeroides R-26: the protein subunits*, Proc. Natl. Acad. Sci. U.S.A. **84** (1987), 6162.

[BBO+83] Bernard R. Brooks, Robert E. Bruccoleri, Barry D. Olafson, Da vid J. States, S. Swaminathan, and Martin Karplus, *CHARMm: a program for macromolecular energy, minimization, and dynamics calculations*, J. Comp. Chem. **4** (1983), no. 2, 187–217.

[Brü90] Axel T. Brünger, *Extension of molecular replacement: A new search strategy based on Patterson correlation refinement*, Acta Cryst. **A46** (1990), 46–57.

[Brü92] Axel T. Brünger, *X-PLOR, version 3.1, a system for X-ray crystallography and NMR*, The Howard Hughes Medical Institute and Department of Molecular Biophysics and Biochemistry, Yale University, 1992.

[BRY88] E.J. Bylina, S.J. Robles, and D.C. Youvan, *Directed mutations affecting the putative bacteriochlorophyll-binding sites in the light-harvesting I antenna of* Rhodobacter capsulatus, Israel Journal of Chemistry **28** (1988), 73.

[BSST87] T. L. Blundell, B. L. Sibanda, M. J. Sternberg, and J. M. Thornton, *Knowledge-based prediction of protein structures and the design of novel molecules*, Nature **326** (1987), 347.

[Bus86] B. Busetta, *Examination of folding patterns for predicting protein topologies.*, Biochimica et Biophysica Acta **870** (1986), 327.

[BVC+93] A. F. Boonstra, R. W. Visschers, F. Calkoen, R. van Grondelle, E. F. van Bruggen, and E. J. Boekema, *Structural characterization of the B800-850 and B875 light-harvesting antenna complexes from Rhodobacter-Sphaeroides by electron microscopy.*, Biochimica et Biophysica Acta **1142** (1993), 181.

[BWS+84] R. A. Brunisholz, V. Wiemken, F. Suter, R. Bachofen, and H. Zuber, *The light-harvesting polypeptides of Rhodospirillum rubrum. II. localisation of the amino-terminal regions of the light-harvesting polypeptides B870-α and B870-β and the reaction-centre subunit L at the cytoplasmic side of the photosynthetic membrane of Rhodospirillum rubrum G-9+*, Hoppe-Seylers Zeitschrift fur Physiologische Chemie **365** (1984), 689.

[CAKF86] F. E. Cohen, R. M. Abarbanel, I. D. Kuntz, and R. J. Fletterick, *Turn prediction in proteins using a pattern-matching approach.*, Biochemistry **25** (1986), 266.

[CCM+87] J.L. Cornette, K.B. Cease, H. Margalit, J.L. Spouge, J.A. Berzofsky, and C. DeLisi, *Hydrophobicity scales and computational techniques for detecting amphipathic structures in proteins*, Journal of Molecular Biology **195** (1987), 659.

[CEHR92] W.A. Cramer, D.M. Engelman, G. Von Heijne, and D.C. Rees, *Forces involved in the assembly and stabilization of membrane proteins*, Faseb Journal **6** (1992), 3397.

[CF78] P.Y. Chou and G.D. Fasman, *Prediction of the secondary structure of proteins from their amino acid sequence.*, Advances in Enzymology and Related Areas of Molecular Biology **47** (1978), 45.

[CSR+92] S. W. Cowan, T. Schirmer, G. Rummel, M. Steiert, R. Ghosh, R. A. Pauptit, J. N. Jansonius, and J. P. Rosenbusch, *Crystal structures explain functional properties of 2 E. coli porins*, Nature **358** (1992), no. 6389, 727–733.

[CSW94] T. Coleman, D. Shalloway, and Z. Wu, *A parallel build-up algorithms for global energy minimizations of molecular clusters using effective energy simulated annealing*, J. of Global Optimization **4** (1994), 171.

[CVF+94] W. Crielaard, R.W. Visschers, G.J. Fowler, R. van Grondelle, K.J. Hellingwerf, and C.N. Hunter, *Probing the B800 bacteriochlorophyll binding site of the accessory light-harvesting complex from* Rhodobacter sphaeroides *using site-directed mutants. I. Mutagenesis, effects on binding, function and electrochromic behaviour of its carotenoids*, Biochim. Biophys. Acta **1183** (1994), 473.

[DEM+85] J. Deisenhofer, O. Epp, K. Mikki, R. Huber, and H. Michel, *Structure of the protein subunits in the photosynthetic reaction center of* Rhodopseudomonas viridis *at 3 Å resolution*, Nature **318** (1985), 618–624.

[DHS84] J. Devereux, P. Haeberli, and O. Smithies, *A comprehensive set of sequence analysis programs for the vax*, Nucleic Acids Research **12** (1984), 387.

[DM89] J. Deisenhofer and H. Michel, *The photosynthetic reaction centre from the purple bacterium Rhodopseudomonas viridis*, EMBO J. **8** (1989), 2149.

[DOR+93] D. Donnelly, J. P. Overington, S. V. Ruffle, J. H. Nugent, and T. L. Blundell, *Modeling $\alpha$-helical transmembrane domains: the calculation and use of substitution tables for lipid-facing residues*, Protein Science **2** (1993), 55.

[Eis84] D. Eisenberg, *Three-dimensional structure of membrane and surface proteins*, Annual Review of Biochemistry **53** (1984), 595.

[ESG86] D. M. Engelman, T. A. Steitz, and A. Goldman, *Identifying nonpolar transbilayer helices in amino acid sequences of membrane proteins*, Ann. Rev. Biophys. Biophys. Chem. **15** (1986), 321–353.

[ESKW84] D. Eisenberg, E. Schwarz, M. Komaromy, and R. Wall, *Analysis of membrane and surface protein sequences with the hydrophobic moment plot*, Journal of Molecular Biology **179** (1984), 125.

[EWTW82] D. Eisenberg, R.M. Weiss, T.C. Terwilliger, and W. Wilcox, *Hydrophobic moments and protein structure*, Faraday Symposia of the Chemical Society **17** (1982), 109.

[Fas89a] G.D. Fasman (ed.), *Prediction of protein structure and the principles of protein conformation*, Plenum, New York, 1989.

[Fas89b] G.D. Fasman, *Protein conformational prediction*, Trends in Biochemical Sciences **14** (1989), 295.

[FSRH94] G.J.S. Fowler, G.D. Sockalingum, B. Robert, and C.N. Hunter, *Blue shifts in bacteriochlorophyll absorbance correlate with changed hydrogen bonding patterns in light-harvesting 2 mutants of* Rhodobacter Sphaeroides *with alterations at $\alpha$-Tyr-44 and $\alpha$-Tyr-45*, Biochemical Journal **299** (1994), 695.

[GD94] C. Geourjon and G. Deleage, *SOPM: a self optimised prediction method for protein secondary structure prediction.*, Protein Engineering **7** (1994), 157.

[GJ79] M.R. Garey and D.S. Johnson, *Computers and intractability: A guide to the theory of np-completeness*, Freeman, San Francisco, 1979.

[GLRM93] L. Germeroth, F. Lottspeich, B. Robert, and H. Michel, *Unexpected similarities of the B800-850 light-harvesting complex from Rhodospirillum molischianum to the B870 light-harvesting complexes from other purple photosynthetic bacteria*, Biochemistry **32** (1993), 5615–5621.

[GOR78] J. Garnier, D.J. Osguthorpe, and B. Robson, *Analysis of the accuracy and implications of simple methods for predicting the secondary structure of globular proteins*, Journal of Molecular Biology **120** (1978), 97.

[HBC+90] R. Henderson, J. M. Baldwin, T. A. Ceska, F. Zemlin, E. Beckmann, and K. H. Downing, *Model for the structure of Bacteriorhodopsin based on high-resolution electron cryo-microscopy*, J. Mol. Biol. **213** (1990), 899–929.

[HC91] A. M. Hawthornthwaite and R. J. Cogdell, *Bacteriochlorophyll binding proteins*, Chlorophylls (Boca Raton) (H. Scheer, ed.), CRC Press, 1991, pp. 493–528.

[HK89] L. H. Holley and M. Karplus, *Protein secondary structure prediction with a neural network*, Proc. Natl. Acad. Sci. USA **86** (1989), 152–156.

[HK91] L. H. Holley and M. Karplus, *Neural networks for protein structure prediction*, Methods in Enzymology **202** (1991), 204.

[HXH+] Xiche Hu, Dong Xu, Kenneth Hamer, Klaus Schulten, Juergen Koepke, and Hartmut Michel, *Predicting the structure of the light-harvesting complex II of Rhodospirillum molischianum*, (Submitted).

[Jäh89] F. Jähnig, *Structure prediction for membrane proteins*, Prediction of protein structure and the principles of protein conformation (New York) (G.D. Fasman, ed.), Plenum, 1989, p. 707.

[JSSB94] M.S. Johnson, N. Srinivasan, R. Sowdhamini, and T.L. Blundell, *Knowledge-based protein modeling*, Critical Reviews in Biochemistry and Molecular Biology **29** (1994), 1.

[KBG95] S. Karrasch, P.A. Bullough, and R. Ghosh, *8.5 Å projection map of the light-harvesting complex I from* Rhodospirillum rubrum *reveals a ring composed of 16 subunits*, EMBO J. **14** (1995), 631.

[KD82] J. Kyte and R. F. Doolittle, *A simple method for displaying the hydropathic character of a protein*, J. Mol. Biol. **157** (1982), 105.

[KGvdB+92] W. Kleinekofort, L. Germeroth, J. van der Broek, D. Schubert, and H. Michel, *The light-harvesting complex II (B800/850) from Rhodospirillum molischianum is an octamer*, Biochimica et Biophysica Acta **1140** (1992), 102–104.

[KL85] L.A. Kuhn and J.S.J. Leigh, *A statistical technique for predicting membrane protein structure*, Biochimica et Biophysica Acta **828** (1985), 351.

[KM] J. Koepke and H. Michel, (To be published).

[Kra91] P. Kraulis, *MOLSCRIPT—a program to produce both detailed and schematic plots of protein structures*, J. Appl. Cryst. **24** (1991), 946–950.

[KS83] W. Kabsch and C. Sander, *Dictionary of protein secondary structure: pattern recognition of hydrogen-bonded and geometrical features*, Bioploymers **22** (1983), 2577.

[KvGH+84] H. J. M. Kramer, R. van Grondelle, C. N. Hunter, W. H. J. Westerhuis, and J. Amesz, *Pigment organization of the B800-850 antenna complex of Rhodopseudomonas sphaeroides*, Biochim. Biophys. Acta **765** (1984), 156–165.

[KWF94] W. Kühlbrandt, D.-N. Wang, and Y. Fujiyoshi, *Atomic model of plant light-harvesting complex by electron crystallography*, Nature **367** (1994), 614.

[Lat85] E. Lattman, *Diffraction methods for biological macromolecules. Use of the rotation and translation functions.*, Methods in Enzymology **115** (1985), 55.

[Lev78] M. Levitt, *Conformational preference of amino acids in globular proteins*, Biochemistry **17** (1978), 4277.

[LP85] D.J. Lipman and W.R. Pearson, *Rapid and sensitive protein similarity searches*, Science **227** (1985), 1435.

[LSBW94] R. Lohmann, G. Schneider, D. Behrens, and P. Wrede, *A neural network model for the prediction of membrane-spanning amino acid sequences*, Protein Science **3** (1994), 1597.

[McR93] D.E. McRee, *Practical protein crystallography*, Academic Press, San Diego, 1993.

[Mic91] H. Michel, *General and practical aspects of membrane protein crystallization*, Crystallization of membrane proteins (Boca Raton, Florida) (H. Michel, ed.), CRC Press, 1991, p. 74.

[MPF+95] G. Mcdermott, S.M. Prince, A.A. Freer, A.M. Hawthornthwalte-Lawless, M.Z. Paplz, R.J. Cogdell, and N.W. Isaacs, *Crystal structure of an integral membrane light-harvesting complex from photosynthetic bacteria*, Nature **374** (1995), 517.

[MWG+86] H. Michel, K. A. Weyer, H. Gruenberg, I. Dunger, D. Oesterhelt, and F. Lottspeich, *The 'light' and 'medium' subunits of the photosynthetic reaction centre from Rhodopseudomonas viridis: Isolation of genes, nucleotide and amino acid sequence*, EMBO J. **5** (1986), 1149.

[NVS91] A. Nayeem, J. Vila, and H.A. Scheraga, *A comparative study of the simulated-annealing and monte carlo-with-minimization approaches to the minimum-energy structure of polypeptides: [met]-enkephalin.*, J. of Computational Chemistry **12** (1991), 594.

[OH94] J. D. Olsen and C. N. Hunter, *Protein structure modelling of the bacterial light-harvesting complex*, Photochem. Photobiol. **60** (1994), 521.

[PA94] B. Persson and P. Argos, *Prediction of transmembrane segments in proteins utilising multiple sequence alignments*, Journal of Molecular Biology **237** (1994), 182.

[PCC92] S. R. Presnell, B. I. Cohen, and F. E. Cohen, *A segment-based approach to protein secondary structure prediction*, Biochemistry **31** (1992), 983.

[PdV90] J. Popot and C. de Vitry, *On the microassembly of integral membrane proteins*, Annual Review of Biophysics and Biophysical Chemistry **19** (1990), 369.

[PdVA94] J.L. Popot, C. de Vitry, and A. Atteia, *Folding and assembly of integral membrane proteins: An introduction*, Membrane protein structure: experimental approaches (New York) (S.H. White, ed.), Oxford University press, 1994, p. 41.

[PE90] J.L. Popot and D.M. Engelman, *Membrane protein folding and oligomerization: the two-stage model*, Biochemistry **29** (1990), 4031.

[Pea90] W.R. Pearson, *Rapid and sensitive sequence comparison with fastp and fasta*, Methods in Enzymology **183** (1990), 63.

[Pop93] J.L. Popot, *Integral membrane protein structure - transmembrane $\alpha$-helices as autonomous folding domains.*, Current Opinion In Structural Biology **3** (1993), 532.

[RA86] J.K. Mohana Rao and P. Argos, *A conformational preference parameter to predict helices in integral membrane proteins*, Biochimica et Biophysica Acta **869** (1986), 197.

[RC93] C.S. Ring and F.E. Cohen, *Modeling protein structures: construction and their applications*, Faseb Journal **7** (1993), 783.

[RDE89] D.C. Rees, L. DeAntonio, and D. Eisenberg, *Hydrophobic organization of membrane proteins*, Science **245** (1989), 510.

[Ros72] M.G. Rossmann (ed.), *The molecular replacement method*, Gordon and Breach, New York, 1972.

[RW88] M.J. Rooman and S.J. Wodak, *Identification of predictive sequence motifs limited by protein structure data base size*, Nature **335** (1988), 45.

[SAL91] G.D. Schuler, S.F. Altschul, and D.J. Lipman, *A workbench for multiple alignment construction and analysis*, Proteins: Structure, Function, and Genetics **9** (1991), 180.

[SB93] A. Sali and T. L. Blundell, *Comparative protein modelling by satisfaction of spatial restraints*, Journal of Molecular Biology **234** (1993), 779.

[SLD+90] J.P. Segrest, H. De Loof, J.G. Dohlman, C.G. Brouillette, and G.M. Anantharamaiah, *Amphipathic helix motif: Classes and properties*, Proteins, Struct. Funct. Genet. **8** (1990), 103.

[SvG91] V. Sundstrom and R. van Grondelle, *Dynamics of excitation energy transfer in photosynthetic bacteria*, Chlorophylls (Boca Raton) (H. Scheer, ed.), CRC Press, 1991, pp. 627–704.

[TEPL94] P. Tuffery, C. Etchebest, J.L. Popot, and R. Lavery, *Prediction of the positioning of the seven transmembrane $\alpha$-helices of bacteriorhodopsin. a molecular simulation study*, Journal of Molecular Biology **236** (1994), 1105.

[TSD+88] Herbert Treutlein, Klaus Schulten, J. Deisenhofer, H. Michel, Axel Brünger, and Martin Karplus, *Molecular dynamics simulation of the primary processes in the photosynthetic reaction center of* Rhodopseudomonas viridis, The Photosynthetic Bacterial Reaction Center: Structure and Dynamics (New York) (J. Breton and A. Verméglio, eds.), NATO ASI Series A: Life Sciences, vol. 149, Plenum, New York, 1988, pp. 139–150.

[VCF+94] R. W. Visschers, W. Crielaard, G. J. Fowler, C. N. Hunter, and R. van Grondelle, *Probing the B800 bacteriochlorophyll binding site of the accessory light-harvesting complex from Rhodobacter sphaeroides using site-directed mutants. II. a low temperature spectroscopy study of structural aspects of the pigment-protein conformation*, Biochim. Biophys. Acta **1183** (1994), 483.

[vGS88] R. van Grondelle and V. Sundstrom, *Excitation energy transfer in photosynthesis*, Photosynthetic Light–Harvesting Systems (Berlin, New York) (H. Scheer, ed.), Walter de Gruyter & Co., 1988, pp. 403–438.

[vH88] G. von Heijne, *Transcending the impenetrable: how proteins come to terms with membranes*, Biochimica et Biophysica Acta **947** (1988), 307.

[vH92] G. von Heijne, *Membrane protein structure prediction—hydrophobicity analysis and the positive-inside rule*, Journal of Molecular Biology **225** (1992), 487.

[vH94a] G. von Heijne, *Decoding the signals of membrane protein sequence*, Membrane protein structure: experimental approaches (New York) (S.H. White, ed.), Oxford University press, 1994, p. 27.

[vH94b] G. von Heijne, *Membrane proteins: from sequence to structure.*, Annual Review of Biophysics and Biomolecular Structure **23** (1994), 167.

[vHM90] G. von Heijne and C. Manoil, *Membrane proteins: from sequence to structure*, Protein Engineering **4** (1990), 109.

[Whi94] S. H. White, *Hydropathy plots and the prediction of membrane protein topology*, Membrane protein structure: experimental approaches (New York) (S. H. White, ed.), Oxford University press, 1994.

[WKN+91] M.S. Weiss, A. Kreusch, U. Nestel, W. Welte, J. Weckesser, and G.E. Schulz, *The structure of porin from* Rhodobacter capsulatus *at 1.8 Å resolution*, FEBS Lett. **280** (1991), 379.

[ZB91] H. Zuber and R.A. Brunisholz, *Structure and function of antenna polypeptides and chlorophyll-protein complexes: Principles and variability*, Chlorophylls (Boca Raton) (H. Scheer, ed.), CRC Press, 1991, pp. 627–692.

[ZBTS87] M. J. Zvelebil, G. J. Barton, W. R. Taylor, and M. J. Sternberg, *Prediction of protein secondary structure and active sites using the alignment of homologous sequences*, J. Mol. Biol. **195** (1987), 957.

[Zub85] H. Zuber, *Structure and function of light-harvesting complexes and their polypeptides*, Photochem. Photobiol. **42** (1985), 821.

[Zub86] H. Zuber, *Structure of light-harvesting antenna complexes of photosynthetic bacteria, cyanobacteria and red algae*, Trends Biochem. Sci. **11** (1986), 414.

[Zub93] H. Zuber, *Structural features of photosynthetic light-harvesting systems*, The Photosynthetic Reaction Center (San Diego) (J. Deisenhofer and J. R. Norris, eds.), Academic Press, 1993, p. 43.

Theoretical Biophysics, Beckman Institute, University of Illinois at Urbana—Champaign, Urbana, IL 61801

*E-mail address*: xhu@ks.uiuc.edu

*E-mail address*: xu@ks.uiuc.edu

*E-mail address*: khamer@ks.uiuc.edu

*E-mail address*: kschulte@ks.uiuc.edu

Max-Planck-Institut für Biochemie, Abteilung Molekulare Membranbiologie, 6000 Frankfurt, Germany

*E-mail address*: koepke@max.mpibp.uni-frankfurt.de

DIMACS Series in Discrete Mathematics
and Theoretical Computer Science
Volume **23**, 1996

# Some Approaches to the Multiple–Minima Problem in Protein Folding

J. Kostrowicki and H. A. Scheraga

ABSTRACT. Once a polypeptide chain is synthesized, it folds spontaneously (in a thermodynamic sense) into the three-dimensional structure of a native protein. Attempts are currently being made to compute this most stable structure as the one of minimum free energy. The conformational energy is expressed as a sum over all pair interactions, and the entropy is computed from the matrix of second derivatives of the conformational energy. Algorithms are available for minimizing the conformational energy, but they lead only to the local minimum that is closest to the starting point, rather than to the global minimum that is required, according to the thermodynamic hypothesis; this is the so-called multiple-minima problem. Various procedures have been developed to surmount this problem. Some of them work efficiently for short linear and cyclic chains, but are not readily extendible to long chains. In order to treat the longer chains that are characteristic of globular proteins, a new procedure (the diffusion equation method) has been introduced.

## 1. Introduction

It is known [**1**] that, after biosynthesis, polypeptide chains fold spontaneously into the three-dimensional structure of a native protein that is required for its biological activity. This experimental observation has given rise to the "thermodynamic hypothesis", viz., that proteins adopt that three-dimensional structure, among many others, for which the free energy of the system (protein plus solvent environment) is a minimum. Consequently, theoreticians have endeavored to compute the most stable structure of a protein of given amino acid sequence with the aid of empirical potential energy functions.

To compute the most stable conformation of any biological macromolecule, it is necessary to (i) generate an arbitrary starting conformation, (ii) compute its conformational energy (with entropy and hydration contributions included), and (iii) locate the global minimum of the conformational energy. Adequate procedures

1991 *Mathematics Subject Classification.* Primary 65K10, 90C30; Secondary 92C05, 92E10.

*Key words and phrases.* Conformational analysis of molecules, global search methods, Self-Consistent Multitorsional Field Method, Diffusion Equation Method, constrained global minimization, end-to-end distance distribution.

This work was supported by NSF Grant No. DMB 90-15815 and by a grant from the Association for International Cancer Research.

are available for steps (i) and (ii) and for minimizing the conformational energy [**2**]. The minimization procedure, however, leads only to the local minimum closest to the starting conformation, rather than to the global minimum; this is the multiple-minima problem [**3, 4**]. It is, therefore, necessary to have efficient methods for searching conformational space to locate the lowest minimum among all those in the whole space. This article summarizes some of the methods that have been developed to facilitate such a search of conformational space. Details of these methods are discussed in reference [**2**].

## 2. Summary of Methods and their Application

The following methods have been developed in our laboratory for searching conformational space.

1. Build-up procedure [**5**]
2. Self-consistent electrostatic field (SCEF) method [**6**]
3. Monte-Carlo-plus-Minimization (MCM) method [**7, 8**]
4. Electrostatically-driven Monte Carlo (EDMC) procedure [**9**]-[**13**]
5. Adaptive importance sampling Monte Carlo procedure [**14**]-[**16**]
6. Relaxation of dimensionality [**17, 18**]
7. Pattern recognition importance-sampling minimization (PRISM) [**19**]
8. Diffusion equation method (DEM) [**20**]-[**23**]
9. Self-consistent multi-torsional field (SCMTF) method [**24**]-[**26**]
10. Minimization-in-a-confined-space approach (CSA) [**27**]-[**29**]
11. Lattice neural network minimization (LNNM) [**30**].

Other methods, such as simulated annealing [**31, 32**], have been proposed for searching conformational space. However, while this method has been shown to be a useful procedure in constrained problems [**33**], it has been found to be inapplicable in unconstrained Monte Carlo calculations, using the pentapeptide Met-enkephalin as an example [**34**].

Examples of applications of these methods are given below. Most of these procedures have been tested first on Met-enkephalin [5, 7, 8, 10, 14-16, 18, 22, 24]. In addition, the build-up procedure has been applied to the 10-residue cyclic peptide gramicidin S [**35**], and to models of collagen [**36**]-[**39**]. The SCEF and EDMC methods have been applied to a-helical structures [**6, 9**], and the MCM, EDMC, adaptive importance sampling, relaxation of dimensionality, DEM, and SCMTF procedures have been applied to various linear oligopeptides. The PRISM procedure has been applied to the 36-residue avian pancreatic polypeptide [**19**]. The confined-space approach (CSA) has been used to unfold and refold bovine pancreatic trypsin inhibitor [**28**]. Finally, the LNNM procedure has been used to fold crambin on a lattice, and to identify chain-folding initiation sites in two globular proteins [**30**].

Efforts are now underway to extend these procedures to larger polypeptides and proteins. Some details of some of these methods are provided below.

## 3. The Self-consistent Multitorsional Field (SCMTF) Method

The SCMTF method [**24**] is based on the idea that the ground-state wave function $\Psi$ of a system of nuclei in a molecule spreads over the entire potential energy surface, irrespective of the number of potential wells. Moreover, the maximum of $|\Psi|^2$ should lie close to the global minimum of the potential energy. The method treats the dihedral angles, $\theta_i$, of the polypeptide as independent and, by application of the variational principle, leads to a set of N coupled *one*-dimensional Schrödinger equations, one for each $\theta_i$, to obtain the corresponding $\phi_i(\theta_i)$, where $\Psi = \prod_i \phi_i(\theta_i)$ and $N$ is the number of dihedral angles.

$$\hat{H}_i\phi_i = \varepsilon_i\phi_i \ (i = 1, \dots, N) \tag{1}$$

with

$$\hat{H}_i = \hat{T}_i + \hat{V}_i^{eff}(\theta_i) \tag{2}$$

$\hat{T}_i$ is the kinetic energy operator, and $\hat{V}_i^{eff}$ is the potential energy operator. Each equation describes the variations of a single dihedral angle in the average field of the others. The Hamiltonian of eq. 1 is

$$\hat{H}_i = \frac{-\hbar^2}{2I_i}\frac{\partial^2}{\partial\theta_i^2} + \hat{V}_i^{eff}(\theta_i) \tag{3}$$

where $I_i$ is a moment of inertia, and the effective potential $\hat{V}_i^{eff}(\theta_i)$ depends on the mean field created by averaging over the other dihedral angles, $\theta_l(l \neq i)$, according to the probability density distribution $\rho_l = |\phi_l|^2$. In order to calculate $\hat{V}_i^{eff}$ , a Monte Carlo procedure is used, i.e.,

$$\hat{V}_i^{eff}(\theta_i) \cong \frac{1}{M^c}\sum_m^{M^c} V(\theta_1^{m^*}, \cdots, \theta_{i-1}^{m^*}, \theta_i, \theta_{i+1}^{m^*}, \cdots, \theta_N^{m^*}) \tag{4}$$

where the summation extends over $M^c$ locally minimized trial points in the (N-1)-dimensional space of all dihedral angles $\theta_l$ except $\theta_i$. First, each point $\theta^m$ in the space is selected randomly according to some preassumed one-dimensional distribution $\rho_l^0$. Then, whenever a $\theta^m$ is chosen, the potential energy $V(\theta)$ is minimized with respect to *all* $\theta$'s. The minimization gives the new $\theta^{m^*}$, which is then used in eq. [4]. Solution of eqs. 1 gives a new set of $\phi_l$ and, therefore, a new set of $\rho_l$. The procedure is repeated iteratively until self-consistency of the $\rho_l$ distributions is achieved. The ECEPP potential (Empirical Conformational Energy Program for Peptides) is used for $V(\theta)$. The SCMTF method has been applied to Met-enkephalin [**24**], icosalanine [**25**], and the 20-residue membrane-bound portion of melittin [**26**].

## 4. The Diffusion Equation Method (DEM)

The DEM is based on the use of the diffusion equation to deform the complex energy hypersurface in successive stages so that higher-energy minima disappear, and only a descendant of the global minimum remains. A reversal of the deformation procedure then recovers the global minimum of the original potential function [**20**]. Therefore, to deform the surface of a multidimensional function $f(x)$ the deformed surface $F$ at any stage of deformation ("time") is obtained by solving the equation

$$\Delta F = \frac{\partial F}{\partial t} \tag{5}$$

where $\Delta$ is the Laplacian operator in the variables $\mathbf{x}$ [**20**].

An analytical solution of eq. 5 can be obtained in a variety of ways, e.g., by evaluating the Fourier-Poisson integral

$$F(\mathbf{x},t) = (4\pi t)^{-d/2} \int_{\Re^d} f(\mathbf{y}) \exp\left[-\frac{1}{4t} \parallel \mathbf{x}-\mathbf{y} \parallel^2\right] d\mathbf{y} \tag{6}$$

if the function $f(\mathbf{y})$ is expressed as a sum of Gaussians,

$$f(\mathbf{y}) = \exp\left[-\sum_{k=1}^{d} a_k y_k^2\right] \tag{7}$$

The solution is

$$F(\mathbf{x},t) = \prod_{k=1}^{d} (1+4a_k t)^{-\frac{1}{2}} \exp\left[-\frac{a_k x_k^2}{1+4a_k t}\right] \tag{8}$$

The theory for the development of the DEM is given in reference [**20**]. An alternative derivation, starting from the Schrödinger equation, is given below.

The Schrödinger equation may be written in the form

$$-\Delta\Psi(R) + V(R)\Psi(R) = E\Psi(R) \tag{9}$$

where $R$ is the vector of Cartesian coordinates of all atoms:

$$R \equiv \begin{bmatrix} r^1 \\ \vdots \\ r^n \end{bmatrix} \tag{10}$$

We seek an approximation of $\Psi$ in the form of $\Phi_{\rho^1,\ldots,\rho^n}(R) = \prod_{j=1}^{n} \phi_{\rho^j}(r^j)$, using the variational principle, with parameters $\rho^1, \ldots, \rho^n$. The function $\phi_{\rho^j}(r^j)$ is assumed to have the Gaussian form

$$\phi_\rho^j(r^j) \equiv a e^{-b\|r^j-\rho^j\|^2} \tag{11}$$

where $b$ is a fixed constant, and $a = (2b/\pi)^{3/4}$ is a normalization factor. In the variational method, we minimize $\langle\Phi_\rho| - \Delta + V|\Phi_\rho\rangle$ with respect to $\rho^j$. Since $\Delta = \sum_j \Delta_j$, where $\Delta_j \equiv \frac{\partial^2}{\partial x_j^2} + \frac{\partial^2}{\partial y_j^2} + \frac{\partial^2}{\partial z_j^2}$, the mean value of the kinetic energy operator, $-\Delta$, is the sum of atomic contributions:

$$\langle\Phi_\rho|\Delta_j|\Phi_\rho\rangle = \int_{\Re^{3n}} \Phi_\rho(R)\Delta_j\Phi_\rho(R)dR = \int_{\Re^3} \phi_{\rho^j}(r^j)\Delta_j\phi_{\rho^j}(r^j)dr^j \tag{12}$$

The integral (12) does not depend on the actual position $\rho^j$ of the center of the Gaussian $\theta_{\rho^j}(r^j)$ of eq. 11 (this can be seen by substituting $\sigma^j = r^j - \rho^j$, and integrating over $\sigma^j$ instead of $r^j$). Therefore, the minimum of $\langle\Phi_\rho| - \Delta + V|\Phi_\rho\rangle$ with respect to $\rho$ has the same position as the minimum of $\langle\Phi_\rho|V|\Phi_\rho\rangle$., where

$$\langle \Phi_\rho | V | \Phi_\rho \rangle = a^{2n} \int_{\Re^{3n}} V(R) e^{-2b\|R-\rho\|^2} dR \tag{13}$$

The integral (13) can be viewed as the Fourier-Poisson integral (6) for a diffusion equation with $V(R)$ as the boundary condition, viz.,

$$\varepsilon(\rho, t) = (2\sqrt{\pi t})^{-3n} \int_{\Re^{3n}} V(R) e^{-\frac{1}{4t}\|R-\rho\|^2} dR \tag{14}$$

The new variable $t$ (time) is related to the coefficient $b$ ($1/4t = 2b$), and the normalization factor $(2\sqrt{\pi t})^{-3n} = a^{2n}$ . The function $\varepsilon(\rho, t)$ satisfies the diffusion equation: (15)

$$\Delta \varepsilon(\rho, t) = \frac{\partial}{\partial t} \varepsilon(\rho, t) \tag{15}$$

with the boundary condition

$$\varepsilon(\rho, 0) = V(\rho) \tag{16}$$

The DEM has been applied to a variety of simple mathematical functions [**20**], and to a series of clusters of Lennard-Jones particles [**21**]. In the latter application, the Lennard-Jones potential function was expressed as a sum of Gaussians. Calculations were carried out for various cluster sizes $n$ = 5,6,7,...,55. For $n = 55$, there are $\approx 10^{45}$ local minima, the global minimum being the Mackay icosahedron. The global minimum was found by the diffusion equation method [**21**] in $\approx 400$ seconds on one processor of an IBM 3090 supercomputer. The DEM has also been applied to water clusters using the MCY potential to treat water-water interactions [**23**].

We also used the DEM to treat our ECEPP algorithm, and applied it to terminally-blocked alanine and to the pentapeptide Met-enkephalin [**22**]. The DEM found the global minimum for the alanine compound in $< 1$ min and a structure close to the global minimum for the pentapeptide in $\approx 10$ min, using one processor of an IBM 3090 supercomputer. We solved the diffusion equation in the space of Cartesian coordinates of all the atoms of the molecule, and then minimized the solution on the manifold of fixed bond lengths and bond angles, as described in reference [**22**]. This approximation is not entirely satisfactory for minimizing an energy function that is expressed as a sum of pairwise interactions. Each atom-atom pairwise potential is transformed by the diffusion equation into a function that varies as $r^2$ for large times and in the range of interatomic distances r in a molecule. Of course, such a harmonic function has only one minimum for clusters (a collapsed configuration; i.e., $r = 0$) but many minima for constrained problems, i.e., for molecules. Hence, we are implementing an improvement in the DEM in order to assure that only one minimum remains in the deformed surface at to, the time at which the reversing procedure is started. This improvement is necessary because, unlike the calculations on clusters of free Lennard-Jones particles [**21**], proper connexity of the chain must be maintained when dealing with polypeptides. Therefore, for two purposes [(i) maintaining proper connexity, and (ii) assuring that only one minimum remains in the deformed surface at $t_0$], we are testing a modification of the DEM which averages the energy function over conformations that are close to the manifold of fixed bond lengths and bond angles. We construct a linear subspace which is tangent to the manifold at a certain conformation, and average the energy function in this subspace. This procedure is described in the next section.

## 5. Solution of the Diffusion Equation in Dihedral Angle Space

Smoothing the conformational (potential) energy $V$ by diffusing it in the space $\Re^{3n}$ of Cartesian coordinates of all atoms $R$ has some significant drawbacks. First, even for a very large time, for which the energy $V$ as a function of $R$ has only one minimum, it may have several constrained minima, i.e., $V[R(\theta)]$ may have minima for several vectors $\theta$ of dihedral angles. Second, and more important, the smoothed value $\varepsilon$ of the energy $V$ for a given time $t$, for conformation $R_o$,

$$\varepsilon(R_o, t) = (2\sqrt{\pi t})^{-3n} \int_{\Re^{3n}} V(\rho) \exp\left[-\frac{1}{4t} \parallel \rho - R_o \parallel^2\right] d\rho \tag{17}$$

is an average over states (geometries) $\rho$ which might significantly violate bond lengths and bond angles. The energy $V(R)$ can have very low values for certain "cluster" geometries which cannot be adopted by the molecule because the interatomic distances for these geometries depart significantly from the correct bond lengths and bond angles. These unphysical states perturb the averaging process on the manifold.

The ideal approach would be to smooth the energy on the manifold. One way of doing so is to solve the diffusion equation in dihedral angle variables. This, however, poses serious mathematical difficulties. For this reason, we designed a method which confines the averaging process to a close neighborhood of the manifold and is practically feasible.

To find the smoothed energy at a point $R_o$, we construct a linear subspace, tangent to the manifold at $R_o$:

$$R = R_o + \frac{\partial R}{\partial \theta}\alpha \tag{18}$$

Then, the energy $V$ is smoothed over states in the tangent space of dimension $S$:

$$\varepsilon_{R_o}(t) = (2\sqrt{\pi t})^{-S} \int_{\Re^S} V\left(R_o + \frac{\partial R}{\partial \theta}\alpha\right) e^{-\frac{1}{4t}\|\alpha\|^2} d\alpha \tag{19}$$

For nonbonded interactions expressed in terms of Gaussians, the integral (19) can be evaluated from an analytical formula.

The quality of this approximation depends on the time $t$, which is a measure of the averaging radius. For small $t$, only geometries close to $R_o$ contribute significantly to the integral (19) (small $\alpha$) and, for small $\alpha$, the tangent space lies very close to the manifold so that an average over states in this space is a good approximation to an average over states on the manifold. This implies that the tangent linear space technique should not be used for large times.

For large values of $t$, we use a different method, based on a Fourier expansion of the energy as a function of the dihedral angles:

$$V(\theta) = \sum_{m_1,\dots,m_S} a_{m_1,\dots,m_S} \exp\left(i\sum_j m_j\theta_j\right) \tag{20}$$

The solution to the diffusion equation in dihedral angles for $V(\theta)$ of eq. (20) as a boundary condition is:

$$\varepsilon(\theta,t) = \sum_{m_1,\dots,m_S} a_{m_1,\dots,m_S} \exp\left[-\left(\sum_j m_j^2\right)t + i\sum_j m_j\theta_j\right] \tag{21}$$

For large times $t$, only the first few terms in the expansion of eq. (21) make a significant contribution. The large-time minimum of $\varepsilon(\theta,t)$ is the unique minimum of a truncated series:

$$F(\theta) \equiv \sum_l A_l e^{i\theta_l} \tag{22}$$

where $A_l \equiv a_{m_1,\dots,m_S}$ with $m_j = 0$ for $j \neq l$ and $m_l = 1$.

The Fourier coefficients $A_l$ are multidimensional integrals:

$$A_l = (2\pi)^{-S} \int_{[0,2\pi]^S} e^{-i\theta_l} V(\theta)d\theta \tag{23}$$

These integrals can be evaluated efficiently for energy as a sum of pairwise interactions $\nu$ depending only on interatomic distances $r$:

$$A_l = (2\pi)^{-S} \int_{[0,2\pi]^S} e^{-i\theta_l} \nu[r(\theta)]d\theta \tag{24}$$

The integration in eq. (24) may be divided into two stages: geometrical and physical:

$$\int e^{-i\theta_l}\nu[r(\theta)]d\theta = \int \nu(r)\left[\int e^{-i\theta_l}\frac{d\theta}{dr}\right]dr \tag{25}$$

The "geometric" integral $D_l(r) \equiv \int e^{-i\theta_l}\frac{d\theta}{dr}$ (where the integration extends over geometries having the same distance $r$) can be evaluated by an efficient recursive algorithm described elsewhere [40]. Having calculated $D_l(r)$, one can evaluate the one-dimensional "physical" integral $\int \nu(r)D_l(r)dr$ even by a numerical quadrature.

The algorithm presently used for the conformational analysis of molecules is as follows:

1. Find the long-time unique minimum of $F(\theta)$ eq. (22).
2. Proceed with the time-reversing algorithm for the tangent- space smoothed energy, starting from the geometry from step 1 and an intermediate time $t_o$.

The method is currently being implemented for large polypeptides. Initial tests on small peptides (Met-enkephalin) show that the new approach finds better minima than the previous one.

## Acknowledgments

Many of the computations were carried out on the IBM 3090 and 9000 supercomputers of the Cornell National Supercomputer Facility (CNSF), a resource of the Cornell Theory Center, which receives major funding from the National Science Foundation and the IBM Corporation, with additional support from New York State and members of its Corporate Research Institute.

## References

[1] C. B. Anfinsen, E. Haber, M. Sela, and F. H. White, Jr., *Proc. Natl. Acad. Sci.*, U.S.A., **47**, 1309-1314 (1961).

[2] H. A. Scheraga - Predicting three-dimensional structures of oligopeptides, in Reviews in Computational Chemistry, Vol. 3, Eds. K. B. Lipkowitz and D. B. Boyd, VCH Publ., New York, pp. 73-142 (1992).

[3] K. D. Gibson and H. A. Scheraga - The multiple-minima problem in protein folding, in "Structure & Expression: Vol. 1: From Proteins to Ribosomes", Eds. R.H. Sarma & M.H. Sarma, Adenine Press, Guilderland, N.Y., p. 67-94 (1988).

[4] H. A. Scheraga - Calculations of stable conformations of polypeptides, proteins, and protein complexes, *Chemica Scripta*, **29A**, 3-13 (1989).

[5] M. Vasquez and H. A. Scheraga - Use of buildup and energy-minimization procedures to compute low-energy structures of the backbone of enkephalin, *Biopolymers*, **24**, 1437-1447 (1985).

[6] L. Piela and H. A. Scheraga - On the multiple-minima problem in the conformational analysis of polypeptides. I. Backbone degrees of freedom for a perturbed ahelix, *Biopolymers*, **26**, S33-S58 (1987).

[7] Z. Li and H. A. Scheraga - Monte Carlo-minimization approach to the multiple-minima problem in protein folding, *Proc. Natl. Acad. Sci.*, U.S.A., **84**, 6611-6615 (1987).

[8] Z. Li and H. A. Scheraga - Structure and free energy of complex thermodynamic systems, *J. Molec. Str. (Theochem)*, **179**, 333-352 (1988).

[9] D. R. Ripoll and H. A. Scheraga - On the multiple-minima problem in the conformational analysis of polypeptides. II. An electrostatically driven Monte Carlo method-tests on poly(Lalanine), *Biopolymers*, **27**, 1283-1303 (1988).

[10] D. R. Ripoll and H. A. Scheraga - The multiple-minima problem in the conformational analysis of polypeptides. III. An electrostatically driven Monte Carlo method; tests on enkephalin, *J. Protein Chem.*, **8**, 263-287 (1989).

[11] D. R. Ripoll and H. A. Scheraga - On the multiple-minima problem in the conformational analysis of polypeptides. IV. Application of the electrostatically driven Monte Carlo method to the 20-residue membrane-bound portion of melittin, *Biopolymers*, **30**, 165-176 (1990).

[12] D. R. Ripoll, M. J. Vasquez and H. A. Scheraga - The electrostatically driven Monte Carlo method: Application to conformational analysis of decaglycine, *Biopolymers*, **31**, 319-330 (1991).

[13] D. R. Ripoll, L. Piela, M. Vasquez and H. A. Scheraga - On the multiple-minima problem in the conformational analysis of polypeptides. V. Application of the self-consistent electrostatic field and the electrostatically driven Monte Carlo methods to bovine pancreatic trypsin inhibitor, Proteins: Structure, Function, and Genetics, **10**, 188-198 (1991).

[14] G. H. Paine and H. A. Scheraga - Prediction of the native conformation of a polypeptide by a statistical-mechanical procedure. I. Backbone structure of enkephalin, *Biopolymers*, **24**, 1391-1436 (1985).

[15] G. H. Paine and H. A. Scheraga - Prediction of the native conformation of a polypeptide by a statistical-mechanical procedure. II. Average backbone structure of enkephalin, *Biopolymers*, **25**, 1547-1563 (1986).

[16] G. H. Paine and H. A. Scheraga - Prediction of the native conformation of a polypeptide by a statistical-mechanical procedure. III. Probable and average conformations of enkephalin, *Biopolymers*, **26**, 1125-1162 (1987).

[17] E. O. Purisima and H. A. Scheraga - An approach to the multiple-minima problem by relaxing dimensionality, *Proc. Natl. Acad. Sci.*, U.S.A., **83**, 2782-2786 (1986).

[18] E. O. Purisima and H. A. Scheraga - An approach to the multiple-minima problem in protein folding by relaxing dimensionality. Tests on enkephalin, *J. Mol. Bio.*, **196**, 697-709 (1987).

[19] M. H. Lambert and H. A. Scheraga - Pattern recognition in the prediction of protein structure, *J. Comput. Chem.*, **10**, 770-797, 798-816, 817-831 (1989).

[20] L. Piela, J. Kostrowicki and H. A. Scheraga - The multiple-minima problem in the conformational analysis of molecules. Deformation of the potential energy hypersurface by the diffusion equation method, *J. Phys. Chem.*, **93**, 3339-3346 (1989).

[21] J. Kostrowicki, L. Piela, B. J. Cherayil and H. A. Scheraga - Performance of the diffusion equation method in searches for optimum structures of clusters of Lennard-Jones atoms, *J. Phys. Chem.*, **95**, 4113-4119 (1991).

[22] J. Kostrowicki and H. A. Scheraga - Application of the diffusion equation method for global optimization to oligopeptides, *J. Phys. Chem.*,**96**, 7442-7449 (1992).

[23] R. J. Wawak, M. M. Wimmer and H. A. Scheraga - Application of the diffusion equation method of global optimization to water clusters, *J. Phys. Chem.*, **96**, 5138-5145 (1992).

[24] K. A. Olszewski, L. Piela and H. A. Scheraga - Mean field theory as a tool for intramolecular conformational optimization. 1. Tests on terminally-blocked alanine and Met-enkephalin, *J. Phys. Chem.*, **96**, 4672-4676 (1992).

[25] K. A. Olszewski, L. Piela and H. A. Scheraga - Mean Field Theory as a tool for intramolecular conformational optimization. 2. Tests on the homopolypeptides decaglycine and icosalanine, *J. Phys. Chem.*, **97**, 260-266 (1993).

[26] K. A. Olszewski, L. Piela and H. A. Scheraga - Mean Field Theory as a tool for intramolecular conformational optimization. 3. Test on melittin,*J. Phys. Chem.*, **97**, 267-270 (1993).

[27] M. H. Hao, S. Rackovsky, A. Liwo, M. R. Pincus and H. A. Scheraga - Effects of compact volume and chain stiffness on the conformations of native proteins, *Proc. Natl. Acad. Sci.*, U.S.A., **89**, 6614-6618 (1992).

[28] M. H. Hao, M. R. Pincus, S. Rackovsky and H. A. Scheraga - Unfolding and refolding of the native structure of bovine pancreatic trypsin inhibitor studied by computer simulations, *Biochemistry*, **32**,9614-9631 (1993).

[29] A. Liwo, M. R. Pincus, R. J. Wawak, S. Rackovsky and H. A. Scheraga - Prediction of protein conformation on the basis of a search for compact structures; test on avian pancreatic polypeptide, *Protein Science*, **2**, 1715-1731 (1993).

[30] A. A. Rabow and H. A. Scheraga - Lattice neural network minimization-application of neural network optimization for locating the global-minimum conformation of proteins, *J. Mol. Biol.*, **232**, 1157-1168 (1993).

[31] S. Kirkpatrick, C. D. Gelatt, Jr. and M. P. Vecchi, *Science*, **220**, 671-680 (1983).

[32] D. Vanderbilt and S. G. Louie, *J. Comput. Phys.*, **56**, 259-271 (1984).

[33] M. Nilges, A. M. Gronenborn, A. T. Brunger and G. M. Clore, *Protein Eng.*, **2**, 27-38 (1988).

[34] A. Nayeem, J. Vila and H. A. Scheraga - A comparative study of the simulated-annealing and Monte Carlo with minimization approaches to the minimum-energy structures of polypeptides: [Met]Enkephalin, *J. Comput. Chem.*, **12**, 594-605 (1991).

[35] M. Dygert, N. G_ and H. A. Scheraga - Use of a symmetry condition to compute the conformation of gramicidin S, *Macromolecules*, **8**, 750-761 (1975).

[36] M. H. Miller and H. A. Scheraga - Calculation of the structures of collagen models. Role of interchain interactions in determining the triplehelical coiled-coil conformation. I. Poly(glycyl-prolyl- prolyl), *J. Polymer Sci.: Polymer Symposia*, No. 54, p. 171-200 (1976).

[37] M. H. Miller, G. Némethy and H. A. Scheraga - Calculation of the structures of collagen models. Role of interchain interactions in determining the triplehelical coiled coil conformation. 2. Poly(glycyl-prolyl-hydroxyprolyl), *Macromolecules*, **13**, 470-478 (1980).

[38] M. H. Miller, G. Némethy and H. A. Scheraga - Calculation of the structures of collagen models. Role of interchain interactions in determining the triplehelical coiled coil conformation. 3. Poly(glycyl-prolyl-alanyl). *Macromolecules*, **13**, 910-913 (1980).

[39] 523. G. Némethy, M. H. Miller and H. A. Scheraga - Calculation of the structures of collagen models. Role of interchain interactions in determining the triple-helical coiled-oil conformation. 4. Poly(glycyl-alanyl-prolyl). *Macromolecules*, **13**, 914-919 (1980).

[40] J. Kostrowicki and H. A. Scheraga - New method for calculation of the exact end-to-end distance distributions for the freely-rotating chain, *Computational Polymer Science*, in press.

*E-mail address*: jarek@scheraga2.tn.cornell.edu
*E-mail address*: has5@cornell.edu

BAKER LABORATORY OF CHEMISTRY, CORNELL UNIVERSITY, ITHACA, NEW YORK 14853–1301

DIMACS Series in Discrete Mathematics
and Theoretical Computer Science
Volume **23**, 1996

# A Deterministic Global Optimization Approach for the Protein Folding Problem

C.D. MARANAS, I.P. ANDROULAKIS, AND C.A. FLOUDAS*

ABSTRACT. A deterministic global optimization algorithm is proposed for locating the global minimum potential energy conformations of oligopeptide chains. The ECEPP/3 detailed potential energy model is utilized to model the energetics of the atomic interactions. The minimization of the total potential energy is formulated on the set of peptide dihedral angles. Based on previous work on the microcluster and molecular structure determination, a procedure for deriving convex lower bounding functions for the total potential energy function is utilized which involves a number of important properties. The global optimization algorithm $\alpha$**BB** which has been shown to be $\epsilon$–convergent to the global minimum potential energy conformation through the solution of a series of nonlinear convex optimization problems is utilized. The ECEPP/3 potential model is interfaced with $\alpha$**BB** in the program **GLOFOLD**, and provisions have been made to accommodate user specified partitioning of the dihedral angles into three sets. The first one (i.e., global variables), consists of dihedral angles where branching occurs. The second set (i.e., local variables) includes the dihedral variables where branching is not necessary. The third set, (i.e., fixed variables) includes the dihedral angles which are kept fixed. The proposed deterministic global optimization is applied on a number of oligopeptide folding problems.

## 1. Introduction

The *protein folding* problem is one of the most challenging problems in biochemistry. Predicting how a protein would fold is of paramount academic and industrial interest. Many products of the rapidly developing biotechnology industry are novel proteins. It is already possible to design genes to direct the synthesis of such proteins. Yet failure to fold properly is a common production

1991 *Mathematics Subject Classification.* Primary 90B80, 90C20, 90C35, 90C27; Secondary 65H20, 65K05.

*Key words and phrases.* global optimization, protein folding, convex lower bounding, $\alpha$**BB**˙
This research was partially funded by the NSF grants CBT–8857013, CTS–9221411, the Air Force Office of Scientific Research, Exxon Co., and Mobil Co.

* Author to whom correspondence should be addressed.

concern. It is possible nowadays to produce proteins with a given amino acid sequence and therefore, knowledge of how the protein would fold would allow one to predict and fine–tune its chemical and biological properties. This would greatly simplify the tasks of interpreting data collected by the human genome project, understanding the mechanisms of hereditary and infectious diseases, designing drugs with specific therapeutical properties, and growing biological polymers with specific material properties.

From a chemical point of view, a *protein* is essentially a polymer chain composed by a sequence of various amino acid residues connected with peptide bonds. Proteins in living cells are composed of only 20 different amino acid residues. The general form of these amino acid residues is shown in Figure 1. The form of the side chain $R$ (e.g., methyl, butyl, benzoic, etc.) defines all different amino acid residues. The chemical structure of a protein is illustrated in Figure 2.

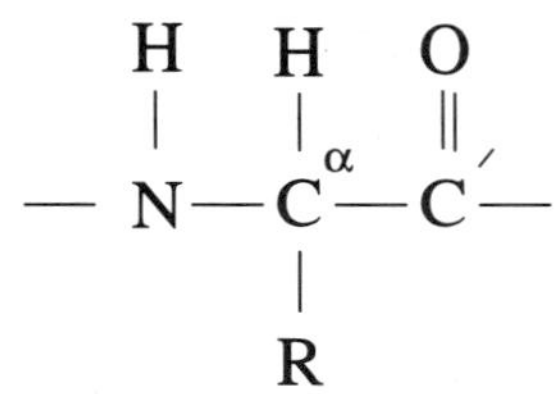

FIGURE 1. Amino acid residue with side chain R.

Note that, the side groups $R_n$ vary from one residue to the other. Also $E_{amino}$, $E_{carboxyl}$ are the amino and carboxyl end groups respectively. The repeating unit -NC$_\alpha$C'- connected with peptide bonds defines the *backbone* of the protein. Although, it appears linear in Figure 2, covalent bond angle requirements and interatomic forces bend and twist the chain in a way characteristic for each protein. The protein chain "curls up" into a *unique* three–dimensional geometric conformation called the *folded state* of the protein. It is exactly this configuration which defines the shape of the protein surface as well as the particular chemically active groups present on the surface which in turn determine the biological function of the protein. Predicting this energetically most favorable conformation based solely on the atomic interactions is the objective of this work.

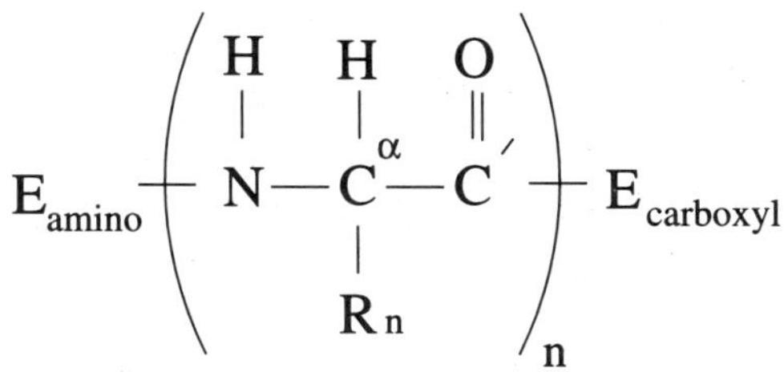


FIGURE 2. Chemical structure of proteins.

In other words, given the *primary structure* of a protein (i.e., residue sequence and type) predict its *tertiary structure* (i.e., 3–D conformation). In the next section, a mathematical description of the protein folding problem is provided.

## 2. Mathematical Description

The geometry of a protein can be fully described by assigning a three–dimensional *coordinate vector*,

$$r_i = \begin{pmatrix} x_i \\ y_i \\ z_i \end{pmatrix} \quad i = 1, \ldots, N$$

which specifies the position of each atom $i = 1, \ldots, N$ in the protein molecule. The *bond vector* between two atoms (i,j) connected with a covalent bond is defined as:

$$r_{ij} = \begin{pmatrix} x_j - x_i \\ y_j - y_i \\ z_j - z_i \end{pmatrix}$$

The corresponding *bond length* is then equal to the Euclidean distance between atoms i and j,

$$|r_{ij}| = \sqrt{(x_j - x_i)^2 + (y_j - y_i)^2 + (z_j - z_i)^2}.$$

The *covalent bond angle* $\theta_{ijk}$ formed by the two adjacent bond vectors $r_{ij}$ and $r_{jk}$ can be computed by the following formulae (See Figure 3).

$$\cos(\theta_{ijk}) = \frac{r_{ij} \cdot r_{jk}}{|r_{ij}|\,|r_{jk}|}, \quad \sin(\theta_{ijk}) = \frac{r_{ij} \times r_{jk}}{|r_{ij}|\,|r_{jk}|}.$$

Here $r_{ij} \cdot r_{jk}$ is the *dot product* of the bond vectors $r_{ij}$ and $r_{jk}$,

$$r_{ij} \cdot r_{jk} = (x_j - x_i)(x_k - x_j) + (y_j - y_i)(y_k - y_j) + (z_j - z_i)(z_k - z_j)$$

and $r_{ij} \times r_{jk}$ is the *cross product*,

$$r_{ij} \times r_{jk} = \begin{pmatrix} (y_j - y_i)(z_k - z_j) - (z_j - z_i)(y_k - y_j) \\ (z_j - z_i)(x_k - x_j) - (x_j - x_i)(z_k - z_j) \\ (x_j - x_i)(y_k - y_j) - (y_j - y_i)(x_k - x_j) \end{pmatrix}.$$

The *dihedral angle* $\omega_{ijkl} \in [-180^o, 180^o]$ or the complementary *torsion angle* $\phi_{ijkl} = \omega_{ijkl} - 180^o$ measure the relative orientation of two adjacent covalent angles $\theta_{ijk}$ and $\theta_{jkl}$ (see Figure 3). It is defined as the angle between the normals through the planes defined by atoms $i, j, k$ and $j, k, l$ respectively, and can be calculated from the following relations:

$$\cos(\omega_{ijkl}) = \frac{(r_{ij} \times r_{jk}) \cdot (r_{jk} \times r_{kl})}{|r_{ij} \times r_{jk}|\,|r_{jk} \times r_{kl}|},$$

$$\sin(\omega_{ijkl}) = \frac{(r_{kl} \times r_{ij}) \cdot r_{jk}\,|r_{jk}|}{|r_{ij} \times r_{jk}|\,|r_{jk} \times r_{kl}|}$$

Instead of specifying the coordinate vector for all atoms in a protein molecule, one can specify all bond lengths, covalent bond angles and dihedral angles. Under biological conditions, the bond lengths and bond angles are fairly rigid and thus can be assumed to be fixed at their equilibrium values. Under this assumption, the dihedral angles along the backbone fully determine the geometric shape of the folded protein.

The names of the dihedral angles of a folded protein chain follow a standard nomenclature. The dihedral angle between the normals of the planes formed by atoms $C'_{i-1}N_iC_{\alpha,i}$ and $N_iC_{\alpha,i}C'_i$ respectively is called $\phi_i$ where $i-1$ and $i$ are two adjacent amino acid residues. The one defined by planes $R_iC_{\alpha,i}C'_i$ and $C_{\alpha,i}C'_iN_{i+1}$ respectively is called $\psi_i$ where $i$ and $i+1$ are two adjacent amino acid residues. Also $\omega_i$ is the dihedral angle defined by the planes $C_{\alpha,i}C'_iN_{i+1}$ and $C'_iN_{i+1}C_{\alpha,i+1}$. The letter $\chi$ is utilized to denote the dihedral angles which are associated with the side groups $R_i$. Also the letter $\theta$ is used to name the dihedral angles associated with the two end groups. Figure 4 pictorially illustrates these conventions.

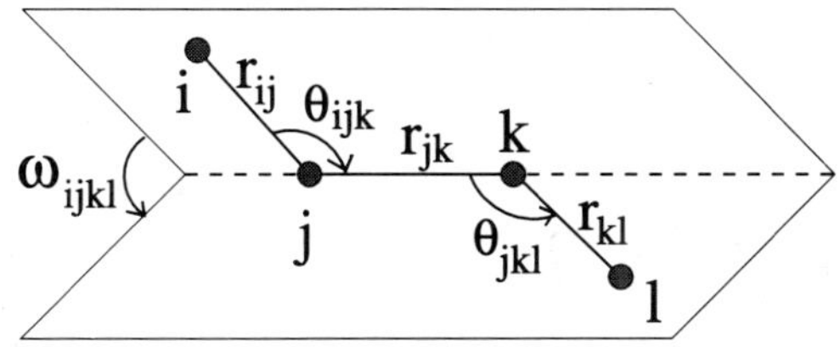


FIGURE 3. Bond vectors, covalent bond angles and dihedral angle

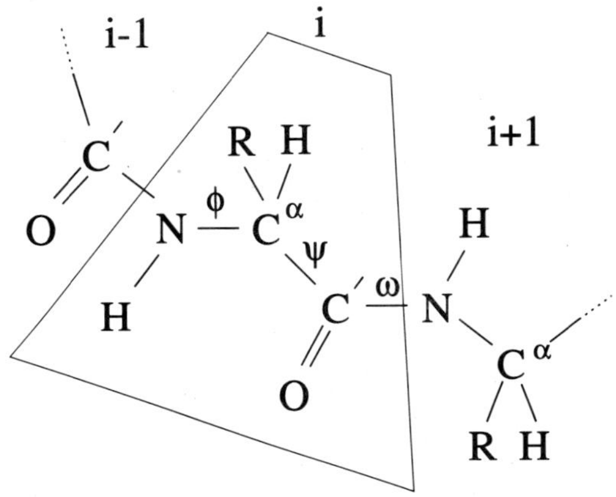


FIGURE 4. Dihedral angles in a protein

## 3. Potential Energy Model

Molecular mechanics calculations employ an empirically derived set of potential energy contributions for approximating these atomic interactions. This set of potential energy contributions, called the *force field*, contains adjustable parameters that are selected in a such a way as to provide the best possible agreement with experimental data. The main assumption introduced in molecular mechanics is that every parameter is associated with *a specific interaction* rather than a specific molecule *(transferability assumption)*. These parameters are bond lengths; covalent bond angles; bond stretching, bending, or rotating constants; non–bonded atom interaction constants, etc. Thus, whenever a specific interaction is present, the same value for the parameter can be used even if this interaction occurs in different molecules [**4**]. Note that experimental results provide sufficient evidence that this is a reasonable assumption in most cases. Many different models have been proposed for approximating the *force field*, and some of the most popular ones are: ECEPP [**10, 11, 12**], MM2 [**1**], ECEPP/2 [**15**], CHARMM [**3**], AMBER [**18**], GROMOS87 [**17**], MM3 [**2**], and ECEPP/3 [**14**].

In this work the ECEPP/3 potential model is utilized. In this potential model, it is assumed that the covalent bond lengths and angles are fixed at their equilibrium values, and thus the protein conformation is only a function of the dihedral angles. This implies that ECEPP/3 accounts for only energy interaction terms which depend on the dihedral angles. The conformational energy is treated as the sum of electrostatic, nonbonded, hydrogen bond and torsional contributions, plus an additional loop closing potential if the polypeptide contains one or more intramolecular disulfide bonds. Also the fixed internal conformational energy of the pyrolidine ring is added for each propyl or hydroxyprolyl residue contained in the polypeptide. The first three energy contributions are computed for each atom pair (i,j) whose interatomic distance is a function of at least one dihedral angle. This set of atomic pairs is denoted as $\mathcal{P}$ and includes the atomic pairs which are separated by at least two other atoms.

The electrostatic energy $U_{ES}$ is computed for each atomic pair $(i,j) \in \mathcal{ES} = \mathcal{P}$ as a Coulomb potential interaction between two atom–centered monopole partial charges $q_i$ and $q_j$ where $D$ is the dielectric constant.

$$U_{ES} = \sum_{(i,j)\in\mathcal{ES}} \frac{q_i q_j}{D\,|r_{ij}|}$$

A modified Lennard–Jones 12–6 potential is used to approximate the nonbonded interaction energies between atomic pairs $(i,j) \in \mathcal{NB}$ [**14**]. The set $\mathcal{NB}$ contains all atomic pairs $\mathcal{P}$ except for hydrogen bonding pairs.

$$U_{NB} = \sum_{(i,j)\in\mathcal{NB}} F\frac{A_{k(i)k(j)}}{|r_{ij}|^{12}} - \frac{C_{k(i)k(j)}}{|r_{ij}|^{6}}$$

Here $k(i)$ returns the atom type of atom i in the protein chain. The coefficients $A_{k(i)k(j)}, C_{k(i),k(j)}$ are assigned specific values for each combination of atom types $k(i)$ and $k(j)$. F is assigned a value of 0.5 for 1–4 interactions and 1.0 for for 1–5+ interactions. An atom pair interaction is defined as 1–4 when the distance between the interacting atoms is a function of only one intervening dihedral angle. Any other interactions $(i,j) \in \mathcal{P}$ is considered to be 1–5+.

Hydrogen–bond interactions are the ones between designated donor and acceptor atoms. The donors (H) are amine, amide, hydroxyl or carboxyl acid hydrogens and the acceptors (X) are uncharged ring nitrogens, amide nitrogens, or hydroxyl ester, carbonyl, or carboxylic acid oxygens. A 12–10 potential function is used to model the hydrogen bond interactions.

$$U_{HX} = \sum_{(i,j)\in\mathcal{HX}} F\frac{A'_{k(i),k(j)}}{|r_{ij}|^{12}} - \frac{B'_{k(i),k(j)}}{|r_{ij}|^{10}}$$

Note that $\mathcal{P} = \mathcal{NB} \cup \mathcal{HX}$.

Torsional energy terms are included in the potential energy model to bring the experimental and computed rotational barrier into agreement. These terms are computed for all $\omega$ dihedral angles and for some designated side–chain $\chi$ and end group $\theta$ dihedral angles, but not for any $\phi$ and $\psi$ angles. Let $\mathcal{TOR}$ be the set of dihedral angles for which a torsional term is calculated. The form of the potential function used is:

$$U_{TOR} = \sum_{k\in\mathcal{TOR}} \frac{U_{o,k}}{2}\left(1 + c_k \cos n_k t_k\right)$$

Here $U_{o,k}$ is the difference between the experimental barrier and the one calculated from the electrostatic, nonbonded, and hydrogen–bond potential functions, $t_k$ is the value of the $k^{th}$ dihedral angle for which a torsional term is included, $n_k$ gives the symmetry of the barrier, and $c_k \in \{-1, 1\}$ defines the sign for the cosine term.

The cystine loop–closing energy $U_{LOOP}$ and torsional energy $U_{CYST}$ are computed as the sums of terms for all disulfide bonds in the peptide. The loop–closing potential penalizes any deviation of the interatomic distances $S_iS_j$, $C_i^\beta\ S_j$, and $C_j^\beta\ S_i$ from their experimentally observed values (see Figure 5). Let $\mathcal{SS}$ be the set of all disulfide bonds in the peptide. Then $U_{LOOP}$ is defined as:

$$U_{LOOP} = \sum_{(i,j)\in\mathcal{SS}} B\left[\left(r_{S_iS_j} - r^o_{S_iS_j}\right)^2 + \left(r_{S_iC_j^\beta} - r^o_{S_iC_j^\beta}\right)^2 + \left(r_{S_jC_i^\beta} - r^o_{S_jC_i^\beta}\right)^2\right]$$

Note that $r^o_{S_iS_j} = 2.04A^o$ is the experimentally observed disulfide bond distance and $r^o_{S_iC_j^\beta} = r^o_{S_jC_i^\beta} = 3.052A^o$ are the distances between $S_{i/j}$ and $C^\beta_{j/i}$ so as the covalent bond angles $\theta_{C_i^\beta S_iS_j}$ and $\theta_{S_iS_jC_j^\beta}$ assume the experimentally observed value of $104^o$. B is a penalty parameter assigned the value $B = 100\ kcal/mol\text{Å}^2$. Torsional contributions to $U_{CYST}$ from the angles $C_i^\alpha C_i^\beta S_iS_j$ and $S_iS_jC_j^\beta C_j^\alpha$ are calculated as mentioned above with $U_o = 1.5 kcal/mol$. The contribution from the dihedral angle $C_i^\beta S_i S_j C_j^\beta$ is computed as a penalty term on the interatomic distance $r_{C_i^\beta C_j^\beta}$ according to the relation:

$$U_{SS} = A \sum_{(i,j) \in \mathcal{SS}} \left( r_{C_i^\beta C_j^\beta} - r^o_{C_i^\beta C_j^\beta} \right)^2$$

The penalty parameter A is set equal to 10 $kcal/mol\text{Å}^2$ and the experimentally observed interatomic distance $r^o_{C_i^\beta C_j^\beta}$ to 3.855Å.

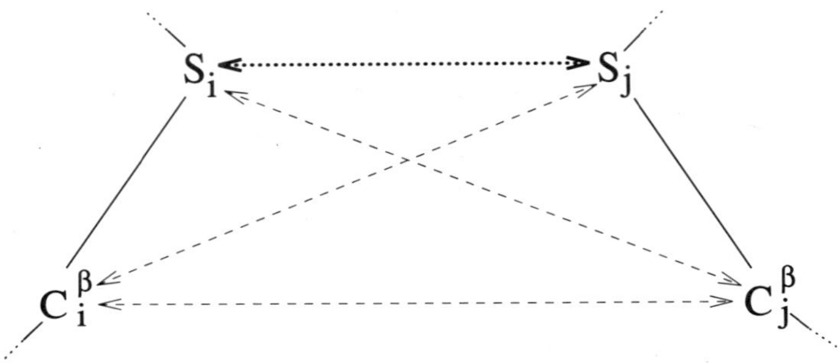


FIGURE 5. Disulfide bonding

Finally, the additional potential energy term $U_{PRO}$ is added to account for the internal conformation energy of proline and hydroxyproline residues. This internal conformation energy depends on whether the peptide bond is on *cis* or *trans* configuration. The total potential energy of the peptide chain, in the context of the potential model ECEPP/3, can then be written as the sum of a number of different interaction and correction terms.

$$U = U_{ES} + U_{NB} + U_{HX} + U_{TOR} + U_{LOOP} + U_{CYST} + U_{PRO}$$

Note that all these terms are functions of the dihedral angles.

## 4. Problem Formulation

The potential energy minimization problem can be formulated as a nonconvex nonlinear optimization problem. Let $i = 1, \ldots, N_{RES}$ be an indexed set describing the sequence of amino acid residues in the peptide chain. This implies that there are $\phi_i, \psi_i, \omega_i,\ i = 1, \dots, N_{RES}$ dihedral angles along the backbone of the peptide chain. Also let $k = 1, \dots, K^i$ denote the dihedral angles of the side group in the $i^{th}$ residue and $j = 1, \dots, J^N$ the dihedral angles of the amino end group and $j = 1, \dots, J^C$ of the carboxyl end group respectively. This defines the side group dihedral angles $\chi_i^k,\ i = 1, \dots, N_{RES},\ k = 1, \dots, K^i$ and the amino $\theta_j^N,\ j = 1, \dots, J^N$ and carboxyl $\theta_j^C,\ j = 1, \dots, J^C$ end group dihedral angles respectively. Based on these definitions the potential model minimization energy problem can be formulated as follows:

$$\min \quad U(\phi_i, \psi_i, \omega_i, \chi_i^k, \theta_j^N, \theta_j^C)$$

$$\begin{aligned}
\text{subject to} \quad -\pi &\leq \phi_i \leq \pi, \ i = 1, \dots, N_{RES} \\
-\pi &\leq \psi_i \leq \pi, \ i = 1, \dots, N_{RES} \\
-\pi &\leq \omega_i \leq \pi, \ i = 1, \dots, N_{RES} \\
-\pi &\leq \chi_i^k \leq \pi, \ i = 1, \dots, N_{RES}, \ k = 1, \dots, K^i \\
-\pi &\leq \theta_j^N \leq \pi, \ j = 1, \dots, J_N \\
-\pi &\leq \theta_j^C \leq \pi, \ j = 1, \dots, J_C
\end{aligned}$$

Here $U$ is the expression for the total potential energy as a function of the peptide dihedral angles. The specific expressions comprising $U$ have been described in detail in the previous section. Note that $U$ is a nonconvex function of these dihedral angles involving numerous local minima even for small peptide systems. These local minima correspond to metastable states of the polypeptide chain. A single global minimum defines the energetically most favorable peptide conformation. A plethora of different methods has been proposed for finding this conformation [**16**]. Most methods attempt to locate this point by tracing, deterministically or stochastically, single or multiple paths on the potential energy surface conjecturing that some of them will converge to the global minimum potential energy point. A review on these methods can be found in [**8**]. The key limitation of these methods is that the obtained conformations depend heavily on the supplied initial conformation expressing the bias of the researcher towards which is the most appropriate conformation. This is why, in practice, many trial geometries need to serve as initial points in an attempt to lessen the initial point dependence. However, there is no guarantee that important conformations are not overlooked. The need for a method that can guarantee convergence to the global minimum potential energy conformation motivated our initial effort to introduce such a method for microclusters [**6, 7**], and small acyclic molecules [**8, 9**] allowing for nonbonded atomic pair interactions. The approach, $\alpha$**BB** has been extended to constrained optimization problems in [**5**]. In this paper, the approach is extended to peptide systems interacting with realistic potential energy models (i.e., ECEPP/3). In the next section, a brief description of $\alpha$**BB** customized for the protein folding problem, is provided.

## 5. Global Optimization

The deterministic branch and bound type global optimization algorithm $\alpha$**BB** [**8**, **5**] is utilized which brackets the global minimum solution by constructing converging lower and upper bounds. These bounds are successively refined by iteratively partitioning the initial feasible region into many subregions. Upper bounds to the global minimum can be obtained by local minimizations of U. Lower bounds are obtained by minimizing a convex function L which is always less than the original nonconvex function U. This function L can be constructed by augmenting U through the addition of a convex separable quadratic term for each dihedral angle.

$$
\begin{aligned}
L = U + \alpha \{ \quad & \sum_{i=1}^{N_{RES}} \left(\phi_i^L - \phi_i\right)\left(\phi_i^U - \phi_i\right) + \\
& \sum_{i=1}^{N_{RES}} \left(\psi_i^L - \psi_i\right)\left(\psi_i^U - \psi_i\right) + \\
& \sum_{i=1}^{N_{RES}} \left(\omega_i^L - \omega_i\right)\left(\omega_i^U - \omega_i\right) + \\
& \sum_{i=1}^{N_{RES}} \sum_{k=1}^{K^i} \left(\chi_i^{k,L} - \chi_i^k\right)\left(\chi_i^{k,U} - \chi_i^k\right) + \\
& \sum_{j=1}^{J^N} \left(\theta_j^{N,L} - \theta_j^N\right)\left(\theta_j^{N,U} - \theta_j^N\right) + \\
& \sum_{j=1}^{J^C} \left(\theta_j^{C,L} - \theta_j^C\right)\left(\theta_j^{C,U} - \theta_j^C\right) \}
\end{aligned}
$$

Note that $\phi_i^L, \psi_i^L, \omega_i^L, \chi_i^{k,L}, \theta_j^{N,L}, \theta_j^{C,L}$ and $\phi_i^U, \psi_i^U, \omega_i^U, \chi_i^{k,U}, \theta_j^{N,U}, \theta_j^{C,U}$ are lower and upper bounds respectively on the dihedral angles $\phi_i, \psi_i, \omega_i, \chi_i^k, \theta_j^N, \theta_j^C$. Also $\alpha$ is a nonnegative parameter which must be greater or equal to the negative one half of the minimum eigenvalue of $U$ inside the current dihedral angles rectangle. Qualitatively, the effect of adding this extra term to $U$ is to make $L$ convex by overpowering the nonconvexity characteristics of $U$ with the addition of the term $2\alpha$ to all of its eigenvalues. This function $L$, defined inside some rectangular region, involves a number of important properties which enable us to construct a global optimization algorithm for finding the global minimum of $U$ in the space defined by the dihedral angles. These properties, whose proof is given in [**8**], demonstrate that (i) $L$ is always a valid underestimator of $U$; (ii) $L$ matches $U$ at all corner points of the box constraints; (iii) $L$ is convex; (iv) the maximum separation between $L$ and $V$ is *bounded* and proportional to $\alpha$ and to the square of the diagonal of the current box constraints; and (v) the underestimators $L$ constructed over supersets of the current set are always *less tight* than the underestimator $U$ constructed over the current box constraints for every point within the current box constraints.

Based on these properties a deterministic branch and bound type global optimization algorithm is proposed for locating the global minimum potential energy of $U$ by constructing converging lower and upper bounds. The approach is implemented in the **GLOFOLD** package. Qualitatively, the steps of the approach are as follows:

**Step 1** An upper bound on the global minimum solution of $U$ is obtained by minimizing $U$ with a local solver (i.e., MINOS [**13**]). The current best upper bound is updated to be the minimum over the stored ones.

**Step 2** The current rectangle is partitioned in two by bisecting along the longest side.

**Step 3** The convex function $L$ is minimized inside both resulting subrectangles. If the solutions are less than the current best upper bound they are stored, otherwise they are discarded (fathoming).

**Step 4** The rectangle involving the minimum solution for $\min L$ is selected for further partitioning and the corresponding solution is erased from the lower bounds stack.

**Step 5** If the current best upper and lower bounds are within $\epsilon$ then terminate, otherwise continue with Step 1.

The approach is shown in [**8**] to terminate in a finite number of iterations to an $\epsilon$–global minimum solution.

## 6. Implementation: GLOFOLD

The proposed approach has been interfaced with ECEPP/3 and implemented in C, in the program **GLOFOLD**. A schematic diagram of the interface between ECEPP/3 and $\alpha$**BB** is shown in Figure 6.

The peptide dihedral angles are partitioned into three sets. The first one (i.e., global variables), consists of dihedral angles where branching occurs. The second set (i.e., local variables) includes the dihedral variables where branching is not performed. The third set, (i.e., fixed variables) includes the dihedral angles for which there exists sufficient (experimental) evidence for keeping them fixed.

The information required by the user, in the current implementation of **GLOFOLD**, is provided in two files. The first one, required by ECEPP/3, contains information about the sequence and number of the amino acid residues and the type of the end groups. Also dihedral angles are initialized and output file numbers are assigned. The second file contains information related with the global optimization phase. In particular, (i) number of dihedral angles, (ii) convergence tolerances, (iii) type of starting point, (iv) lower/upper bounds on dihedral vari-

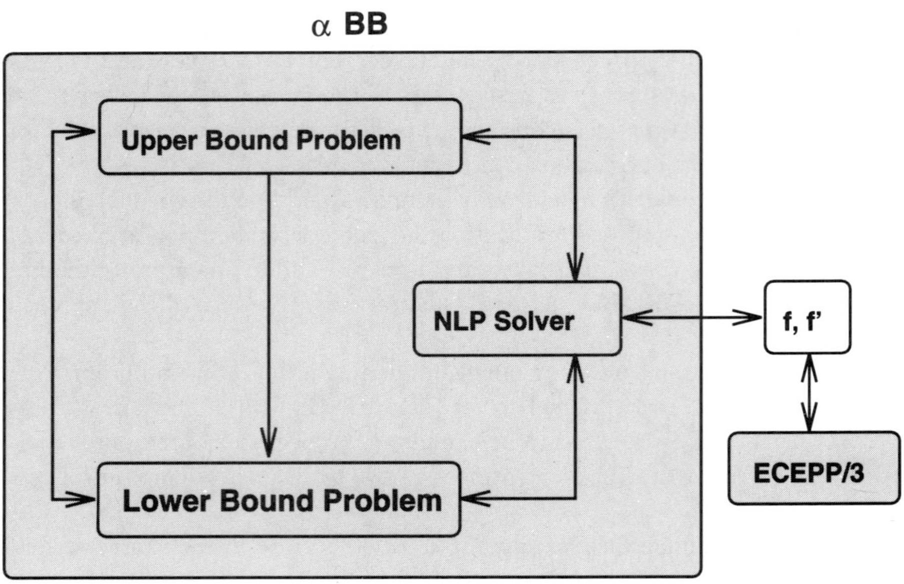


FIGURE 6. Interface between ECEPP/3 and $\alpha$**BB**

ables, (v) values for $\alpha$ parameter, and (vi) variables where branching occurs are required. In addition to the ability of **GLOFOLD** to locate the global minimum total potential energy oligopeptide conformation, low energy protein conformations close to the global minimum one can be obtained with multi-start local optimizations initiated from the obtained solutions for the lower bounding problems. Furthermore, PDB format files are created for all solutions which can be readily interfaced with graphics programs.

## 7. Computational Studies

**GLOFOLD** has been tested on two classes of oligopeptide folding problems. The selected relative convergence tolerance is $10^{-2}$ and computational requirements in seconds are reported for an HP–730 workstation. First, the method was applied on all 20 naturally occurring amino acids. The amino end group used was acetyl, and the carboxyl end group was methyl. Note that all dihedral angles were treated as global variables except for the three $\theta$ angles in the end groups which were treated as local variables. The results are summarized in Table 1. The CPU time required is presented for each residue, as well as the average $< CPU >$ time needed for residues of the same size. The computational effort compares favorably with that reported in [**19**] employing a simulated annealing implementation and fewer dihedral angles. The computational effort increases, as expected, as a function of the number of global variables $n$, but it stays under the $n^3$ curve (see Figure 7). Even though in [**19**] a different potential model (i.e. AMBER) was utilized the large differences in computational requirements are quite suggestive about the relative computational efficiencies.

| Amino acid | # Dihedrals | Energy | Iter | CPU | $< CPU >$ |
|---|---|---|---|---|---|
| Pro | 5 | -19.81 | 28 | 6 | **6** |
| Gly | 6 | -6.33 | 67 | 14 | **14** |
| Ala | 7 | -5.18 | 141 | 55 | |
| Cys | 7 | -5.84 | 142 | 45 | **50** |
| His | 8 | -8.92 | 298 | 173 | |
| Phe | 8 | -8.43 | 298 | 169 | |
| Ser | 8 | -7.86 | 184 | 102 | |
| Trp | 8 | -9.56 | 306 | 227 | **167** |
| Asn | 9 | -22.95 | 345 | 220 | |
| Asp | 9 | -20.05 | 452 | 239 | |
| Thr | 9 | -9.59 | 285 | 208 | |
| Tyr | 9 | -8.48 | 753 | 506 | |
| Val | 9 | -4.19 | 644 | 387 | **312** |
| Gln | 10 | -18.99 | 601 | 460 | |
| Glu | 10 | -15.87 | 640 | 386 | |
| Ile | 10 | -2.54 | 388 | 352 | |
| Leu | 10 | -5.72 | 1123 | 613 | |
| Met | 10 | -6.91 | 1284 | 641 | **480** |
| Lys | 11 | -7.98 | 922 | 1070 | **1070** |
| Arg | 13 | -31.84 | 1000 | 1660 | **1660** |

TABLE 1. Naturally occurring amino acids

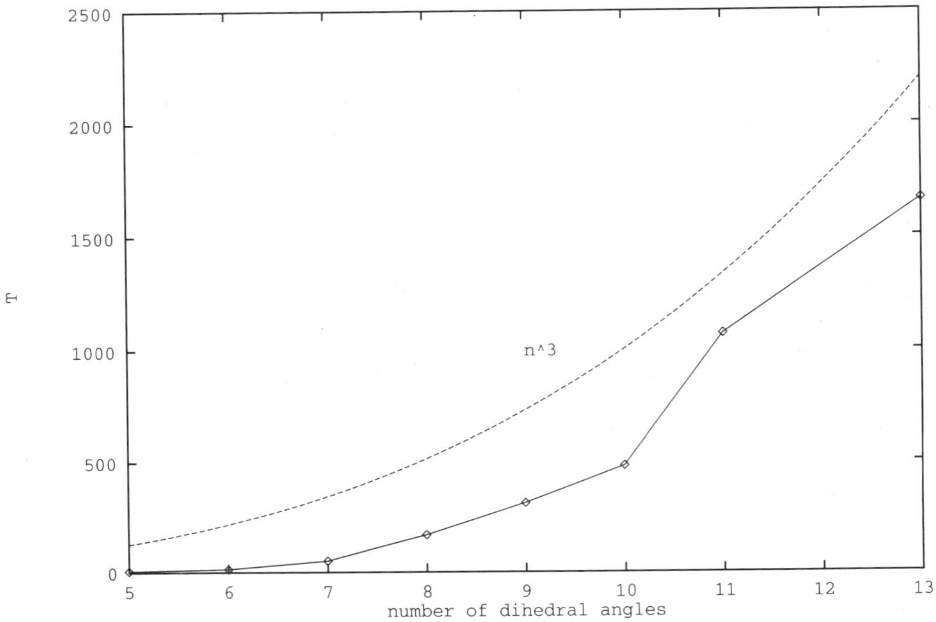


FIGURE 7. CPU times as a function of the number of global dihedral angles.

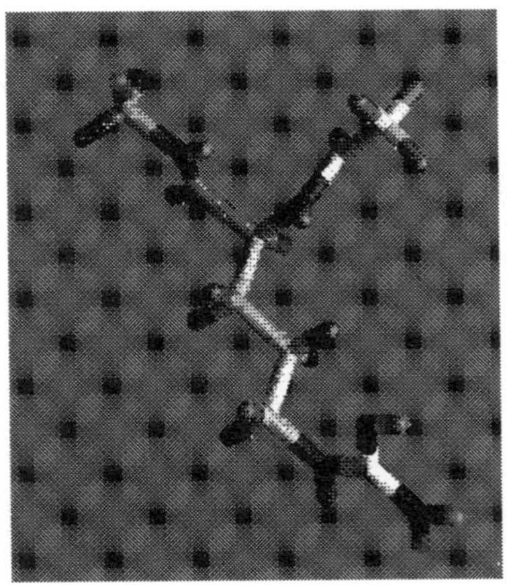

FIGURE 8. NaN'm–arginine.

| No. | Name |
|---|---|
| 1 | NaN'm – L-Alanyl-L-proline(D) |
| 2 | NaN'm – L-Alanyl-L-proline(U) |
| 3 | NaN'm – L-Prolyl(D)-L-proline(D) |
| 4 | NaN'm – L-Prolyl(D)-L-proline(U) |
| 5 | NaN'm – L-Prolyl(U)-L-proline(D) |
| 6 | NaN'm – L-Prolyl(U)-L-proline(U) |

TABLE 2. Oligopeptide names

| Residue | $\theta_1^1$ | $\theta_1^2$ | $\phi_1$ | $\psi_1$ | $\omega_1$ | $\chi_1$ | $\chi_2$ | $\phi_1$ | $\psi_1$ | $\omega_1$ | $\theta_2^1$ | $U^*$ |
|---|---|---|---|---|---|---|---|---|---|---|---|---|
| $1^{known}$ | 180.0 | 180 | -149.0 | 67 | 177 | 180 | – | -68.8 | 79 | 180 | 180 | -17.39 |
| $1^{found}$ | 60.0 | -180 | -149.0 | 67 | 177 | 178 | – | -68.8 | 79 | 179 | 60 | -17.39 |
| $2^{known}$ | 180 | 180 | -150.0 | 77 | 178 | 180 | – | -53.0 | -38 | 180 | 180 | -17.08 |
| $2^{found}$ | -180 | 180 | -150.0 | 77 | 178 | -61 | – | -53.0 | -38 | -180 | -180 | -17.08 |
| $3^{known}$ | 180 | 180 | -68.8 | 162 | 176 | – | – | -68.8 | 81 | 178 | 180.0 | -32.33 |
| $3^{found}$ | 60 | 180 | -68.8 | 162 | 176 | – | – | -68.8 | 80 | 178 | 60.0 | -32.34 |
| $4^{known}$ | 180 | 176 | -53.0 | 157 | 177 | – | – | -68.8 | 81 | 178 | 180 | -31.38 |
| $4^{found}$ | 55 | 176 | -53.0 | 157 | -179 | – | – | -68.8 | 81 | 178 | 180 | -31.38 |
| $5^{known}$ | 180 | 179 | -68.8 | 169 | 181 | – | – | -53.0 | -38 | 179 | 180 | -31.30 |
| $5^{found}$ | -63 | 179 | -68.8 | 169 | 177 | – | – | -53.0 | -38 | 179 | 180 | -31.30 |
| $6^{known}$ | 180 | 174 | -53.0 | 167 | 179 | – | – | -53.0 | -37 | 179 | 180 | -30.22 |
| $6^{found}$ | -53 | 174 | -53.0 | 167 | -179 | – | – | -53.0 | -37 | -179 | 60 | -30.22 |

TABLE 3. Oligopeptide results

The second set of examples involves a number of oligopeptides listed in Table 2. The results are summarized in Table 3. Dihedral angle and energy values are reported at the global minimum solution. Note that in all cases the obtained solution is at least as good as the best over the ones reported in the literature. In case (3) a slightly improved solution is found, over the best known so far. The computational performance and efficiency of a simulated annealing implementation was next studied on the oligopeptide examples of Table 3. The parameters employed in the simulated annealing implementation follow the suggestions of [**19**] and are shown in Table 4. Nine different combinations of Markov chain lengths and number of successive annealing steps are considered (see Table 4). Note that MC is the Markov chain length and NT the number of successive annealing steps. Furthermore, the initial temperature was $T_0 = 5.0$, the maximum allowable step was set to $\delta = 90^o$ and the cooling schedule factor was $\beta = 0.9$. The results in Table 5 indicate that the $\alpha$**BB** algorithm always locates the global minimum conformation and requires less CPU time than the simulated annealing implementation. On top of that, the failure rate (numbers in parentheses) for the simulated annealing is very high. Notably, in certain

| MC | NT | | |
|---|---|---|---|
| | 25 | 50 | 100 |
| 100 | 1 | 2 | 3 |
| 250 | 4 | 5 | 6 |
| 500 | 7 | 8 | 9 |

TABLE 4. Parameters used in different simulated annealing runs

| | | $\alpha$BB | Simulated Annealing | | | | | | | | |
|---|---|---|---|---|---|---|---|---|---|---|---|
| Example | D.A. | | 1 | 2 | 3 | 4 | 5 | 6 | 7 | 8 | 9 |
| | | cpu | cpu (%F) | cpu (%F) | cpu (%F) | cpu (%F) | cpu (%F) | cpu (%F) | cpu (%F) | cpu (%F) | cpu (%F) |
| 1 | 4 | 29.8 s | 8.26 s (100) | 16.6 s (100) | 33.0 s (100) | 20.7 s (100) | 41.3 s (100) | 82.5 s (90) | 41.2 s (90) | 82.5 s (90) | 165. s (90) |
| 2 | 4 | 40.6 s | 8.20 s (90) | 16.6 s (100) | 33.0 s (90) | 20.6 s (100) | 41.5 s (90) | 82.7 s (90) | 41.3 s (90) | 84.4 s (70) | 164. s (90) |
| 3 | 5 | 60.5 s | 9.60 s (100) | 19.4 s (60) | 39.1 s (80) | 24.2 s (90) | 48.6 s (100) | 97.1 s (90) | 49.0 s (60) | 97.4 s (100) | 193. s (100) |
| 4 | 5 | 62.2 s | 9.60 s (70) | 19.2 s (80) | 39.1 s (80) | 24.2 s (80) | 48.7 s (90) | 39.1 s (80) | 48.9 s (100) | 97.1 s (100) | 195. s (90) |
| 5 | 5 | 140. s | 9.70 s (90) | 19.3 s (100) | 39.1 s (90) | 24.3 s (90) | 48.8 s (100) | 96.7 s (90) | 48.5 s (90) | 96.9 s (80) | 194. s (90) |
| 6 | 5 | 138. s | 9.70 s (100) | 19.4 s (100) | 38.7 s (100) | 24.2 s (100) | 48.9 s (90) | 96.5 s (80) | 48.6 s (90) | 96.7 s (90) | 193. s (70) |
| | | | 88 | 85 | 83 | 87 | 81 | 82 | 84 | 81 | 79 |

TABLE 5. Comparison of $\alpha$**BB** and simulated annealing

cases this failure rate reaches 100 %.

## 8. Conclusions

A deterministic global optimization method was described for locating the global minimum potential energy conformations of oligopeptide chains based on $\alpha$**BB**. The ECEPP/3 detailed potential energy model was selected to model the energetics of the atomic interactions and the minimization of the total potential energy was formulated on the set of polypeptide dihedral angles. The proposed approach was implemented in C, in the program **GLOFOLD** and provisions were made to accommodate user specified partitioning of the dihedral angles into three sets. The first one (i.e., global variables), consisted of dihedral angles where branching occurs. The second set (i.e., local variables) included the dihedral variables where branching was not necessary. The third set, (i.e., fixed variables) included the dihedral angles which were kept fixed. **GLOFOLD** was applied to a number of oligopeptide folding problems. Computational performance compared favorably with a simulated annealing implementation.

## REFERENCES

1. N.L. Allinger, J. Am. Chem. Soc. **99** (1977), 8127.
2. N.L. Allinger, Y.H. Yuh, and J.-H. Lii, J. Am. Chem. Soc. **111** (1989), 8551.
3. B. Brooks, R. Bruccoleri, B. Olafson, D. States, S. Swaminathan, and M. Karplus, J. Comput. Chem. **8** (1983), 132.
4. A.J. Hopfinger, *Conformational Properties of Macromolecules*, Academic Press, New York, NY, 1973.
5. C.D. Maranas I.P. Androulakis and C.A. Floudas, accepted in J. Global Opt. (1995).
6. C.D. Maranas and C.A. Floudas, J. Chem. Phys. **97** (1992), no. 10, 7667.
7. ______, Ann. Oper. Res. **42** (1993), 85.
8. ______, J. Global Opt. **4** (1994a), 135.
9. ______, J. Chem. Phys. **100** (1994b), no. 2, 1247.
10. F.A. Momany, L.M. Carruthers, R.F. McGuire, and H.A. Scheraga, J. Phys. Chem. **78** (1974a), 1595.
11. F.A. Momany, L.M. Carruthers, and H.A. Scheraga, J. Phys. Chem. **78** (1974b), 1621.
12. F.A. Momany, R.F. McGuire, A.W. Burgess, and H.A. Scheraga, J. Phys. Chem. **79** (1975), 2361.
13. B.A. Murtagh and M.A. Saunders, *MINOS5.0 Users Guide*, Systems Optimization Laboratory, Dept. of Operations Research, Stanford University, CA., 1983, Appendix A: MINOS5.0, Technical Report SOL 83-20.
14. G. Némethy, K.D. Gibson, K.A. Palmer, C.N. Yoon, G. Paterlini, A. Zagari, S. Rumsey, and H.A. Scheraga, J. Phys. Chem. **96** (1992), 6472.
15. G. Némethy, M.S. Pottle, and H.A. Scheraga, J. Phys. Chem. **89** (1983), 1883.
16. G. Xue P.M. Pardalos and D. Shalloway, J. Global Opt. **4** (1994), no. 2, 117.
17. W.F. van Groningen and H.J.C. Berendsen, *GROMOS*, Groningen Molecular Simulation, Groningen, The Netherlands, 1987.
18. S. Weiner, P. Kollmann, D. Nguyen, and D. Case, J. Comput. Chem. **7** (1986), 230.
19. S.R. Wilson and W. Cui, Biopolymers **29** (1990), 225.

(C.D. Maranas) DEPARTMENT OF CHEMICAL ENGINEERING, PRINCETON UNIVERSITY, PRINCETON, NJ 08544–5263 USA

*E-mail address*, C.D. Maranas: kmaranas@titan.Princeton.EDU

(I.P. Androulakis) DEPARTMENT OF CHEMICAL ENGINEERING, PRINCETON UNIVERSITY, PRINCETON, NJ 08544–5263 USA

*E-mail address*, I.P. Androulakis: androula@titan.Princeton.EDU

(C.A. Floudas) DEPARTMENT OF CHEMICAL ENGINEERING, PRINCETON UNIVERSITY, PRINCETON, NJ 08544–5263 USA

*E-mail address*, C.A. Floudas: floudas@titan.Princeton.EDU

DIMACS Series in Discrete Mathematics
and Theoretical Computer Science
Volume **23**, 1996

# $\mathcal{E}$-OPTIMAL SOLUTIONS TO DISTANCE GEOMETRY PROBLEMS VIA GLOBAL CONTINUATION

**Jorge J. Moré and Zhijun Wu**

ABSTRACT. We show that a continuation approach to global optimization with global smoothing techniques can be used to obtain $\varepsilon$-optimal solutions to distance geometry problems. We show that determining an $\varepsilon$-optimal solution is still an NP-hard problem when $\varepsilon$ is small. A discrete form of the Gaussian transform is proposed based on the Hermite form of Gaussian quadrature. We show that the modified transform can be used whenever the transformed functions cannot be computed analytically. Our numerical results show that the discrete Gauss transform can be used to obtain $\varepsilon$-optimal solutions for general distance geometry problems, and in particular, to determine the three-dimensional structure of protein fragments.

## 1 Introduction

Distance geometry problems arise in the interpretation of NMR data and in the determination of protein structures. For a general review of the distance geometry problem and its relationship to macromolecular conformation, see Crippen and Havel [4], Havel [10], Kuntz, Thomason, and Oshiro [17], and Brünger and Nilges [1].

A distance geometry problem is specified by a subset $\mathcal{S}$ of all atom pairs and by the distances $\delta_{i,j}$ between atoms $i$ and $j$ for $(i,j) \in \mathcal{S}$. A solution to

*1991 Mathematics Subject Classification.* Primary: 49M37, 65D32, 68Q22, 68Q25, 92C40, 92E10.

*Key Words and Phrases.* Global optimization, NP-complete problems, continuation methods, Gaussian quadratures, distance geometry problems, macromolecular modeling.

This work was supported by the Mathematical, Information, and Computational Sciences Division subprogram of the Office of Computational and Technology Research, U.S. Department of Energy, under Contract W-31-109-Eng-38 and by the Argonne Director's Individual Investigator Program.

the distance geometry problem is a set of positions $x_1, \ldots, x_m$ in $\mathbb{R}^3$ such that

$$\|x_i - x_j\| = \delta_{i,j}, \qquad (i,j) \in \mathcal{S}. \tag{1}$$

Usually, $\mathcal{S}$ is sparse; in other words, only a small subset of distances is known.

In practice, lower and upper bounds on the distances are specified instead of their exact values. The distance geometry problem with lower and upper bounds is to find a set of positions $x_1, \ldots, x_m$ such that

$$l_{i,j} \le \|x_i - x_j\| \le u_{i,j}, \qquad (i,j) \in \mathcal{S}, \tag{2}$$

where $l_{i,j}$ and $u_{i,j}$ are lower and upper bounds on the distance constraints, respectively.

The distance geometry problem (1) is computationally intractable in general because the restriction of the distance geometry problem to a one-dimensional space is equivalent to the set partition problem, which is known to be NP-complete [6]. Even stronger results have been obtained showing that $k$-dimensional distance geometry problems are strongly NP-hard for all $k \ge 1$. For detailed proofs of these results, see Saxe [22]. One of the purposes of this paper is to study the complexity of solving the bounded distance geometry problem (2).

The distance geometry problem, exact or bounded, can be formulated as a global optimization problem. The objective function for the distance geometry problem is defined so that the constraints are satisfied at a global minimizer of the problem. Special optimization techniques for this class of problems have been developed by Crippen and Havel [4], Havel [10], Hendrickson [11, 12], Glunt, Hayden, and Raydan [8, 9], and Moré and Wu [19]. In particular, Moré and Wu [19] used global smoothing techniques and a continuation approach to solve distance geometry problems.

We extend the work of Moré and Wu [19], by showing that the global continuation approach can also be used to obtain $\varepsilon$-optimal solutions to the distance geometry problem, that is, a set of positions $x_1, \ldots, x_m$ such that

$$\Big| \|x_i - x_j\| - \delta_{i,j} \Big| \le \varepsilon, \qquad (i,j) \in \mathcal{S}, \tag{3}$$

where $\varepsilon$ is a positive number. An $\varepsilon$-optimal solution to the distance geometry problem is all that can be expected in practice, with the choice of $\varepsilon > 0$ dictated by the accuracy in the data. An $\varepsilon$-optimal solution can also be useful when the exact solution does not exist because of small errors in the data. The latter case can happen, for example, if the triangle inequality

$$\delta_{i,j} \le \delta_{i,k} + \delta_{k,j}$$

is violated for atoms $\{i, j, k\}$ because of possible inconsistency in the experimental data.

We study problem (3) as a special case of the bounded distance geometry problem (2). Our approach is to formulate problem (2) in terms of finding the global minimum of the function

$$f(x) = \sum_{i,j \in \mathcal{S}} h_{i,j}(x_i - x_j), \tag{4}$$

where

$$h_{i,j}(x) = \min{}^2 \left\{ \frac{\|x\|^2 - l_{i,j}^2}{l_{i,j}^2}, 0 \right\} + \max{}^2 \left\{ \frac{\|x\|^2 - u_{i,j}^2}{u_{i,j}^2}, 0 \right\}. \tag{5}$$

Clearly, $x = \{x_1, \ldots, x_m\}$ solves problem (2) if and only if $x$ is a global minimizer of $f$ and $f(x) = 0$ (see Crippen and Havel [4]).

In the continuation approach to global optimization an integral transformation is used to smooth the original function into a function with fewer minimizers. An optimization algorithm is then used to trace the minimizers of the transformed functions back to the original function. The smoothing transformation was first proposed and studied in the diffusion equation method for protein conformation by Scheraga and coworkers [21, 14, 15, 16, 23]. Similar transformations have been used in the packet annealing algorithm of Shalloway [25, 24], and in the algorithms used by Coleman, Shalloway, and Wu [2, 3] for molecular conformation problems. Recent development of this approach from a mathematical and computational point of view can be found in Wu [28] and Moré and Wu [19]. A general review on this approach can also be found in Pardalos, Shalloway, and Xue [20].

A major issue in applying the continuation method to problem (2) is that the potential function (4) cannot be transformed analytically. While the function can certainly be transformed by using standard techniques for numerical integration, the cost of evaluating the transformed functions would be prohibitive. In this paper we introduce a Gauss-Hermite discrete transformation that can be evaluated at a reasonable cost and is applicable to a wide class of functions, including (4).

Our main concern is with the determination of $\varepsilon$-optimal solutions to the exact distance geometry problem, that is, vectors $x \in \mathbb{R}^n$ that satisfy (3). Complexity issues are examined in Section 2, where we show that determining $\varepsilon$-optimal solutions to distance geometry problems is an NP-hard problem when the atoms are restricted to $\mathbb{R}$.

In Section 3 we define the Gaussian transform and show how the Gaussian transform of (4) can be computed in terms of one-dimensional integrals. We then turn to computational issues. In particular, Section 4 introduces the discrete Gauss-Hermite transform as an approximation to the Gaussian transform, while Section 5 contains numerical results for a continuation method based on this transformation. We consider a set of model distance geometry problems, as well as the problem of determining the three-dimensional structure of a 63-atom protein fragment. Although preliminary, these results show that continuation algorithms based on smoothing techniques can solve these global optimization problems with a high degree of reliability.

## 2 Approximate Solutions

The solution of the exact distance geometry problem (1) may not exist because, for example, the constraints may not be consistent. Even if the constraints are consistent, a small perturbation may render the constraints inconsistent. Since experimental data is not 100% accurate, it is quite reasonable to search for approximate solutions to the exact distance geometry problem.

In this section we analyze the complexity of determining the approximate solution to the exact distance geometry problem (1). Any $x \in \mathbb{R}^n$ that satisfies (3) is an $\varepsilon$-optimal solution to th exact distance geometry problem. We wish to show that if $\varepsilon$ is small, problem (3) is just as difficult as the original exact distance geometry problem. We prove this result by restricting all the atoms to lie in $\mathbb{R}$, and all distances to be positive integers. The result can be extended to higher dimensions, but we do not pursue these extensions.

Given $n$ positive integers $a_i \in \mathbf{Z}^+$, the standard set partition problem is to find a partition $S_1$, $S_2$ of $\{1, \ldots, n\}$ such that

$$\sum_{i \in S_1} a_i = \sum_{i \in S_2} a_i .$$

This problem is one of the basic NP-hard problems (see Garey and Johnson [6]). We now generalize this problem by allowing approximate solutions. Given $n$ positive integers $a_i \in \mathbf{Z}^+$ and $\varepsilon > 0$, the $\varepsilon$-set partition problem is to find a partition $S_1(\varepsilon)$, $S_2(\varepsilon)$ of $\{1, \ldots, n\}$, and numbers $a_i(\varepsilon) \in \mathbb{R}$ such that

$$|a_i(\varepsilon) - a_i| \le \varepsilon, \qquad \left| \sum_{i \in S_1(\varepsilon)} a_i(\varepsilon) - \sum_{i \in S_2(\varepsilon)} a_i(\varepsilon) \right| \le \tfrac{1}{2}. \tag{6}$$

Clearly, any solution to the standard set partition problem is a solution of the $\varepsilon$-set partition problem for any $\varepsilon > 0$. This seems to suggest that the $\varepsilon$-set partition problem may be easier than the standard set partition problem. However, we now show that obtaining a solution to the $\varepsilon$-set partition problem is still NP-hard when $\varepsilon$ is small.

**Theorem 2.1** *If $a_i(\varepsilon) \in \mathbb{R}$ solves the $\varepsilon$-set partition problem and $n\varepsilon < \frac{1}{2}$, then*

$$\sum_{i \in S_1(\varepsilon)} a_i = \sum_{i \in S_2(\varepsilon)} a_i$$

**Proof.** The triangle inequality and the second inequality in (6) show that

$$\left| \sum_{i \in S_1(\varepsilon)} a_i - \sum_{i \in S_2(\varepsilon)} a_i \right| \le \sum_{i \in S_1(\varepsilon)} |a_i(\varepsilon) - a_i| + \sum_{i \in S_2(\varepsilon)} |a_i(\varepsilon) - a_i| + \tfrac{1}{2}.$$

Now use the first inequality in (6), and recall that $S_1$, $S_2$ is a partition of $\{1, \ldots, n\}$ to obtain that

$$\left| \sum_{i \in S_1(\varepsilon)} a_i - \sum_{i \in S_2(\varepsilon)} a_i \right| \le n\varepsilon + \tfrac{1}{2} < 1.$$

The sums in the above inequality add integers, and thus the sums must be integers. Since the difference between these two sums is less than one, they must be equal. This is the desired result. ∎

An immediate consequence of Theorem 2.1 is that the $\varepsilon$-set partition problem is NP-hard. We now show that determining an $\varepsilon$-optimal solution to the distance geometry problem in $\mathbb{R}$ is NP-hard. The proof shows that any $\varepsilon$-set partition problem can be directly reduced to a distance geometry problem with $n+1$ atoms in $\mathbb{R}$.

**Theorem 2.2** *Determining an $\varepsilon$-optimal solution to the distance geometry problem in $\mathbb{R}$ is NP-hard.*

**Proof.** Given an instance of the $\varepsilon$-set partition problem, consider the distance geometry problem with $n+1$ atoms in $\mathbb{R}$ with

$$\delta_{i,i+1} = a_i, \quad 1 \le i \le n, \qquad \delta_{1,n+1} = \tfrac{1}{2}.$$

If $x_1, \ldots, x_{n+1}$ is an $\varepsilon$-optimal solution to this distance geometry problem, then

$$\Big| |x_i - x_{i+1}| - \delta_{i,i+1} \Big| \le \varepsilon, \qquad |x_1 - x_{n+1}| \le \tfrac{1}{2}.$$

In particular, if we define $a_i(\varepsilon) = |x_i - x_{i+1}|$, then $|a_i(\varepsilon) - a_i| \le \varepsilon$. Now note that

$$\left| \sum_{i=1}^{n} (x_i - x_{i+1}) \right| = |x_1 - x_{n+1}| \le \tfrac{1}{2} \tag{7}$$

and that, if $S_1(\varepsilon) = \{i : x_i > x_{i+1}\}$ and $S_2(\varepsilon) = \{i : x_i \le x_{i+1}\}$, then (7) implies that

$$\left| \sum_{i \in S_1(\varepsilon)} |x_i - x_{i+1}| - \sum_{i \in S_2(\varepsilon)} |x_i - x_{i+1}| \right| \le \tfrac{1}{2}.$$

Since $a_i(\varepsilon) = |x_i - x_{i+1}|$, we have shown that the $a_i(\varepsilon)$ solve the $\varepsilon$-set partition problem. ∎

# 3 Global Smoothing

In the continuation approach to global optimization, the objective function is gradually transformed into a smoother function with fewer local minimizers. An optimization algorithm is then applied to the transformed function, tracing the minimizers back to the original function. In this section we define the Gaussian transform and describe some of the interesting properties of this transformation. The emphasis is on computational considerations; motivation and additional details can be found in Wu [28] and Moré and Wu [19].

The Gaussian transform depends on a parameter $\lambda$ that controls the degree of smoothing. The original function is obtained if $\lambda = 0$, while smoother functions are obtained as $\lambda$ increases.

**Definition 3.3** *The Gaussian transform $\langle f \rangle_\lambda$ of a function $f : \mathbb{R}^n \mapsto \mathbb{R}$ is*

$$\langle f \rangle_\lambda(x) = \frac{1}{\pi^{n/2}\lambda^n} \int_{\mathbb{R}^n} f(y) \exp\left( -\frac{\|y-x\|^2}{\lambda^2} \right) dy. \tag{8}$$

The value $\langle f \rangle_\lambda(x)$ is an average of $f$ in a neighborhood of $x$, with the relative size of this neighborhood controlled by the parameter $\lambda$. The size of the neighborhood decreases as $\lambda$ decreases so that when $\lambda = 0$, the neighborhood is the center $x$. The Gaussian transform $\langle f \rangle_\lambda$ can also be viewed as the expectation value of $f$ with respect to the Gaussian density function

$$\rho_\lambda(y) = \frac{1}{\pi^{n/2}\lambda^n} \exp\left( -\frac{\|y\|^2}{\lambda^2} \right).$$

We could have used other density functions in the definition of the transform, but Gaussian density function has stronger smoothing properties.

In principle the computation of the Gaussian transform requires the evaluation of $n$-dimensional integrals, but for many functions that arise in practice, it is possible to compute the Gaussian transform explicitly, or in terms of one-dimensional integrals. In particular, if the function is *decomposable*, that is, if the function $f : \mathbb{R}^n \mapsto \mathbb{R}$ can be written in the form

$$f(x) = \sum_{k=1}^{m} f_k(x), \qquad f_k(x) = \prod_{j=1}^{n} f_{k,j}(x_j),$$

for some set of functions $\{f_{k,j}\}$, where $f_{k,j} : \mathbb{R} \mapsto \mathbb{R}$, then a computation shows that

$$\langle f \rangle_\lambda(x) = \sum_{k=1}^{m} \left( \prod_{j=1}^{n} \langle f_{k,j} \rangle_\lambda(x_j) \right).$$

Thus, computing $\langle f \rangle_\lambda$ for a decomposable function requires the computation of only the one-dimensional integrals for each $\langle f_{k,j} \rangle_\lambda$. As we shall see in the next section, we avoid computing one-dimensional integrals by using special-purpose quadratures.

For the distance geometry problem, as well as for other problems in macromolecular conformation, we are interested in transforming a class of functions defined in terms of the distances between pairs of atoms. Given functions $h_{i,j} : \mathbb{R}^p \mapsto \mathbb{R}$ of the distance between atoms $i$ and $j$, we consider the potential function

$$f(x) = \sum_{i,j \in \mathcal{S}} h_{i,j}(x_i - x_j), \tag{9}$$

where $\mathcal{S}$ is some subset of all pairs of atoms, and $x_i \in \mathbb{R}^p$ is the position of the $i$-th atom. In general we are concerned with three-dimensional problems where $p = 3$.

The following result of Moré and Wu [19] shows that computing the Gaussian transform of (9) requires only the Gaussian transform of $h_{i,j}$.

**Theorem 3.4** *If $f : \mathbb{R}^n \mapsto \mathbb{R}$ and $h : \mathbb{R}^p \mapsto \mathbb{R}$ are related by*

$$f(x) = h(P^T x),$$

*for some matrix $P \in \mathbb{R}^{n \times p}$ such that $P^T P = \sigma^2 I$, then*

$$\langle f \rangle_\lambda(x) = \langle h \rangle_{\sigma\lambda}(P^T x).$$

Theorem 3.4 reduces the computation of the Gaussian transform of the mapping $f$, which is defined on $\mathbb{R}^n$, to the computation of the Gaussian transform of $h$, which is defined on $\mathbb{R}^p$. As an application of this result, note that

$$\langle f \rangle_\lambda(x) = \sum_{i,j \in \mathcal{S}} \langle h_{i,j} \rangle_{\sqrt{2}\lambda}(x_i - x_j)$$

is the Gaussian transform of the potential function defined by (9). In this case $f$ is defined on $\mathbb{R}^{pn}$, but $h_{i,j}$ is defined on $\mathbb{R}^p$.

In some application we can compute the Gaussian transform $\langle h_{i,j} \rangle_\lambda$ explicitly. For example, in the distance geometry problem (1), the function $h_{i,j} : \mathbb{R}^p \mapsto \mathbb{R}$ is of the general form

$$h(x) = \left( \|x\|^2 - \delta^2 \right)^2 .$$

This function is decomposable. Moreover, the Gaussian transform is explicitly given by

$$\langle h \rangle_\lambda(x) = h(x) + [3 + (p-1)]\lambda^2 \|x\|^2 + \tfrac{1}{4} p(p+2)\lambda^4 - p\delta^2\lambda^2.$$

For details of this computation see Section 4 of Moré and Wu [19].

The Gaussian transform of the function defined by (4) requires a different approach since the element functions $h_{i,j}$ in (4) are not decomposable. The key observation is that the element functions $h_{i,j}$ in (4) are of the general form

$$h(x) = \mathcal{H}(\|x\|), \tag{10}$$

where $\mathcal{H} : \mathbb{R} \mapsto \mathbb{R}$ is given by

$$\mathcal{H}(s) = \min^2 \left\{ \frac{s^2 - l^2}{l^2}, 0 \right\} + \max^2 \left\{ \frac{s^2 - u^2}{u^2}, 0 \right\}. \tag{11}$$

We now show that the Gaussian transform of $h : \mathbb{R}^3 \mapsto \mathbb{R}$ can be reduced to the calculation of one-dimensional integrals.

**Theorem 3.5** *Given $\mathcal{H} : \mathbb{R} \mapsto \mathbb{R}$, define $h : \mathbb{R}^3 \mapsto \mathbb{R}$ by (10). If $r = \|x\|$ then*

$$\langle h \rangle_\lambda(x) = \frac{1}{\lambda\sqrt{\pi}r} \int_0^\infty s\mathcal{H}(s) \left[ \exp\left( -\frac{(r-s)^2}{\lambda^2} \right) - \exp\left( -\frac{(r+s)^2}{\lambda^2} \right) \right] ds. \tag{12}$$

**Proof.** First note that Theorem 3.4 shows, in particular, that $\langle h \rangle_\lambda(y) = \langle h \rangle_\lambda(x)$ for any vector $y$ such that $\|y\| = \|x\|$. Thus, without loss of generality, we assume that $x = (0, 0, r)$ since this simplifies the proof. Now note that the definition of $\langle h \rangle_\lambda$ implies that

$$\langle h \rangle_\lambda(x) = \frac{1}{\lambda^3\sqrt{\pi}} \int_{\mathbb{R}^3} \mathcal{H}(\|y\|) \exp\left( -\frac{\|y-x\|^2}{\lambda^2} \right) dy.$$

Making a change to spherical coordinates

$$y = \begin{pmatrix} s\cos\theta \\ s\sin\theta \\ s\cos\varphi \end{pmatrix},$$

and noting that $y^T x = rs\cos\varphi$ when $x = (0, 0, r)$, we obtain that

$$\langle h \rangle_\lambda(x) = \frac{2}{\lambda^3\sqrt{\pi}} \int_0^\infty \int_0^\pi s^2\mathcal{H}(s) \exp\left( -\frac{r^2+s^2}{\lambda^2} \right) \exp\left( -\frac{rs\cos\varphi}{\lambda^2} \right) \sin\varphi \, d\varphi \, ds.$$

A computation now shows that

$$\int_0^\pi \exp\left( -\frac{rs\cos\varphi}{\lambda^2} \right) \sin\varphi \, d\varphi = \frac{\lambda^2}{2rs} \left[ \exp\left( \frac{2rs}{\lambda^2} \right) - \exp\left( -\frac{2rs}{\lambda^2} \right) \right],$$

and this yields the desired result. ∎

Theorem 3.5 can be used to determine the Gaussian transform of $h$, but the computation can be poorly conditioned when $r$ is small. The following result gives an alternative expression that can be used for small $r$.

**Corollary 3.6** *If $\mathcal{H} : \mathbb{R} \mapsto \mathbb{R}$ and $h : \mathbb{R}^3 \mapsto \mathbb{R}$ are as in Theorem 3.5, then*

$$\lim_{x \to 0} \langle h \rangle_\lambda(x) = \frac{4}{\lambda^3 \sqrt{\pi}} \int_0^\infty s^2 \mathcal{H}(s) \exp\left(-\frac{s^2}{\lambda^2}\right) ds.$$

**Proof.** Since $r = \|x\|$, the result follows by taking the limit as $r \to 0$ in (12). ∎

For the general distance geometry problem the function $\mathcal{H}$ is defined by (11), and thus $\mathcal{H}$ is an even function. We now show that Theorem 12 can be simplified in this case.

**Corollary 3.7** *If the mapping $\mathcal{H}$ in Theorem 3.5 is an even function, then*

$$\langle h \rangle_\lambda(x) = \frac{1}{\lambda \sqrt{\pi} r} \int_{-\infty}^{+\infty} s \mathcal{H}(s) \exp\left(-\frac{(r-s)^2}{\lambda^2}\right) ds. \qquad (13)$$

**Proof.** By direct computation,

$$\int_0^\infty s\mathcal{H}(s) \exp\left(-\frac{(r-s)^2}{\lambda^2}\right) ds - \int_0^\infty s\mathcal{H}(s) \exp\left(-\frac{(r+s)^2}{\lambda^2}\right) ds$$

$$= \int_0^\infty s\mathcal{H}(s) \exp\left(-\frac{(r-s)^2}{\lambda^2}\right) ds + \int_{-\infty}^0 s\mathcal{H}(s) \exp\left(-\frac{(r-s)^2}{\lambda^2}\right) ds$$

$$= \int_{-\infty}^{+\infty} s\mathcal{H}(s) \exp\left(-\frac{(r-s)^2}{\lambda^2}\right) ds,$$

which yields the result. ∎

Interestingly enough, Corollary 3.7 shows that the computation of the Gaussian transform of $h$ when $\mathcal{H}$ is an even function reduces to the computation of the transform of the mapping $s \mapsto s\mathcal{H}(s)$.

## 4 Gauss-Hermite Approximation

We consider techniques for the approximation of the Gaussian transform of a function $f : \mathbb{R} \mapsto \mathbb{R}$. These techniques are useful for the computation of the Gaussian transform of a decomposable function defined on $\mathbb{R}^n$, and for the computation of the Gaussian transform of the $3n$-dimensional potential function (9) when the element functions $h_{i,j}$ are of the form (10) with $\mathcal{H}$ an even function.

The change of variables $y = x + \lambda s$ in (8) shows that the Gaussian transform of a mapping $f : \mathbb{R} \mapsto \mathbb{R}$ can be written in the form

$$\langle f \rangle_\lambda(x) = \frac{1}{\sqrt{\pi}} \int_{-\infty}^{+\infty} f(x + \lambda s) \exp(-s^2)\, ds. \tag{14}$$

We can certainly compute $\langle f \rangle_\lambda(x)$ with standard techniques for numerical integration (for example, an adaptive quadrature), but these techniques usually require a large number of function evaluations.

Another approach used to evaluate $\langle f \rangle_\lambda(x)$ is to approximate $f$ with a linear combination of functions $\phi_i$ whose Gaussian transforms can be obtained analytically. If

$$f(x) \approx \sum_{i=1}^{m} \phi_i(x),$$

then the Gaussian transform can be evaluated by

$$\langle f \rangle_\lambda(x) \approx \sum_{i=1}^{m} \langle \phi_i \rangle_\lambda(x).$$

This approach was used in [15] for the Lennard-Jones potential with $\{\phi_i\}$ a set of Gaussian functions. The validity of this approach, however, requires the approximation to be valid over a wide range. Moreover, the approximation must be recalculated if the potential function changes.

We propose an alternative approach to the approximation of the Gaussian transform (14) based on Gaussian quadrature. This approach mixes elements of the numerical integration and function approximation approaches, but does not require the calculation of a different approximation for each application.

Given a positive weight function $w : \mathbb{R} \mapsto \mathbb{R}$, the Gaussian quadrature formula

$$\int_{-\infty}^{+\infty} w(s) f(s) ds \approx \sum_{i=1}^{m} w_i f(s_i), \tag{15}$$

is exact for all polynomials of degree less than $2m$. The weights $w_i$ are the zeros of the $m$-th degree orthogonal polynomial $p_m$ with respect to the weight function $w$; the nodes $s_i$ are also determined by $p_m$. The $w$-orthogonal polynomials can be generated by a three term recurrence relationship. For the weight function $w(s) = \exp(-s^2)$ the $w$-orthogonal polynomials are the Hermite polynomials, which are generated by the recurrence

$$p_{k+1}(x) = x p_k(x) - (\tfrac{1}{2}k) p_{k-1}(x), \qquad k \geq 0,$$

with $p_{-1} \equiv 0$, and $p_0 \equiv 1$. Details about $w$-orthogonal polynomials for various weight functions are discussed in standard numerical analysis textbooks, for example, Stoer and Bulirsch [26] and Kincaid and Cheney [13]. For an advanced treatment of Gaussian quadrature, see Stroud and Secrest [27] and Davis and Rabinowitz [5].

If we use Gaussian quadrature on the integral in (14), we obtain an approximation

$$\langle f \rangle_{G,\lambda}(x) = \frac{1}{\sqrt{\pi}} \sum_{i=1}^{m} w_i f(x + \lambda s_i) \tag{16}$$

to the Gaussian transform of $f$, which we call the Gauss-Hermite transform of $f$. This terminology is appropriate because (16) defines a discrete transformation that shares many of the properties of the standard Gaussian transform. In particular, note that the Gauss-Hermite transform coincides with the Gaussian transform if $f$ is a polynomial of degree less than $2m$.

The weights $w_i$ and nodes $s_i$ in the Gauss-Hermite transform are independent of $f$ and can be found in the literature (for example, Stroud and Secrest [27]) or can be computed with the **gauss** subroutine in ORTHOPOL (Gautschi [7]). The computation of the Gauss-Hermite transform requires $m$ function evaluations.

We are interested in the use of Gaussian quadratures to approximate integrals of the form (13) where the function $\mathcal{H} : \mathbb{R} \mapsto \mathbb{R}$ is given by (11). Error bounds for this type of integrals can be obtained by using the Peano kernel theorem.

**Theorem 4.8** *Let $\langle f \rangle_{G,\lambda}$ be the Gauss-Hermite transform of $f : \mathbb{R} \mapsto \mathbb{R}$. If $f^{(l)}$ is piecewise continuous on $\mathbb{R}$ for some $l \leq 2m$, there is a constant $\mu_l$, independent of $f$, such that*

$$|\langle f \rangle_\lambda(x) - \langle f \rangle_{G,\lambda}(x)| \leq \mu_l \lambda^l \sigma(x),$$

*where*

$$\sigma(x) = \left\{ \int_{-\infty}^{+\infty} \exp(-s^2) \left| f^{(l)}(x + \lambda s) \right|^2 ds \right\}^{1/2}.$$

**Proof.** Since Gaussian quadratures are exact for polynomials of degree less than $2m$, the Peano kernel theorem (for example, Davis and Rabinowitz [5, Section 4.3]) shows that

$$|\langle f \rangle_\lambda(x) - \langle f \rangle_{G,\lambda}(x)| \leq \lambda^l \int_{-\infty}^{+\infty} \exp(-s^2) \left| f^{(l)}(x + \lambda s) \right| \left| K_l(s) \right| ds,$$

where $K_l$ is the Peano kernel. Hence,

$$|\langle f \rangle_\lambda(x) - \langle f \rangle_{G,\lambda}(x)| \leq \mu_l \lambda^l \sigma(x),$$

where $\mu_l = \|K_l\|_2$ is the weighted $L_2$-norm of the kernel. ∎

Theorem 4.8 shows that the Gauss-Hermite transform is likely to be a good approximation to $\langle f \rangle_\lambda$ provided $\lambda < 1$, but that the accuracy is likely to deteriorate if $\lambda > 1$. This is not a serious difficulty because for large $\lambda$ the Gauss-Hermite transform is used to guide an algorithm to a global minimizer, but for small $\lambda$ we work with the original function $f$.

We can apply Theorem 4.8 to the mapping $f(x) = x\mathcal{H}(x)$, where $\mathcal{H} : \mathbb{R} \mapsto \mathbb{R}$ is given by (11). In this case, $f''$ is piecewise continuous, and thus Theorem 4.8 applies with $l = 2$ provided we use $m \geq 1$ nodes in the quadrature. Also note that in this case $f$ is a piecewise polynomial of degree 5, and thus we can expect better results provided we use $m \geq 3$ nodes.

# 5 Numerical Results

We now present an algorithm for determining a solution to the bounded distance geometry problem (2). Our results are restricted to the problem of obtaining $\varepsilon$-optimal solutions to the distance geometry problem. In particular, we present preliminary numerical results on the determination of the three-dimensional structure of a 63-atom protein fragment with experimental data provided by Bruce Hendrickson. All computations were done on the IBM SP parallel system at Argonne National Laboratory with 128 nodes, each node an IBM RS/6000-370 with 128 Mbytes of memory and 125 Mflops of peak performance.

We attack the bounded distance geometry problem (2) by seeking a global minimizer of the function

$$f(x) = \sum_{i,j \in \mathcal{S}} h_{i,j}(x_i - x_j), \tag{17}$$

where

$$h_{i,j}(x) = \min{}^2 \left\{ \frac{\|x\|^2 - l_{i,j}^2}{l_{i,j}^2}, 0 \right\} + \max{}^2 \left\{ \frac{\|x\|^2 - u_{i,j}^2}{u_{i,j}^2}, 0 \right\}. \tag{18}$$

This function is appropriate for the bounded distance geometry problem because $f(x) \geq 0$ for all $x \in \mathbb{R}^n$, and $f(x) = 0$ if and only if $x$ solves (2).

The computation of the Gaussian transform of $f$ follows from the results in Section 3. Theorem 3.4 shows that

$$\langle f \rangle_\lambda(x) = \sum_{i,j \in \mathcal{S}} \langle h_{i,j} \rangle_{\sqrt{2}\lambda}(x_i - x_j).$$

The functions $h_{i,j}$ defined by (18) are not decomposable, but in this case Corollary 3.7 is applicable because

$$h_{i,j}(x) = \mathcal{H}_{i,j}\left(\|x\|\right), \tag{19}$$

where

$$\mathcal{H}_{i,j}(r) = \min^2\left\{\frac{r^2 - l_{i,j}^2}{l_{i,j}^2}, 0\right\} + \max^2\left\{\frac{r^2 - u_{i,j}^2}{u_{i,j}^2}, 0\right\} \tag{20}$$

is an even function. Hence, Corollary 3.7 shows that

$$\begin{aligned} \langle h_{i,j}\rangle_{\sqrt{2}\lambda}(x) &= \frac{1}{\lambda\sqrt{2\pi}\,r}\int_{-\infty}^{+\infty} s\mathcal{H}_{i,j}(s)\exp\left(-\frac{(r-s)^2}{2\lambda^2}\right) ds \\ &= \frac{1}{\sqrt{\pi}\,r}\int_{-\infty}^{+\infty}(r+\sqrt{2}\lambda s)\mathcal{H}_{i,j}(r+\sqrt{2}\lambda s)\exp\left(-s^2\right) ds, \end{aligned}$$

where $r = \|x\|$. In summary, these computations show that

$$\langle f\rangle_\lambda(x) = \sum_{i,j\in\mathcal{S}} \frac{1}{\sqrt{\pi}\,r_{i,j}}\int_{-\infty}^{+\infty}(r_{i,j}+\sqrt{2}\lambda s)\mathcal{H}_{i,j}(r_{i,j}+\sqrt{2}\lambda s)\exp\left(-s^2\right) ds,$$

where $r_{i,j} = \|x_i - x_j\|$, is the Gaussian transform of the function defined by (17), (19), and (20). This expression for the Gaussian transform is valid for any function defined by (17) and (19), for any even function $\mathcal{H}_{i,j}$.

We approximate the Gaussian transform with the Gauss-Hermite transform as described in Section 4. For this problem the Gauss-Hermite transform is

$$\langle f\rangle_{G,\lambda}(x) = \sum_{i,j\in\mathcal{S}} \frac{1}{\sqrt{\pi}\,r_{i,j}}\sum_{k=1}^{n_q} w_k(r_{i,j}+\sqrt{2}\lambda s_k)\mathcal{H}_{i,j}(r_{i,j}+\sqrt{2}\lambda s_k), \tag{21}$$

where $w_k$ and $s_k$ are the weights and nodes for the Gaussian quadrature, and $r_{i,j} = \|x_i - x_j\|$.

We use a simple version of the global continuation algorithm that uses a prespecified set $\lambda_0 > \lambda_1 > \ldots > \lambda_p = 0$ of smoothing parameters.

**Algorithm GMIN**

Choose a random vector $x_0 \in \mathbb{R}^{m\times 3}$.
**for** $k = 0, 1, \ldots, p$
  Determine $x^* =$ **locmin** $(\langle f\rangle_{G,\lambda_k},\ x_0)$.
  Set $x_0 = x^*$.
**end do**

We did not attempt to optimize the choice of smoothing parameters or the starting point for the local minimization procedure **locmin** in **GMIN**. We chose

$$\lambda_k = \lambda_0\left(1 - \frac{k}{p}\right), \qquad 0 \le k \le p,$$

and the starting point of **locmin** as the local minimizer found at the previous iteration. We will show that these choices work remarkably well.

We tested **GMIN** with $\lambda_0 > 0$, but we also tested **GMIN** with a choice of $\lambda_0 = 0$, since for this choice **GMIN** reduces to the use of **locmin** on the original function $f$ from a random starting point. A standard multistart method is obtained if **GMIN** is used from a set of randomly generated starting points with $\lambda_0 = 0$.

The choice of local minimization algorithm has to be done with some care because $\langle f \rangle_{G,\lambda_k}$ is not twice continuously differentiable. The Hessian matrix is discontinuous at points where the argument of $\mathcal{H}_{i,j}$ coincides with either $l_{i,j}$ or $u_{i,j}$. We cannot expect to avoid these discontinuities, in particular, if $l_{i,j}$ or $u_{i,j}$ are close. We used the variable-metric limited-memory code VMLM in MINPACK-2, which is an implementation of the Liu and Nocedal [18] algorithm.

We study problem (3) as a sepcial bounded distance geometry problem. In our numerical results we consider **GMIN** to be successful from a starting point $x_0$ if $x^*$ satisfies

$$\left| \|x_i - x_j\|^2 - \delta_{i,j}^2 \right| \le \varepsilon\, \delta_{i,j}^2, \qquad (i,j) \in \mathcal{S}, \tag{22}$$

for a given $\varepsilon > 0$. This termination test may not be appropriate in practice because smaller distances tend to be determined more accurately. We used **GMIN** with

$$l_{i,j}^2 = \delta_{i,j}^2(1-\varepsilon), \qquad u_{i,j}^2 = \delta_{i,j}^2(1+\varepsilon).$$

In general, $\varepsilon = .10$ is a reasonable value, but we also tested **GMIN** with smaller $\varepsilon$.

We first tested the algorithm with the model distance geometry problem constructed by Moré and Wu [19]. In this problem the molecule has $m = s^3$ atoms located in the three-dimensional lattice

$$\{(i_1, i_2, i_3) : \; 0 \le i_1 < s, \; 0 \le i_2 < s, \; 0 \le i_3 < s\}$$

for some integer $s \ge 1$. The problem is to determine the structure of this molecule if we are given $\delta_{i,j} \in \mathcal{S}$, where

$$\mathcal{S} = \{(i,j) : \; |i-j| \le r\},$$

and $r$ is an integer between 1 and $m$.

Numerical results for obtaining the $\varepsilon$-optimal solution of this problem are presented in Table 1. We used molecules with $(m,r) = (s^3, s^2)$, where $4 \le s \le 7$. For each problem, we ran **GMIN** from 10 randomly chosen starting points with $\lambda_0 = 1$ and $p = 20$ for continuation. The results show that the continuation method found the $\varepsilon$-optimal solutions to the problems successfully. On the other hand, the multistart method failed to find a solution in all cases.

Table 1: Results for the model problem, $\varepsilon = .10$

| $\lambda_0$ | $m = 64,\ r = 16$ | $m = 125,\ r = 25$ | $m = 216,\ r = 36$ | $m = 343,\ r = 49$ |
|---|---|---|---|---|
| 1 | 70% | 30% | 60% | 40% |
| 0 | 0% | 0% | 0% | 0% |

Table 2: Results for the 63-atom protein fragment

| $\lambda_0$ | $\varepsilon = .01$ | $\varepsilon = .02$ | $\varepsilon = .04$ | $\varepsilon = .06$ | $\varepsilon = .08$ | $\varepsilon = .10$ |
|---|---|---|---|---|---|---|
| 1 | 0% | 4% | 30% | 69% | 76% | 96% |
| 0 | 0% | 0% | 8% | 23% | 40% | 66% |

We also tested **GMIN** with a 63-atom protein fragment. For this problem, 236 pairwise distances are specified. Hendrickson [11] showed that the molecule with this set of distance data is nearly rigid and difficult to determine. Our numerical results confirm this finding.

We ran **GMIN** from 100 starting points for $\varepsilon = .01, .02, .04, .06, .08, .10$. In all cases we set $\lambda_0 = 1$ and $p = 20$ for continuation. The results from these experiments, given in Table 2, clearly show that the problem becomes harder as $\varepsilon$ is decreased. The continuation method failed for $\varepsilon = .01$, but was successful for all other larger values of $\varepsilon$. If compared with the multi-start method, note that the continuation method has a significantly higher probability of success.

## Acknowledgment

We thank Bruce Hendrickson for providing the data for the problems of determining the structures of protein fragments.

# References

[1] A. T. Brünger and M. Nilges, *Computational challenges for macromolecular structure determination by X-ray crystallography and solution NMR-spectroscopy*, Q. Rev. Biophys., 26 (1993), pp. 49–125.

[2] T. F. Coleman, D. Shalloway, and Z. Wu, *Isotropic effective energy simulated annealing searches for low energy molecular cluster states*, Comp. Optim. Applications, 2 (1993), pp. 145–170.

[3] ——, *A parallel build-up algorithm for global energy minimizations of molecular clusters using effective energy simulated annealing*, J. Global Optim., 4 (1994), pp. 171–185.

[4] G. M. Crippen and T. F. Havel, *Distance Geometry and Molecular Conformation*, John Wiley & Sons, 1988.

[5] P. J. Davis and P. Rabinowitz, *Methods of Numerical Integration*, Academic Press, second ed., 1984.

[6] M. R. Garey and D. S. Johnson, *Computers and Intractability*, W. H. Freeman, 1979.

[7] W. Gautschi, *Algorithm 726:* ORTHOPOL – *A package of routines for generating orthogonal polynomials and Gauss-type quadrature rules*, ACM Trans. Math. Software, 20 (1994), pp. 21–62.

[8] W. Glunt, T. L. Hayden, and M. Raydan, *Molecular conformation from distance matrices*, J. Comp. Chem., 14 (1993), pp. 114–120.

[9] ——, *Preconditioners for distance matrix algorithms*, J. Comp. Chem., 15 (1994), pp. 227–232.

[10] T. F. Havel, *An evaluation of computational strategies for use in the determination of protein structure from distance geometry constraints obtained by nuclear magnetic resonance*, Prog. Biophys. Mol. Biol., 56 (1991), pp. 43–78.

[11] B. A. Hendrickson, *The molecule problem: Determining conformation from pairwise distances*, Ph.D. thesis, Cornell University, Ithaca, New York, 1991.

[12] ——, *The molecule problem: Exploiting structure in global optimization*, SIAM J. Optimization, (1995). To appear.

[13] D. Kincaid and W. Cheney, *Numerical Analysis*, Brooks/Cole Publishing Company, 1991.

[14] J. Kostrowicki and L. Piela, *Diffusion equation method of global minimization: Performance for standard functions*, J. Optim. Theory Appl., 69 (1991), pp. 269–284.

[15] J. Kostrowicki, L. Piela, B. J. Cherayil, and H. A. Scheraga, *Performance of the diffusion equation method in searches for optimum structures of clusters of Lennard-Jones atoms*, J. Phys. Chem., 95 (1991), pp. 4113–4119.

[16] J. KOSTROWICKI AND H. A. SCHERAGA, *Application of the diffusion equation method for global optimization to oligopeptides*, J. Phys. Chem., 96 (1992), pp. 7442–7449.

[17] I. D. KUNTZ, J. F. THOMASON, AND C. M. OSHIRO, *Distance geometry*, in Methods in Enzymology, N. J. Oppenheimer and T. L. James, eds., vol. 177, Academic Press, 1993, pp. 159–204.

[18] D. C. LIU AND J. NOCEDAL, *On the limited memory BFGS method for large scale optimization*, Math. Programming, 45 (1989), pp. 503–528.

[19] J. J. MORÉ AND Z. WU, *Global continuation for distance geometry problems*, Preprint MCS-P505-0395, Argonne National Laboratory, Argonne, Illinois, 1995.

[20] P. M. PARDALOS, D. SHALLOWAY, AND G. XUE, *Optimization methods for computing global minima of nonconvex potential energy functions*, J. Global Optim., 4 (1994), pp. 117–133.

[21] L. PIELA, J. KOSTROWICKI, AND H. A. SCHERAGA, *The multiple-minima problem in the conformational analysis of molecules: Deformation of the protein energy hypersurface by the diffusion equation method*, J. Phys. Chem., 93 (1989), pp. 3339–3346.

[22] J. B. SAXE, *Embeddability of weighted graphs in k-space is strongly NP-hard*, in Proc. 17th Allerton Conference in Communications, Control and Computing, 1979, pp. 480–489.

[23] H. A. SCHERAGA, *Predicting three-dimensional structures of oligopeptides*, in Reviews in Computational Chemistry, K. B. Lipkowitz and D. B. Boyd, eds., vol. 3, VCH Publishers, 1992, pp. 73–142.

[24] D. SHALLOWAY, *Application of the renormalization group to deterministic global minimization of molecular conformation energy functions*, J. Global Optim., 2 (1992), pp. 281–311.

[25] D. SHALLOWAY, *Packet annealing: A deterministic method for global minimization, application to molecular conformation*, in Recent Advances in Global Optimization, C. Floudas and P. Pardalos, eds., Princeton University Press, 1992, pp. 433–477.

[26] J. STOER AND R. BULIRSCH, *Introduction to Numerical Analysis*, Springer-Verlag, 1980.

[27] A. H. STROUD AND D. SECREST, *Gaussian Quadrature Formulas*, Prentice-Hall, Inc., 1966.

[28] Z. WU, *The effective energy transformation scheme as a special continuation approach to global optimization with application to molecular conformation*, Preprint MCS-P442-0694, Argonne National Laboratory, Argonne, Illinois, 1994.

(Jorge J. Moré) MATHEMATICS AND COMPUTER SCIENCE DIVISION, ARGONNE NATIONAL LABORATORY, ARGONNE, IL 60439, USA
*E-mail address*, Jorge J. Moré: **more@mcs.anl.gov**

(Zhijun Wu) MATHEMATICS AND COMPUTER SCIENCE DIVISION, ARGONNE NATIONAL LABORATORY, ARGONNE, IL 60439, USA
*E-mail address*, Zhijun Wu: **zwu@mcs.anl.gov**

DIMACS Series in Discrete Mathematics
and Theoretical Computer Science
Volume **23**, 1996

# The Design of Chromophore Containing Biomolecules

R. Pachter, Z. Wang, J. A. Lupo, S. B Fairchild, and B. Sennett

ABSTRACT. In this study we present a combination of approaches for structure and properties prediction of polypeptide-bound chromophores for the design of an ordered three-dimensional network of biopolymers with controlled nonlinear optical properties. Structure prediction of smaller biomolecular systems is being achieved by applying the adaptive simulated annealing (ASA) approach. Results for [Met]-Enkephalin indicate that ASA is a robust and efficient algorithm. Further, a computational approach is applied for the prediction of folds of large biomolecular systems, consisting of a neural networks and the double-iterated Kalman Filter (DIKF) technique. Specifically, a neural network learns to predict the spatial proximity of C$\alpha$ atoms, while the DIKF algorithm is employed to elucidate the structure using a data set that includes these pairwise atomic distances, and the distances and angles that define the chemistry and stereochemistry of the amino acids sequence. The results for Crambin demonstrate that such an integrated method is useful for the prediction of the biomolecular structure at an intermediate resolution. Finally, detailed effects of chromophore substitution are studied utilizing molecular dynamics simulations, specifically for an amphipathic coiled coil.

## 1. Introduction

In order to design an ordered network of biopolymers and other macromolecules with controlled nonlinear optical properties [**1, 2, 3**], we use a combination of approaches for structure and properties prediction. Biopolymers of defined secondary and tertiary structure derivatized with chromophores are well suited for use as nonlinear optical materials. Although various polymers have been used as backbones for the design of molecular assemblies with large second-order optical nonlinearities ($\chi^2$), as recently reviewed by Kanis *et al.* [**4**], polypeptides provide the advantage of controlling backbone structure at the tertiary or even quaternary level, thereby providing temporal stability to attached chromophores through packing phenomena. Rod-like coiled coil structures are of particular interest for such an application, providing larger potential for self-assembly [**5**].

The development of optimization techniques, recently reviewed [**6**], for determining structural motifs of such molecular systems, is first required. Indeed, for

1991 *Mathematics Subject Classification.* Primary 65K10, 90C30; Secondary 92C05, 92E10.
*Key words and phrases.* Biopolymers, adaptive simulated annealing, Kalman Filter, neural network, molecular dynamics, chromophores.

structure determination of peptides substituted with chromophores, or for the refinement of larger proteins whose initial fold has been determined, conformational search techniques are being explored. A number of approaches were developed to address the problem of reducing the scope of conformational search, specificallyby imposing constraints or biasing the search towards regions where the lowest energy may lie. Such methods include Monte Carlo (MC) with minimization [**7**], electrostatically driven Monte Carlo [**8**], simulated annealing (SA) [**9, 10**], or dynamic programming [**11**] techniques. However, these approaches are still limited by the high-dimensionality of the problem and low efficiency. The recently developed ASA algorithm [**12**], as described below, utilizing ECEPP/2, has been shown to be faster and more efficient as compared to SA [**13**]. In this paper we outline some of the aspects of the method, and report results for its application to a [Met]-Enkephalin model.

An integrated approach is applied for determining a structure at an intermediate resolution, to be thereafter refined by energy minimization. Specifically, an expert system is used to develop the training set consisting of specific protein structures obtained from the Brookhaven Protein Data Base, with the data files being preprocessed to extract the backbone atomic coordinates and calculate the appropriate torsion angles. A neural network then learns to predict the spatial proximity of $C\alpha$ atoms. The DIKF [**14**] algorithm (coded in PROTEAN2 [**15**]) is subsequently employed to elucidate the structure using a data set that includes these pairwise atomic distances, and the distances and angles that define the molecular structure. The PROTEAN2 code has been ported to the Thinking Machines Inc. CM-5 and the multiprocessor Cray C-90, and timing comparisons were previously described [**16**]. Results show that the neural network constraints set was found to be partially adequate compared to modeling with an exhaustive set of all $C\alpha$ pair distances derived from the crystal structure of Crambin [**17**].

Molecular dynamics simulations based on the coiled coil configuration illustrate the destabilization of the structure on chromophore substitution. Although methods for modeling protein conformation and dynamics in general and polypeptides of defined secondary structure in particular have been carried out [**18**], and the motions of an $\alpha$-helical polypeptide were examined, for example, by molecular and harmonic dynamics [**19**], polypeptide bound chromophores have generally not been modeled. A recent investigation on the influence of the spiropyran to merocyanine ring opening on a model of poly(spiropyran-L-glutamate) [**3**] has shown that such an approach lends support to the experimental observation and yields an understanding of the interactions involved. Results for poly(spiropyran-L-tyrosine) imply similar effects [**20**], namely the importance of chromophore substitution on the amount of helix bending. This destabilization for the coiled coil system is found to be moderate and the interaction between the two segments is still indicated. It is interesting to point out that the individual segments maintain a more stable secondary structure that may be due to relaxation of the tertiary conformation.

## 2. Methods and Computational Details

**2.1. Simulated Annealing.** The cooling schedule for SA can be expressed as:

$$T_k = T_0 e^{-(1-c)k}, \quad (0 < c < 1) \tag{1}$$

being equivalent to $T_{k+1} = T_k c$. Trial and error are applied for the cooling rate and the starting temperature, and a MC simulation is carried out at each temperature step $k$ with temperature $T_k$. Previous modifications that have significantly improved the efficiency of the conformational search include moving dihedral angles in the MC step, adjusting the maximum deviation of variables as the temperature decreases to insure that the acceptance ratio is more than 25%, and treating the variables differently according to their importance in the folding process, *e.g.*, sampling was increased for the backbone dihedral angles relative to those of the side-chains.

### 2.2. Adaptive Simulated Annealing.

2.2.1. *MC Configurations.* In a D-dimensional conformational space with dihedral angles $p^i$ in the range $[A_i, B_i]$ , $p^i_{k+1}$ is a new point about the $k'th$ last saved point (*e.g.*, a local optimum):

$$p^i_{k+1} = p^i_k + y^i(B_i - A_i), \tag{2}$$

where the random variables $y^i$ in $[-1, 1]$ (*non-uniform*) are generated from a random number $u^i$ uniformly distributed in $[0, 1]$ and the 'temperatures' $T_i$.

2.2.2. *Annealing Schedule.* The annealing schedule for from a starting temperature $T_{0i}$ is:

$$T_i(k_i) = T_{0i} exp(-c_i k_i^{1/D}). \tag{3}$$

However, the index for re-annealing the cost function, $k_{cost}$ , is determined by the number of accepted points instead of the number of generated points as used for the other parameters. This choice was made since the Boltzmann acceptance criterion uses an exponential distribution which is not as fat-tailed as the ASA distribution used for the parameters.

2.2.3. *Re-Annealing.* The parameter 'temperatures' may be periodically adaptively re-annealed, or rescaled according to the sensitivity of the cost function to the parameter. At any given annealing time, the temperature range is 'stretched out' over the relative insensitive parameters, thus guiding the search 'fairly' among the parameters. The sensitivity is calculated by:

$$S_i = \frac{\partial E}{\partial p^i}, \tag{4}$$

while the re-annealing temperature is determined by

$$T_i(k') = T_i(k)\frac{S_i}{S_{max}}. \tag{5}$$

In this way, less sensitive parameters anneal faster. This is done approximately every 100 accepted-events. ASA has proven to be more robust than other simulated annealing techniques (*e.g.*, Cauchy annealing where $T_i = T_0/k$, and Boltzmann annealing where $T_i = T_0/lnk$) for complex problems with multiple local minima.

2.2.4. *Potential Energy.* Within the ECEPP approximation the bond lengths and bond angles are kept constant, and the variables are only the dihedral angles. It is given in the form:

$$U = E_{nb} + E_{es} + E_{tor} \tag{6}$$

where $E_{nb}$, $E_{es}$, $E_{tor}$ are non-bonded (Lennard-Jones and hydrogen-bonding), electrostatic, and torsional energies, respectively, expressed as:

$$E_{nb} = \sum_{i \neq j} \epsilon_{ij}[F(\frac{r_{ij}^0}{r_{ij}})^{12} - 2(\frac{r_{ij}^0}{r_{ij}})^6] + \sum_{i' \neq j'} \epsilon_{ij}[5(\frac{r_{ij}^0}{r_{ij}})^{12} - 6(\frac{r_{ij}^0}{r_{ij}})^{10}] \tag{7}$$

$$E_{es} = \sum_{i \neq j} \frac{q_i q_j}{D r_{ij}} \tag{8}$$

$$E_{tor} = \sum_k \frac{A_k}{2}(1 \pm cos\, n\theta_k) \tag{9}$$

$\epsilon_{ij}$ and $r_{ij}^0$ are the potential depth and the position of the minimum of the $12-6$ or $12-10$ pair potentials; $q$ is the partial atomic charge; $A_k$ is the barrier height of rotation around the $kth$ bond; $\theta_k$ is the dihedral angle; $n$ is the $n$-fold degeneracy of the torsional potential; and $F$ is assigned the value 0.5 for interactions between atoms three bonds apart or the value 1 otherwise. The $12-10$ potential models hydrogen-bonds.

The ASA code with the ECEPP/2 potential form was integrated and tested [**21**]. Each ASA calculation is started from random peptide backbone dihedral angles $\phi$, $\psi$, with the $\omega$ dihedral fixed to 180°. The initial temperature was determined by the average energy of $5-10$ random samplings (the typical maximum number of energy function calls was 30,000). The ASA run was terminated when the best energy value for $3-5$ re-annealing cycles was repeated (each cycle generates 100 configurations). Further refinement was obtainedby the Conjugate Gradient Method, so that the energy gradient was less than 0.01. This combination improved the efficiency of the search. A discussion on the optimal control parameters is given elsewhere [**21**].

**2.3. Neural Networks.** Neural networks are trained on the dependence between the amino acid sequence and the distance relation between proximate $C\alpha$'s in the polypeptide backbone. Tertiary structure information is generated in the form of binary distance constraints between the $C\alpha$ atoms, being 1 if their distance is less than a given threshold, and 0 otherwise. The proteins used in the training set were the first 48 of the set collected by Kabsch and Sander [**22**] (total of 8315 residues). A feed forward neural network with one hidden layer was used, and backpropagation was the learning algorithm. Note that the input to the network is a vector describing a section of 31 amino acids wide, while the output results in the predicted binary distance relationships. The components of the output vector are real numbers between $-1$ and 1, to be converted to binary values. Output values above the threshold are converted to 1, while values below that are converted to 0. Note that each residue was represented by its hydrophobicity [**23**] (scaled between $0 \pm 1$). The Chou Fasman [**24**] probability was also used in the encoding scheme. These numbers are assigned by the frequency of residues in the $\alpha$-helical regions for the 20 amino acids relative to their respective average frequency. Additional details are described elsewhere [**25**], as well as results when using the secondary structure information explicitly in the encoding scheme [**26**]. The neural network runs on a Sun SPARC workstation II using the NeuralWorks Professional II/Plus software package [**27**]. The training involved about 250 cycles through the training set.

**2.4. Double Iterated Kalman Filter.** The DIKF approach for structure determination was initially introduced by Altman and Jardetzky [**28**]. Applications of the technique indicated that a known structure can be reproduced to within a small RMS error even when a limited data set is used [**29, 30**]. Several applications using Nuclear Magnetic Resonance (NMR) Nuclear Overhauser Effect (NOE) data utilizing this novel method proved the approach successful [**31, 32, 33**]. Note that the input to PROTEAN2 is automated [**34**]. An outline of the method is given below.

The structural molecular model of a polypeptide consisting of $N$ atoms is described by the mean Cartesian coordinates $x$ and the covariance matrix $C(x)$. The elements of $C$ for any two atoms $i, j$ are symmetric, and the matrices on the diagonal of the covariance matrix $C(x_{ii})$ describe the extent of the $3-D$ uncertainty in the position of the $i-th$ atom, and the covariance between two variables (the non-diagonal elements $C(x_{ij})$ estimates their correlation. Measurements are given by:

$$z = h(x) + v \tag{10}$$

where $z$ is the observed value of the data, $h(x)$ the relationship between the state vector variables and the distance/dihedral angle values, and v the variance of the data. Given this information, a sequential linear estimator for the minimum variance estimate of the state is obtained by the extended Kalman filter for nonlinear measurement functions [**14**]:

$$x(+) = x(-) + K[z - h(x(-))] \tag{11}$$

with

$$C(+) = C(-) - KHC(-) \tag{12}$$

where $(-)$ signifies a previous structural representation to be sequentially updated to $(+)$ . The criterion for the choice of the Kalman estimator gain matrix $K$ , given by:

$$K = C(-)H^T[HC(-)H^T + v]^{-1} \tag{13}$$

is to minimize a weighted scalar sum of the diagonal elements of the error covariance matrix $C$. The term within the inverse in the expression for $K$ represents the variance of the observed measurement:

$$C(v) = C(h(x) + v). \tag{14}$$

The first-order Taylor approximation of $C(h(x))$ is $HCH^T$ where $H$ is the derivative of the data model $h$, and $H^T$ is the transpose of $H$. The derivatives of the distance model are calculated analytically, while the derivatives of the dihedral angle data model are approximated by using a finite difference calculation. Note that in addition to the optimality of $K$ contained in its structure, it can also be seen as the ratio between the uncertainty in the estimate and that of the measurement.

The extended Kalman filter approach is used to obtain higher-order nonlinear filters:

$$x(+)_k = x(-) + K_k[z - [h(x)_{k-1}) + H(x(-) - x(+)_{k-1})]] \tag{15}$$

with a similar expression for $C(+)_k$ for any iteration $k$. This iterative procedure is carried out for each of the distance and dihedral angle constraints. However, since the filter is not optimal in the nonlinear case, residual inaccuracies may still result.

Therefore, the mean positions obtained after all data are introduced are used for another cycle of updating. The covariance matrix is reset to its initial large value in order to allow atoms freedom to move in response to the constraints, and all measurements are re-introduced into the system for each of these doubly-iterated cycles. The successive cycles are repeated until all of the constraints are satisfied to within a pre-set threshold of standard deviations from the error $e$:

$$e = \frac{h(x)_{final} - z}{\sqrt{v}} \tag{16}$$

**2.5. Molecular Dynamics.** The polypeptide bound chromophore models were studied using the CHARMm force-field given by Brooks *et al.* [**35**], implemented in QUANTA, consisting of the following terms [**36**]:

$$\sum_{bonds} K_b(b - b_0)^2 + \sum_{angles} K_\theta(\theta - \theta_0)^2 + \sum_{dihedrals} K_\phi[1 + \cos(n\phi - \delta)] \tag{17}$$

and non-bonded interactions:

$$+ \sum_{non-bonded} \frac{A}{r^{12}} - \frac{C}{r^6} + \frac{q_1 q_2}{4\pi\epsilon r}. \tag{18}$$

The parameters and reference values are described elsewhere [**35**]. Note that CHARMm is parametrized for the chromophore molecules.

The molecular structures were optimized by energy minimization. A heating procedure to 300 K was then employed by assigning and periodically adjusting random velocities to the atoms, and subsequently an equilibration was carried out at constant temperature by scaling the velocities. After equilibration a molecular dynamics simulation was performed using the Verlet algorithm [**37**] for 0.1 ns at 300 K, using a 0.5 fs integration time step since SHAKE was not employed, a cutoff distance of 15 Å, and a smoothing function was also employed. All computations were performed with a solvent-averaged potential of mean force implied by setting the dielectric constant to the distance of the nonbonded pairs [**38**].

## 3. Results and Discussion

In order asses the ASA algorithm, it has been applied for the structure determination of a [Met]-Enkephalin model, a small pentapeptide defined by the five-amino-acid sequence $Tyr - Gly - Gly - Phe - Met$. Note that the neutral $-NH_2$ and $-COOH$ end groups were substituted at the $N-$ and $C-$ termini of the model [**39**]. The results of the best structure obtained with ECEPP/2 are: $\phi_1$=-87.3; $\psi_1$=152.7; $\omega_1$=181.3; $\phi_2$=-158.6; $\psi_2$=69.2; $\omega_2$=176.0; $\phi_3$=68.3; $\psi_3$=-91.5; $\omega_3$=176.5; $\phi_4$=-83.5; $\psi_4$= -31.5; $\omega_4$=183.5; $\phi_5$=—80.8; $\psi_5$= 143.7; $\omega_5$=181.5; $\chi_1^1$ =-180.3; $\chi_1^2$=66.3; $\chi_1^3$=-42.1; $\chi_4^1$=179.4; $\chi_4^2$=-80.3; $\chi_5^1$=-65.1; $\chi_5^2$=-179.0; $\chi_5^3$=-179.3; and $\chi_5^4$=-59.9. The energy of this configuration is -10.3 kcal/mol after CGM refinement with $\omega$ not fixed. These results are consistent with previous works [**40**], and are currently being refined by using the ECEPP/2-84 version [**21, 41**].

The trained neural network was applied to determine the distance constraints for the amino acid sequence for Crambin (which was not contained in the training set). The results were quantified by the number of correct predictions, the overall success rate and by the correlation coefficient [**42**]. The success rate is the number of correct predictions relative to the total number of predictions. The correlation coefficient takes into account the number of over- and under- predicted cases, and is a better estimate of prediction success. In our case the correct predictions/total = 1121/1380; positive prediction success ratio = 246/378; negative prediction ratio = 889/1002; success rate = 0.81, and the correlation coefficient = 0.52. The neural network results were used for the application with the DIKF. The Crambin model contains 327 atoms and the starting structure assumes an extended configuration. The distance constrains include the bond distances and distances implied by bond angles with a low variance in the range of 0.1 $\AA^2$ - 0.3 $\AA^2$, with a total of 915 constraints, as well as the non-bonded distances derived from the NN applications (assigned 5 Å) with a large variance of 1.1 $\AA^2$, resulting in a total of 131 constraints. 82 Dihedral angle constraints are also included. Good convergence was obtained. A comparison with the experimental X-ray structure results in total all-atom average Root Mean Square (RMS) deviation of 6.3 Å for the backbone (note the RMS deviation between the initial and X-ray structures is *ca.* 35 Å). This result in being further refined using the adaptive simulated annealing approach. This additional minimization and also results for BPTI are to be reported elsewhere [**46**]. These results indicate that such an integrated approach may be useful for an intermediate biomolecular structure prediction, to be further refined by energy minimization and molecular dynamics.

Underivatized **(I)** and derivatized amphipathic coil coiled polypeptide models with the chromophore disperse orange 3 **(II)** (Figure 1) were studied, consisting of two identical $\alpha$-helical segments, maintaining the sequence $Y-G-E-L-E-E-K-(L-E-E-L-E-E-K)_2-L-E-E$. This suggested sequence is expected to assume a coiled coil motif, which describes the self-assembly of $\alpha$-helices [**43**], that would otherwise tend to be unstable in an aqueous environment [**44**]. The orientation of the residues, with the leucine amino acids occupying the $a$ and $d$ positions, enables the formation of a leucine zipper [**45**], where interdigitation of the hydrophobes in an aqueous environment stabilizes the super helix. Moreover, electrostatic interactions between the glutamate and lysine side chains at positions $e$ and $g$, further stabilize the dimer, and it is therefore hypothesized that the ammonium groups of the lysine side chains and the carboxyl groups of the glutamate side chains will form salt bridges. The positioning of the residues adjacent to the hydrophobic face of the $\alpha$-helices facilitates the formation of the leucine zipper by constraining the orientation of the segments *via* their interaction. The $\alpha$-helices are parallel and in register, with a required deformation in the torsional values along the backbones to maintain a super-coiled conformation. The chain lengths were chosen so as to limit the size of the computational problem, while for synthesis a polyheptad is required to induce a coiled coil motif [**46**]. The $N-$ and $C-$ segments termini were blocked by acetyl and amide groups, respectively, to remove any zwitterionic contribution to destabilization of the secondary structure.

In order to gain an insight of the influence of the side-chain substituted chromophore on the coiled coil configuration, the degree of secondary structure maintained by each segment in the coiled coil and the degree of superhelicity maintained by both segments were examined. Preliminary molecular dynamics simulation results have shown [**26**] that structure **II** exhibits a larger variation in the RMS difference between the initial and last conformation generated during the molecular dynamics simulation, particularly 3.2 Å for structure **II**, as compared to 1.2 Å for structure **I**. Further, the mean $\phi, \psi$ backbone dihedral angles of structures **I** and **II** show that the distributions of the dihedral angles assume the values of an $\alpha$-helical configuration, but the larger mean fluctuations are exhibited by **I**, and are representative of residues throughout the structure. A greater degree of $\alpha$-helical structure for the segments of **II** is found, which is further proven by an examination of hydrogen bonding patterns in those systems.

Intrachain hydrogen bonding of residues $(i, i+4)$ ($\delta_i$, approximately 3.04 Å in Figure 2), where the error bars represent the standard deviation of each distribution, show that **II** has a more uniform occurrence of intrachain hydrogen bonding along the segment. This is in contrast to **I**, where several residues within the chain exhibit a large $\delta_i$. The lack of hydrogen bonding at residues 17, 36 and 37 may be attributed to conformational constraints imposed by the tertiary configuration. Furthermore, structure **I** exhibits frequent interchain hydrogen bonding, while structure **II** lacks any interchain hydrogen bonding on one face of the superhelical structure. This can be seen as an opening of the super helix, which exposes the hydrophobic residues. The lack of superhelical stability results in improved $\alpha$-helical stability of the segments, as they are less constrained by the affinity of the lysine and glutamate side chains.

A further description of the coiled coil superstructure is given by the superhelical angle $\chi_s$ defining the angle between two lines that are described by an RMS fit to the set of averaged Cartesian coordinates of each segment. $\chi_s$ defines the mean angle of incidence between the two $\alpha$-helices, with an ideal value near 18 ° **[46]** . Structure **I** exhibits a superhelical angle of incidence with a mean value of 20 °, while **II** maintains a $\chi_s$ value of 44 °. This demonstrates the destabilization of the coiled coil super structure when substituted with the chromophore. The proximity of the chromophore and $\alpha$-helical segment is described by $\chi_c$, that is the angle between two lines defined by an RMS fit to the averaged coordinates of disperse orange 3 and segment 1 of structure **II**, and is calculated in this case as a parallel position.

In summary, the design of biopolymer-based nonlinear optical materials clearly requires first the determination of the biomolecular structure, to be followed by the prediction of the chromophore effects. This study describes the derivation of an initial folding pattern based only on the knowledge of the primary structure and specified metric constraints. The results for Crambin demonstrate that this integrated approach is useful for an intermediate tertiary structure prediction. Furthermore, an adaptive simulated annealing approach for structure prediction of polypeptides may offer the capability for refinement of large structures or prediction of smaller biomolecular configurations. Detailed effects of the chromophore substitution are then studied utilizing molecular dynamics simulations. The results suggest that the biopolymer will exhibit rigid-rod packing and thus provide for the temporal stability of the attached chromophore necessary for enhanced $\chi^2$ nonlinear optical effects. Synthesis efforts lend further support to this premise **[5]**, indicating that the structure has not fully moved away from the $\alpha$-helical form.

## 4. Figures

Figure 1: The disperse orange 3 chromophore is bound to the side chain of glutamic acid

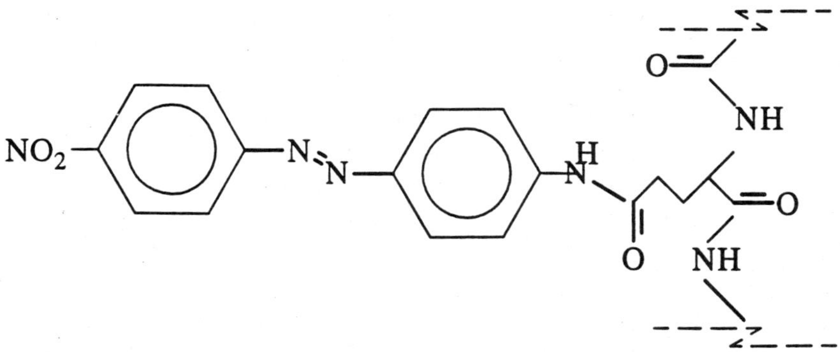

Figure 2: Intrachain hydrogen $(i, i+4)$ bonding for **I** and **II**

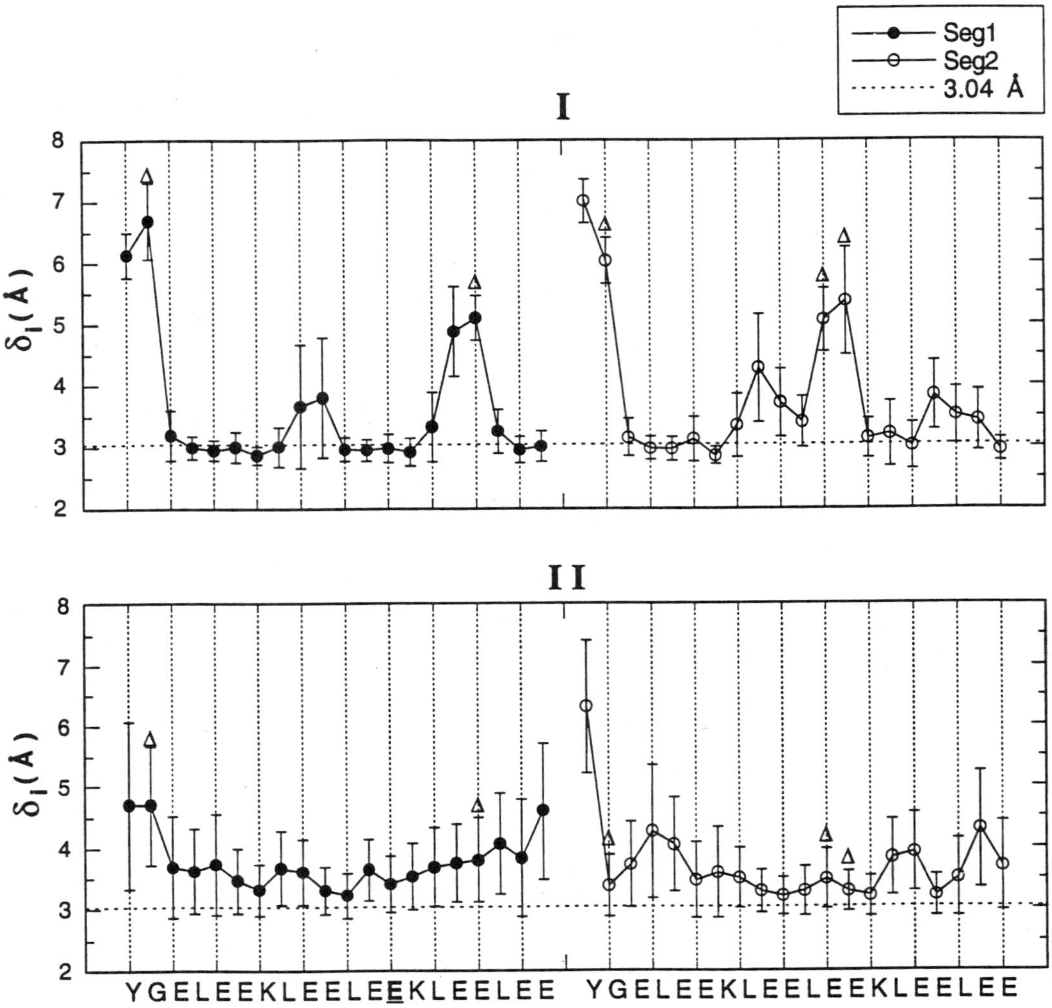

## References

[1] B. F. Levine, C. G. Bethea, E. Wasserman, and L. Leenders, J. Chem. Phys. **68** (1978) 5042; T. Ishii, T. Wada, A. Garito, H. Sasabe, and A. Yamada, Mat. Res. Soc. Symp. Proc. **175** (1990) 129.

[2] T. M. Cooper, L. V. Natarajan, R. Strasser, R. Pachter and R. L. Crane, ACS Pol. Prep. **33** (1992) 1018; T. M. Cooper, K. Obermeier, L. Natarajan, R. Crane, Photochem. Photobiol. **55** (1992) 1.

[3] R. Pachter, T. M. Cooper, L.V. Natarajan, K. Obermeier, R. L. Crane and W. W. Adams, Biopolymers **32** (1992) 1129.

[4] D. R. Kanis, M. A. Ratner, and T. J. Marks, Chem. Rev. **94** (1994) 159.

[5] T. M. Cooper, unpublished results.

[6] M. Vasquez, G. Nemethy and H. A. Scheraga, Chem. Rev. **94** (1994) 2183.

[7] Z. Li and H. A. Scheraga, J. Mol. Structure (THEOCHEM) **179** (1988) 333.

[8] D. R. Ripoll, M. Vasquez, and H. A. Scheraga, Biopolymers **30** (1990) 165; D. R. Ripoll and H. A. Scheraga, Biopolymers **31** (1991) 319.

[9] Y. Okamoto, T. Kikuchi and H. Kawai, Chemistry Letters (The Chemical Society of Japan) (1992) 1275.

[10] A. Nayeem, J. Vila and H. A. Scheraga, J. Comp. Chem. **12** (1991) 594.

[11] S. Vajda and C. Delisi, Biopolymers **29** (1990) 1755.

[12] L. Ingber, Mathl. Comp. Modeling **12** (1989) 967.

[13] L. Ingber and B. Rosen, Mathl. Comp. Modeling **16** (1992) 87.

[14] A. Gelb, In: *'Applied Optimal Estimation'*, MIT Press, 1984.

[15] R. B. Altman, R. Pachter, E. A. Carrara, and O. Jardetzky, QCPE **10** (1990) 596 (2).

[16] J. A. Lupo, R. Pachter, S. B. Fairchild, and W. W. Adams, In: *'Parallel Computing in Computational Chemistry'*, to be published.

[17] M. M. Teeter, Proc. Nat. Acad. Sci. USA **81** (1984) 6014.

[18] H. A. Scheraga, Chemica Scripta **29A** (1989) 3.

[19] D. Perahia, R. Levy, and M. Karplus, Biopolymers **29** (1990) 645.

[20] R. Pachter and T. M. Cooper, unpublished results.

[21] Z. Wang and R. Pachter, manuscript in preparation.

[22] W. Kabsch and C. Sanders, FEBS Letters **155** (1983) 179.

[23] G. Wilcox, M. Poliac, and M. Liebman, Tetrahedron Computer Methodology **3** (1990) 196.

[24] P. Y. Chou and G. D. Fasman, Advances in Enzymology **47** (1978) 45.

[25] S. B. Fairchild, R. Pachter, and R. Perrin, Mathematica Journal, in press.

[26] R. Pachter, S. B. Fairchild, J. A. Lupo, R. L. Crane, and W. W. Adams, In: *'Biomolecular Materials by Design'* (Alper, M., Bayley, H., Kaplan, D., Navia, M. eds.) **330** (1994) 71.

[27] NeuralWare, Inc. Penn Center West, Bldg IV, Pittsburgh, PA, 15276.

[28] R. B. Altman, R. Pachter, and O. Jardetzky, Methods in Enzymology **177** (1989) 218.

[29] R. B. Altman, R. Pachter, and O. Jardetzky, In: *'Protein Structure and Engineering'*, (O. Jardetzky, Ed.), Plenum Press, New York, 1989, pp79.

[30] R. Pachter, R. B. Altman, and O. Jardetzky, J. Magn. Reson. **89** (1990) 578.

[31] C. H. Arrowsmith, R. Pachter, R. B. Altman, S. B. Iyer, and O. Jardetzky, Biochemistry **29** (1990) 6332; C. H. Arrowsmith, R. Pachter, R. B. Altman, and O. Jardetzky, FEBS Eur. J. Biochemistry **202** (1991) 53.

[32] R. B. Altman, C.H. Arrowsmith, R. Pachter, and O. Jardetzky, In: *'Computational Aspects of the Study of Biological Macromolecules by NMR Spectroscopy'*, (J. C. Hoch, Ed.), Plenum Press, New York 1991, pp375; R. B. Altman, R. Pachter, and O. Jardetzky, Applied Magn. Reson. **4** (1993) 441.

[33] R. Pachter, R. B. Altman, J. Czaplicki, and O. Jardetzky, J. Magn. Reson. **92** (1991) 648.

[34] R. Pachter, Programs: PROPARE, PROCON, PROCOPY, unpublished.

[35] B. R. Brooks, R. E. Bruccoleri, B. D. Olafson, and D. J. States, S. Swaminathan, and M. Karplus, J. Comp. Chem. **4**(1983) 187.

[36] Quanta: Release 3.0, Polygen Corporation, 1990.

[37] L. Verlet, Physical Review **159** (1967) 98.

[38] J. A. McCammon, S. C. Harvey, *'Dynamics of Proteins and Nucleic Acids'* (Cambridge University Press, New York, 1987).

[39] J. Bascle, T. Garel, H. Orland and B. Velikson, Biopolymers **33** (1993) 1843.

[40] S. R. Wilson and W. Cui, Biopolymers **29** (1990) 225.

[41] S. Fossey, private communication.

[42] N. Qian and T. Sejnowski, J. Mol. Biol. **202** (1988) 868.

[43] J. A. Talbot, R. S. Hodges, Acc. Chem. Res. **15** (1982) 224.

[44] C. Cohen, D. Parry, Trends Biochem. Sci. **11** (1986) 245.

[45] E. K. O'Shea, R. Rheba, P. S. Kim, Science **243** (1989) 538.

[46] R. Pachter, S. B. Fairchild, and J. A. Lupo, manuscript in preparation.

[47] S. Y. M. Lau, A. K. Taneja, R. S. Hodges, J. Biol. Chem. **259** (1984) 13253.
[48] F. H. C. Crick, Acta Cryst. **6** (1953) 689.

*E-mail address*: pachterr@ml.wpafb.af.mil
*E-mail address*: wangz@ml.wpafb.af.mil
*E-mail address*: lupoja@ml.wpafb.af.mil

MATERIALS DIRECTORATE, WRIGHT LABORATORY WL/MLPJ, BLDG. 651, 3005 P STREET STE 1, WRIGHT-PATTERSON AIR FORCE BASE, OH 45433-7702.

DIMACS Series in Discrete Mathematics
and Theoretical Computer Science
Volume **23**, 1996

# Molecular Structure Determination by Convex Global Underestimation of Local Energy Minima

A.T. Phillips, J.B. Rosen, V.H. Walke

ABSTRACT. The determination of a stable molecular structure can often be formulated in terms of calculating the global (or approximate global) minimum of a potential energy function. Computing the global minimum of this function is very difficult because it typically has a very large number of local minima which may grow exponentially with molecule size. The optimization method presented involves collecting a large number of conformers, each attained by finding a local minimum of the potential energy function from a random starting point. The information from these conformers is then used to form a convex quadratic global underestimating function for the potential energy of all known conformers. This underestimator is an $L_1$ approximation to all known local minima, and is obtained by a linear programming formulation and solution. The minimum of this underestimator is used to predict the global minimum for the function, allowing a localized conformer search to be performed based on the predicted minimum. The new set of conformers generated by the localized search serves as the basis for another quadratic underestimation step in an iterative algorithm. This algorithm has been used to determine the structures of homopolymers of length $n \leq 30$ with no sidechains. While it is estimated that there are $O(3^n)$ local minima for a chain of length $n$, this method requires $O(n^4)$ computing time on average. It is also shown that the global minimum potential energy values lie on a concave quadratic curve for $n \leq 30$. This important property permits estimation of the minimum energy for larger molecules, and also can be used to accelerate the global minimization algorithm.

## 1. Introduction

An important class of difficult global minimization problems arise as an essential feature of molecular structure calculations. The determination of a stable molecular structure can often be formulated in terms of calculating the global (or approximate global) minimum of a potential energy function (see [8]). Computing the global minimum of this function is very difficult because it typically has a very large number of local minima which may grow exponentially with molecule size.

---

1991 Mathematics Subject Classification. Primary 65K05; Secondary 90C26, 90C90.

The first author was supported by a grant from the Minnesota Supercomputer Institute; the second author was supported by the AFOSR grant 91-0147; the third author was supported by the Trident Scholar Program at the United States Naval Academy.

One such application is the well known protein folding problem. It is widely accepted that the folded state of a protein is completely dependent on the one-dimensional linear sequence (i.e. "primary" sequence) of amino acids from which the protein is constructed: external factors, such as enzymes, present at the time of folding have no effect on the final, or native, state of the protein. This led to the formulation of the protein folding problem: given a known primary sequence of amino acids, what would be its native, or folded, state in three-dimensional space.

Recently, several successful predictions of folded protein structures have been made and announced before the experimental structures were known (see [2], [9]). While most of these have been made with a blend of a human expert's abilities and computer assistance, fully automated methods have shown promise for producing previously unattainable accuracy.

These machine based prediction strategies attempt to lessen the reliance on experts by developing a completely computational method. Such approaches are generally based on two assumptions. First, that there *exists* a potential energy function for the protein; and second that the folded state corresponds to the structure with the lowest potential energy (minimum of the potential energy function) and is thus in a state of thermodynamic equilibrium. This view is supported by in vitro observations that proteins can successfully refold from a variety of denatured states. Evolutionary theory also supports a folded state at a global energy minimum. Protein sequences have evolved under pressure to perform certain functions, which for most known occurrences requires a stable, unique, and compact structure. Unless specifically required for a certain function, there was no biochemical need for proteins to hide their global minimum behind a large kinetic energy barrier. While kinetic blocks may occur, they should be limited to special proteins developed for certain functions (see [1]).

## 2. A Simplified Model for Computational Methods

Unfortunately, finding the "true" energy function of a molecular structure, if one even exists, is virtually impossible. For example, with proteins ranging in size up to 1,053 amino acids (a collagen found in tendons), exhaustive conformational searches will never be tractable. One possible way of finding the global energy minimum is to use a simplified model of the molecular structure. By using a simplified model, the complexity of the problem formulation could be reduced to an acceptable level for optimization techniques. Care must be taken, however, to insure that the error included in such an approximation does not drive the computational solution too far away from the true native state.

One possible approximation method, which can be applied to molecules that form a linear chain, represents the molecular structure as a string of beads where the position of each bead is defined by its location relative to the previous three beads in the sequence (see [4]). In this model, the chain monomers come in two forms, H (hydrophobic) and P (polar), where the H-type monomers exhibit a strong pairwise attraction, and hence the lowest free energy is obtained by those conformations with the greatest number of HH "contacts" (see [4], [10]). One significant advantage of this formulation of the folding problem is that it allows the model to take advantage of known scientific knowledge about the chemical structure of sequences of molecules. The use of knowledge such as the Ramachandran plot (see [7]), which specifies the allowable angles between consecutive amino acids in proteins, greatly simplifies the problem. This paper will examine one such method for the global optimization of a potential energy function.

### 3. Potential Energy Function

Molecular structure information is generally given in terms of internal molecular coordinates which consist of bond lengths, bond angles, and the mainchain backbone torsion angles. For example, a protein consists of $n$ amino acids in a "primary" sequence $a_1, a_2, \ldots, a_n$, where $a_i$ represents the $i^{th}$ amino acid in the sequence. For every consecutive pair of amino acids, $a_{i-1}$ and $a_i$, $l_i$ represents the distance by which they are separated. For every three consecutive amino acids, $a_{i-2}$, $a_{i-1}$, and $a_i$, the bond angle, $\theta_i$, represents the position of the third amino acid with respect to the line containing the previous two amino acids. Similarly, for every four consecutive amino acids, $a_{i-3}$, $a_{i-2}$, $a_{i-1}$, and $a_i$, the torsion angle, $\varphi_i$, represents the relative position of the fourth amino acid, $a_i$, with respect to the plane containing the previous three amino acids (see Figure 3.1).

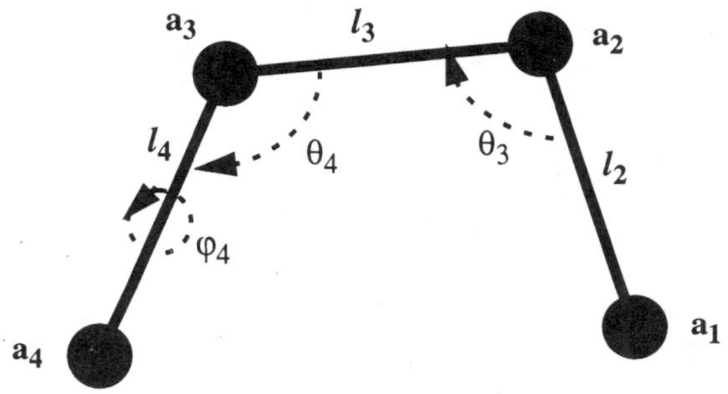


**Figure 3.1 Conformation for a Four Bead Sequence**

Empirical representations of the protein's potential energy include energy terms to represent chemical bonds, angles, and torsions, as well as non-bonded interactions between amino acids farther apart in the chemical structure. This simplified definition resembles a physical model in which beads (e.g. amino acids) are connected by springs (first term of (1)) at a distance of $l_i$. Bond angles, $\theta_i$, which are determined by a sequence of three beads $(a_{i-2}, a_{i-1}, a_i)$ are maintained by similar "springs" (the second term of (1)). Torsion angles, $\varphi_i$, are modeled by a trigonometric based penalty (third term of (1)). Such a formulation is often called a molecular mechanics potential. One often used formula for the overall potential $U$ is:

$$(1) \qquad U = \sum_i \frac{1}{2} k_l (l_i - l_0)^2 + \sum_i \frac{1}{2} k_\theta (\theta_i - \theta_0)^2 + \sum_i \frac{V_{\bar{n}}}{2} [1 + \cos(\bar{n}\varphi_i + \delta)] + \sum_{i,l} \varepsilon_{i,l} \left( \left( \frac{\sigma_{i,l}}{r_{i,l}} \right)^{12} - 2H_{i,l} \left( \frac{\sigma_{i,l}}{r_{i,l}} \right)^6 \right)$$

where $k_l$ and $k_\theta$ are the bond stretching and angle bending force constants. In the third term of the equation, $V_n$ is the $n$-fold torsional constant with a phase shift of $\delta$. This term provides for preferred torsion angles within the molecular structure. For example, if $n = 3$ and the phase shift $\delta$ is zero, preferred torsion angles (configurations resulting in a small penalty) can be found at 60°, 180°, and 300°. The constants $l_0$ and $\theta_0$ represent the preferred bond length and bond angle for each sequence of consecutive beads.

The fourth term of the (1), $\sum_{i,l} \left( (\sigma_{i,l}/r_{i,l})^{12} - 2H_{i,l} (\sigma_{i,l}/r_{i,l})^6 \right)$, is known as the Lennard-Jones pairwise potential (see Figure 3.2). This term defines the potential energy contributions of all beads separated by more than two along the primary chain. In (1), $\varepsilon_{i,l}$ and $\sigma_{i,l}$ are the Lennard-Jones coefficients, which are constants defined by the relationships

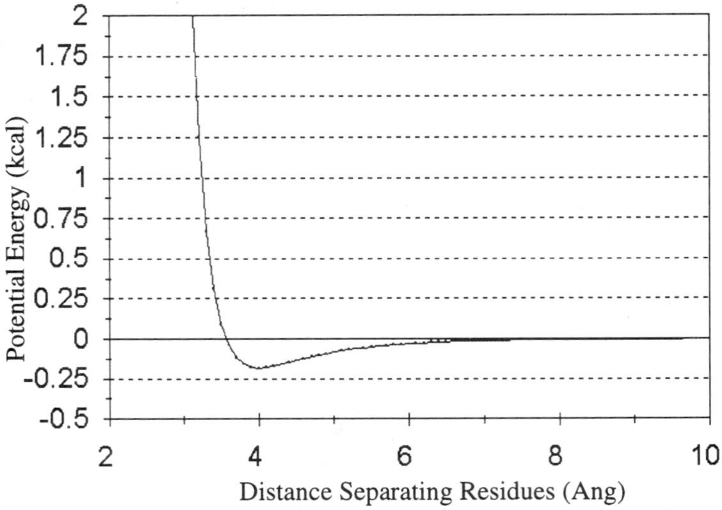


**Figure 3.2 Lennard-Jones Pairwise Potential** ($\varepsilon$=0.181, $\sigma$=4.0)

between the two specific beads (e.g. amino acids) involved. The terms involving $r_{i,l}$ in the Lennard-Jones expression represent the Euclidean distance between the beads $a_i$ and $a_l$ (see [3], [5], [8]). The constant $H_{i,l} = 1$ if beads $a_i$ and $a_l$ are both H-type (attractive monomers), and hence both a repulsive force (ensuring that the chain is "self-avoiding") and an attractive force (since the beads are H-H) are added to the potential energy. On the other hand, $H_{i,l} = 0$ if the beads $a_i$ and $a_l$ are H-P, P-H, or P-P pairs, so that the Lennard-Jones contribution to the total potential energy is just the repulsive force that ensures self-avoidance.

## 4. Coordinate Representations

Potential energy functions for molecular conformations usually involve computing Euclidean distances between molecules. For example the potential function in (1) contains the Lennard-Jones pairwise potential, $\sum_{i,l}\left((\sigma_{i,l}/r_{i,l})^{12} - 2H_{i,l}(\sigma_{i,l}/r_{i,l})^{6}\right)$, which depends on the lengths between all beads separated by more than two residues along the primary chain. Unfortunately, computing distances using bond lengths, bond angles, and torsion angles is extremely difficult. Since optimization methods require this computation to be executed often, it is desirable to convert the representation of the problem into cartesian coordinates. Once the conversion to three dimensional cartesian coordinates has been completed, the length between the residues can be found in $O(n^2)$ time. However, the conversion itself becomes very complex since every set of three residues creates *its own reference frame* for bond lengths, bond angles, and torsion angles.

To perform the conversion from internal molecular coordinates (bond lengths, bond angles, and torsion angles) to a cartesian representation, the first three beads in the sequence can be fixed, without loss of generality, as depicted in Figure 4.1. The first bead, $a_1$, is fixed at the origin, (0,0,0). The second bead, $a_2$, is positioned at $(-l_2,0,0)$, a distance from $a_1$ equal to the bond length, $l_2$, along the negative x axis. The location of the third bead, $a_3$, is fixed at $(l_3\cos(\theta_3)-l_2, l_3\sin(\theta_3), 0)$. The fourth and any other beads constituting the primary sequence are found using the cartesian representation associated with the pre-

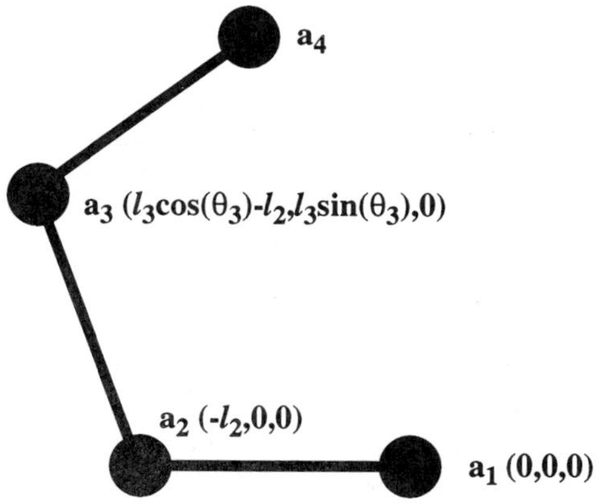


**Figure 4.1 Four Bead Sequence (Cartesian Coordinates)**

vious three beads, and the bond length, bond angle, and torsion angle associated with the bead under consideration. The cartesian representation of this and all following beads in the sequence can be computed in a variety of ways, two of which are presented in [11] and [12]. Appendix A (Section 12) presents a discussion of both the computations required to perform the transformations, and also a derivation of an analytic form of the gradient of the potential function $U$.

## 5. *n*-Chain Homopolymer Problem

One simple application of the "folding" problem, which can be modeled as a string of beads, is the *n-chain homopolymer problem*. Homopolymers can be modeled in manner similar to proteins: a string of beads ($a_i$, where $i = 1,...,n$), positioned by their respective bond lengths, $l_i$, bond angles, $\theta_i$, and torsion angles, $\varphi_i$. However, each "residue" in this case can be considered *identical* (i.e. all of type H). Furthermore, it is well known that in the case of hydrocarbons, these homopolymers have "preferred" bond lengths $l_0 = 1.526$ Ang and bond angles $\theta_0 = 109.47°$ (1.91 rad), and hence this information can be used in some optimization techniques to reduce the number of problem parameters.

Heptane is an $n$-chain hydrocarbon homopolymer of length seven ($n = 7$). It provides a simple, tractable problem for which the conformational space can (almost) be exhaustively explored. While this structure reflects great simplifications over the general molecular conformation problem, its complexity must not be underestimated.

The potential energy function is given by (1), where $r_{i,l}$ is the distance between beads, in this case hydrocarbons, separated by two or more residues along the central chain, $k_l =$ 310.0 kcal/Ang$^2$, $k_\theta = 40.0$ kcal/rad$^2$, $V_n = 1.3$ kcal, $\varepsilon_{i,l} = 0.181$ kcal, $\sigma_{i,l} = 4.0$ Ang, $l_0 =$ 1.526 Ang, $\theta_0 = 109.47°$ (1.91 rad), and $H_{i,l} = 1$ for all pairs $i,l$. Notice that $l_0$ and $\theta_0$ represent the preferred bond angles and bond lengths and $\bar{n}$ is set to three, providing three preferred torsion angles at 60°, 180°, and 300°, while $\delta$=0°. Since all hydrocarbons can be considered identical, only one value of the $\varepsilon_{i,l}$ and $\sigma_{i,l}$ is required for all $i$ and $l$.

Despite these simplifications, the conformation problem remains very difficult. Since torsion angles have three preferred positions (60°,180°,300°), $3^{n-3}$, or 81, local minima are expected (recall that three of the torsion angles are fixed). Tests on the potential function support this estimate: by starting from 1000 random starting points, 77 local minima were discovered. However, merely selecting preferred angles does not guarantee that the initial conformer is also a local minimum. Notice in Table 5.1 that the local minimum found by

**Table 5.1 Selected Minima of Heptane**

| $U(\phi)$ | Initial Position, $\phi^{(0)}$ | | | | Minimized Position, $\phi^*$ | | | |
|---|---|---|---|---|---|---|---|---|
| -0.339727 | 180 | 180 | 180 | 180 | 180.000000 | 180.000000 | 180.000000 | 180.000000 |
| -0.215609 | -60 | 60 | -60 | 60 | -62.657878 | 179.600918 | -179.857154 | 179.956953 |
| -0.215524 | -60 | 180 | 180 | 180 | -62.706316 | 179.627701 | 180.119720 | 180.000000 |
| -0.104118 | 60 | 60 | -60 | 60 | 62.550498 | 180.191468 | -179.798849 | 62.592457 |
| 2.097992 | 180 | -60 | 60 | 60 | 179.378780 | -79.278332 | 79.236470 | 62.423103 |
| 4.384072 | 180 | -60 | 60 | -60 | 178.741629 | -93.723491 | 69.284953 | -94.158124 |

starting at (-60,60,-60,60) results in the same local minimum as the point with initial torsion angles (-60,180,180,180). The global minimum for heptane corresponds to the point with all torsion angles equal to 180 (see Table 5.1 and Figure 5.1).

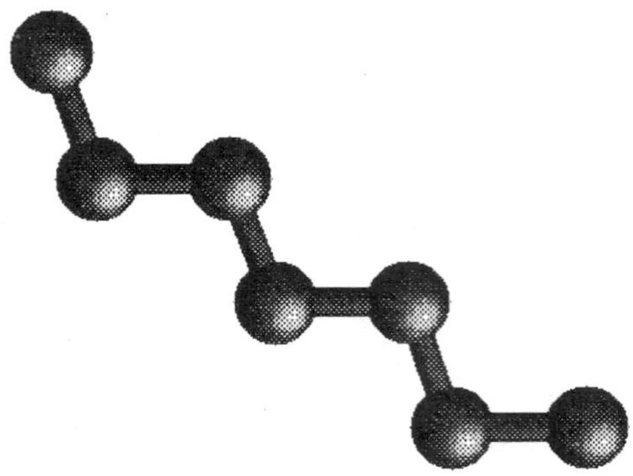

**Figure 5.1 Global Minimum Conformation, $\phi^*$, for Heptane**

While it may seem that 77 local minima is relatively small, larger $n$-chain hydrocarbons produce an exponentially increasing number (i.e. $O(3^n)$) of local minima. A molecule with as few as 12 beads has 19,683 possible preferred angle configurations, with most of these likely to produce a local minimum. It can clearly be seen that finding the global minimum for chains of even moderate length (e.g. $n = 20$ or more) is intractable with exhaustive methods. The next few sections discuss a method which attempts to find the global solution without resorting to an exponential dependence on problem size.

## 6. A Convex Global Underestimator

One possibility for aiding the search for the global minimum of the molecule's potential energy function is to use a global underestimator to localize the search in the region of the global minimum. This new method is designed to fit all known local minima with a convex function which underestimates all of them, but which differs from them by the minimum possible amount in the discrete $L_1$ norm. The minimum of this underestimator is used to predict the global minimum for the function, allowing a localized conformer search to be performed based on the predicted minimum. A new set of conformers generated by the localized search can then serve as a basis for another quadratic underestimation. After several repetitions, the global minimum can be found with reasonable assurance.

The use of an underestimating function allows the translation of a *very* complex function into a simple underestimator. If the underestimator is well suited to the problem (i.e. provides accurate predictions for the global minimum), immense savings in time can be achieved. The presence of quadratic terms in the length and bond angle portions of the molecular energy function support the use of a convex quadratic to provide a suitable approximation for this problem.

This method is presented in terms of a differentiable function $F(\phi)$, where $\phi \in \mathbf{R}^{\tau}$ ($\tau$ represents the number of torsion angles), and where $F(\phi)$ has many local minima. Since the bond lengths and bond angle terms in the potential function carry severe quadratic penalties, they can be assumed to be fixed for the purposes of the global underestimator. Therefore, $\phi$ is a vector of torsion angles of length $\tau$, where $\tau = n - 3$ and $n$ is the number of residues (recall that the first three are fixed by definition). To begin the iterative process, a set of $k \geq 2\tau+1$ distinct local minima are computed. This can be done with relative ease by using an efficient unconstrained minimizer, starting with a large enough set of points chosen at random in an initial hyperrectangle $H\phi$, which is assumed to enclose the torsion angle space.

Assuming that $k \geq 2\tau+1$ local minima $\phi^{(j)}$, for $j=1,\ldots,k$, have been computed, a convex quadratic underestimator function $\Psi(\phi)$ is now fitted to these local minima so that it underestimates all the local minima, and normally interpolates $F(\phi^{(j)})$ at $2\tau+1$ points. This is accomplished by determining the coefficients in the function $\Psi(\phi)$ so that

$$\delta_j = F(\phi^{(j)}) - \Psi(\phi^{(j)}) \geq 0 \tag{2}$$

for $j=1,\ldots,k$, and where $\sum_{j=1}^{k} \delta_j$ is minimized. That is, the difference between $F(\phi)$ and $\Psi(\phi)$ is minimized in the discrete $L_1$ norm over the set of $k$ local minima $\phi^{(j)}$, $j=1,\ldots,k$. The underestimating function $\Psi(\phi)$ is given by

$$\Psi(\phi) = c_0 + \sum_{i=1}^{\tau} \left( c_i \phi_i + \frac{1}{2} d_i \phi_i^2 \right). \tag{3}$$

Note that $c_i$ and $d_i$ appear linearly in the constraints of (2) for each local minimum $\phi^{(j)}$. Convexity of this quadratic function is guaranteed by requiring that $d_i \geq 0$ for $i=1,\ldots,\tau$. Other linear combinations of convex functions could also be used, but this quadratic function is the simplest.

Additionally, in order to guarantee that $\Psi(\phi)$ attains its global minimum $\Psi_{\min}$ in the hyperrectangle $H\phi = \{\phi_i : 0 \leq \underline{\phi}_i \leq \phi_i \leq \bar{\phi}_i \leq 2\pi\}$, the additional set of constraints are imposed on the coefficients of $\Psi(\phi)$:

$$c_i + \underline{\phi}_i d_i \leq 0 \text{ and } c_i + \bar{\phi}_i d_i \geq 0 \text{ for } i=1,\ldots,\tau. \tag{4}$$

Note that the satisfaction of (4) implies that $c_i \leq 0$ and $d_i \geq 0$ for $i=1,\ldots,\tau$.

## 7. Constructing the Underestimating Function

The unknown coefficients $c_i$, $i=0,\ldots,\tau$, and $d_i$, $i=1,\ldots,\tau$, can be determined by a linear program which may be considered to be in the dual form. For reasons of efficiency, the equivalent primal of this problem is actually solved, as described below. The solution to this primal linear program provides an optimal dual vector, which immediately gives the underestimating function coefficients $c_i$ and $d_i$. Since the convex quadratic function $\Psi(\phi)$ gives a global approximation to the local minima of $F(\phi)$, then its easily computed global minimum function value $\Psi_{\min}$ is a good candidate for an approximation to the global minimum of the correct energy function $F(\phi)$.

An efficient linear programming formulation and solution satisfying (2), (3), and (4) will now be summarized. Let $f^{(j)} = F(\phi^{(j)})$, for $j=1,...,k$, and let $f \in \mathbf{R}^k$ be the vector with elements $f^{(j)}$. Also let $\omega^{(j)} \in \mathbf{R}^\tau$ be the vector with elements $1/2(\phi^{(j)}{}_i)^2$, $i=1,...,\tau$, and $e_k \in \mathbf{R}^k$ be the vector of ones. Now define the following two matrices $\Phi \in \mathbf{R}^{(\tau+1)\times k}$ and $\Omega \in \mathbf{R}^{\tau\times k}$:

$$(5) \qquad \Phi = \begin{bmatrix} e_k^T \\ \phi^{(1)}\,\phi^{(2)}\dots\phi^{(k)} \end{bmatrix}, \quad \Omega = \begin{bmatrix} \omega^{(1)}\,\omega^{(2)}\dots\omega^{(k)} \end{bmatrix}.$$

Finally, let $c \in \mathbf{R}^{\tau+1}$, $d \in \mathbf{R}^\tau$, and $\delta \in \mathbf{R}^k$ be the vectors with elements $c_i$, $d_i$, and $\delta_i$, respectively. Then (2), (3), and (4) can be restated as the linear program (with variables $c$, $d$, and $\delta$):

$$\underset{c,\,d,\,\delta}{\text{minimize}} \quad e_k^T\delta$$

$$(6) \qquad \text{subject to} \begin{bmatrix} \Phi^T & \Omega^T & 0 \\ -\Phi^T & -\Omega^T & -I_k \\ I'_\tau & \underline{D} & 0 \\ -I'_\tau & -\bar{D} & 0 \end{bmatrix} \begin{bmatrix} c \\ d \\ \delta \end{bmatrix} \le \begin{bmatrix} f \\ -f \\ 0 \\ 0 \end{bmatrix}$$

where $\underline{D} = diag(\underline{\phi}_1,\underline{\phi}_2,...,\underline{\phi}_\tau)$, $\bar{D} = diag(\bar{\phi}_1,\bar{\phi}_2,...,\bar{\phi}_\tau)$, $I_k$ is the identity matrix of order $k$, and $I'_\tau$ is the $\tau$x$(\tau+1)$ "augmented" matrix $[0 : I_\tau]$ where $I_\tau$ is the identity matrix of order $\tau$.

Since the matrix in (6) has more rows than columns $2\cdot(k+\tau)$ rows and $k+2\tau+1$ columns, where $k \ge 2\tau+1$), it is computationally more efficient to consider it as a dual problem, and to solve the equivalent primal. After some simple transformations, this primal problem reduces to:

$$\underset{y_1,\,y_2,\,y_3}{\text{minimize}} \quad f^Ty_1 - f^Te_k$$

$$(7) \qquad \text{subject to} \begin{bmatrix} \Phi & I'^T_\tau & -I'^T_\tau \\ \Omega & \underline{D} & -\bar{D} \end{bmatrix} \begin{bmatrix} y_1 \\ y_2 \\ y_3 \end{bmatrix} = \begin{bmatrix} \Phi e_k \\ \Omega e_k \end{bmatrix}, \quad y_1, y_2, y_3 \ge 0$$

which has only $2\tau+1$ rows and $k+2\tau \ge 4\tau+1$ columns, and the obvious initial feasible solution $y_1 = e_k$ and $y_2 = y_3 = 0$. Furthermore, since the first of the $2\tau+1$ constraints in (7) in fact requires that $e_k^Ty_1 = 1$, then the function $f^Ty_1 - f^Te_k$ is also bounded below, and so this primal linear program always has an optimal solution. This optimal solution gives the values of $c$, $d$, and $\delta$ via the dual vectors, and also determines which values of $f^{(j)}$ are interpolated by the potential function $\Psi(\phi)$. That is, the basic columns in the optimal solution to (7) correspond to the conformations $\phi^{(j)}$ for which $f^{(j)} = \Psi(\phi^{(j)})$.

Note that once an optimal solution to (7) has been obtained, the addition of new local minima is very easy. It is done by simply adding new columns to $\Phi$ and $\Omega$, and therefore to the constraint matrix in (7). The number of primal rows remains fixed at $2\tau+1$, independent of the number $k$ of local minima.

## 8. Iterative Approximation to the Global Minimum

The convex quadratic underestimating function $\Psi(\phi)$ determined by the values $c \in \mathbf{R}^{\tau+1}$ and $d \in \mathbf{R}^{\tau}$ now provides a global approximation to the local minima of $F(\phi)$, and its easily computed global minimum point $\phi_{\min}$ is given by $(\phi_{\min})_i = -c_i/d_i$, $i=1,\ldots,\tau$, with corresponding function value $\Psi_{\min}$ given by $\Psi_{\min} = c_0 - \sum_{i=1}^{\tau} c_i^2/(2d_i)$ . The value $\Psi_{\min}$ is a good candidate for an approximation to the global minimum of the correct energy function $F(\phi)$, and so $\phi_{\min}$ can be used as an initial starting point around which additional configurations (i.e. local minima) should be generated. These local minima are added to the constraint matrix in (7) and the process is repeated. Before each iteration of this process, it is necessary to reduce the volume of the hyperrectangle $H\phi$ over which the new configurations are produced so that a tighter fit of $\Psi(\phi)$ to the local minima "near" $\phi_{\min}$ is constructed.

The rate and method by which the hyperrectangle size is decreased, and the number of additional local minima computed at each iteration will be determined by computational testing. But clearly the method depends most heavily on computing local minima quickly and on solving the resulting linear program efficiently to determine the approximating function $\Psi(\phi)$ over the current hyperrectangle.

If $E_c$ is a cutoff energy, then one means for decreasing the size of the hyperrectangle $H\phi$ at any step is to let $H\phi = \{\phi\colon \Psi(\phi) \le E_c\}$. To get the bounds of $H\phi$, consider $\Psi(\phi) \le E_c$ where $\Psi(\phi)$ satisfies (3). Then limiting $\phi_i$ requires that

$$\sum_{i=1}^{\tau}\left(c_i\phi_i + \frac{1}{2}d_i\phi_i^2\right) \le E_c - c_0 .$$

As before, the minimum value of $\Psi(\phi)$ is attained when $\phi_i = -c_i/d_i$, $i=1,\ldots,\tau$. Assigning this minimum value to each $\phi_i$, except $\phi_k$, then results in

$$c_k\phi_k + \frac{1}{2}d_k\phi_k^2 \le E_c - c_0 + \frac{1}{2}\sum_{i \ne k} c_i^2/d_i \equiv \beta_k .$$

The lower and upper bounds on $\phi_k$, $k=1,\ldots,\tau$, are given by the roots of the quadratic equation

$$c_k\phi_k + \frac{1}{2}d_k\phi_k^2 = \beta_k . \tag{8}$$

Hence, these bounds can be used to define the new hyperrectangle $H\phi$ in which to generate new configurations.

Clearly, if $E_c$ is reduced, the size of $H\phi$ is also reduced. At every iteration the predicted global minimum value $\Psi_{\min}$ satisfies $\Psi_{\min} \le F(\phi^*)$, where $\phi^*$ is the smallest known local minimum conformation. Therefore, $E_c = F(\phi^*)$ is often a good choice. If at least one improved point $\phi$, with $F(\phi) < F(\phi^*)$, is obtained in each iteration, then the search domain $H\phi$ will strictly decrease at each iteration, and may decrease substantially in some iterations.

## 9. The Algorithm

Based on the results of the previous three sections, a general method for computing the global, or near global, energy minimum of the potential energy function $U$ can now be described. Recall that $\phi \in \mathbf{R}^\tau$ where $\tau$ represents the number of torsion angles, and the bond lengths and bond angle are assumed to be fixed for the purposes of the global underestimator.

1. Compute $k \geq 2\tau+1$ distinct local minima $\phi^{(j)}$, for $j=1,...,k$, of the function $F(\phi)$.
2. Compute the convex quadratic underestimator function

$$\Psi(\phi) = c_0 + \sum_{i=1}^{\tau} \left( c_i\phi_i + \frac{1}{2}d_i\phi_i^2 \right)$$

by solving the linear program

$$\underset{y_1,\, y_2,\, y_3}{\text{minimize}} \quad f^T y_1 - f^T e_k$$

$$\text{subject to} \quad \begin{bmatrix} \Phi & I'^T_\tau & -I'^T_\tau \\ \Omega & D & -D \end{bmatrix} \begin{bmatrix} y_1 \\ y_2 \\ y_3 \end{bmatrix} = \begin{bmatrix} \Phi e_k \\ \Omega e_k \end{bmatrix}, \quad y_1, y_2, y_3 \geq 0$$

The optimal solution to this linear program gives the values of $c$ and $d$ via the dual vectors.

3. Compute the predicted global minimum point $\phi_{\min}$ given by $(\phi_{\min})_i = -c_i/d_i$, $i=1,...,\tau$, with corresponding function value $\Psi_{\min}$ given by $\Psi_{\min} = c_0 - \sum_{i=1}^{\tau} c_i^2/(2d_i)$ .
4. If $\phi_{\min} = \phi^*$, where $\phi^* = \text{argmin}\{F(\phi^{(j)}), j=1,2,...\}$ is the best local minimum found so far, then stop and report $\phi^*$ as the approximate global minimum conformation.
5. Reduce the volume of the hyperrectangle $H\phi$ over which the new configurations will be produced, and remove all columns from $\Phi$ and $\Omega$ which correspond to the conformations which are excluded from $H\phi$.
6. Use $\phi_{\min}$ as an initial starting point around which additional local minima $\phi^{(j)}$ of $F(\phi)$ (restricted to the hyperrectangle $H\phi$) are generated. Add these new local minimum conformations as columns to the matrices $\Phi$ and $\Omega$.
7. Return to step 2.

The number of new local minima to be generated in step 6 is unspecified since there is currently no theory to guide this choice. In general, a value exceeding $2\tau+1$ would be required for the construction of another convex quadratic underestimator in the next iteration (step 2). In addition, the means by which the volume of the hyperrectangle $H\phi$ is reduced in step 5 may vary. One such method, presented in section 8, would use the two roots of (8) to define the new bounds of $H\phi$. Another method would be simply to use $H\phi = \{\phi_i : (\phi_{\min})_i - \delta_i \leq \phi_i \leq (\phi_{\min})_i + \delta_i, i=1,...,\tau\}$ where $\delta_i = |(\phi_{\min})_i - (\phi^*)_i|$, $i=1,...,\tau$.

## 10. Computational Results

The computational results presented in this section were obtained on the Cray Y-MP C90 and Cray T3D systems (at the Minnesota Supercomputer Institute) using the network PVM message passing system for communication between the C90 and T3D. Steps 1 and 6 of the algorithm (presented in section 9) were performed in parallel on 8 processors of the T3D, while the remaining steps of the algorithm were done sequentially on the C90.

Using this system, solutions were obtained for the $n$-chain hydrocarbon problem with lengths 4 through 22, 25, and 30. As an example, a solution for the 7 bead hydrocarbon heptane was obtained in only 8 seconds after 3 iterations (a total of 180 local minimizations were performed). Table 10.1 summarizes the minimization of the potential function for heptane (an entry of NC indicates "no change" between the predicted and computed torsion angles in that step). Notice that the first iteration reports the current "best" conformation (corresponding to the lowest potential energy value) to be approximately the set of torsion angles (180°,180°,300°,180°). Recall that the bond lengths and bond angles are

**Table 10.1 Solution Summary for Heptane**

| | Initial Position, $\phi^{(0)}$ | Randomize over Radius | New Global Minimum (25 random points) | Predicted Global Minimum | Randomize over Radius | New Global Minimum (20 random points) | Predicted Global Minimum | Randomize over Radius | New Global Minimum (10 random points) | Predicted Global Minimum |
|---|---|---|---|---|---|---|---|---|---|---|
| $\varphi_1$ | 20 | 180 | 180.12 | 195.45 | 20 | 180.01 | NC | 4 | 180.02 | NC |
| $\varphi_2$ | 330 | 180 | 179.53 | 187.96 | 15 | 179.99 | NC | 4 | 179.99 | 179.88 |
| $\varphi_3$ | 278 | 180 | 297.50 | 163.06 | 140 | 180.06 | 180.14 | 5 | 180.12 | NC |
| $\varphi_4$ | -60 | 180 | 179.63 | 195.98 | 20 | 179.97 | 179.88 | 5 | 179.91 | NC |
| F($\phi$) | | | -.2967 | | | -.3398 | | | -.3398 | |
| | | | | | | | | | $\phi$* | |

considered fixed for the purposes of the global underestimator. However, the global underestimator predicts a conformation close to (180°, 180°, 180°, 180°). After the second iteration, with H$\phi$ defined by $\{175.45° \le \phi_1 \le 215.45°, 172.96° \le \phi_2 \le 203.96°, 23.06° \le \phi_3 \le 303.06°, 175.98° \le \phi_4 \le 215.98°\}$, we find that a better function value can be realized at this conformation. The underestimator and the current global minimum now reside at almost identical points. The third iteration confirms (180°, 180°, 180°, 180°) as the global minimum. This result has been shown through enumeration to be the correct global minimum conformation for the given potential energy function (the solution is shown in Figure 5.1).

As the problem size $n$ increases, the number of local minima increases exponentially at the rate of $O(3^n)$ as can be seen in Figure 10.1. With exponential increases in the number of local minima, exhaustive searches for the global minimum quickly become computationally intractable. The time required for exhaustive techniques is driven by the number of local minimizations needed to generate and "relax" each possible conformation. For example, there are $3^{12}$ = 531,441 possible conformations for the 15 bead problem, and hence, to test every possible local minimum conformation is expected to take over 100 hours on a Cray Y-MP C90!

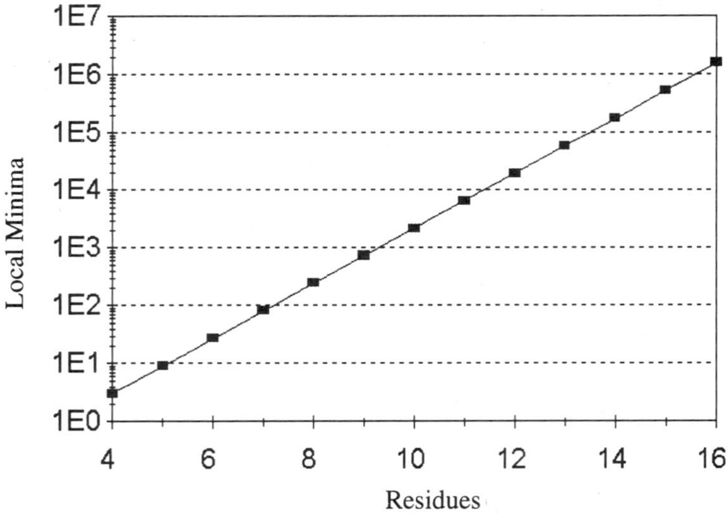


**Figure 10.1 Total Number of Local Minima as a Function of the Number of Residues**

Through the use of the global underestimator we can avoid such prohibitive numbers of calculations; however, the underestimator still remains dependent on the accumulation of a large number of local minima. In contrast to enumeration, the underestimator method requires a much less dramatic increase in the number of computations. Since the total number of underestimation steps needed to find a global minimum remains close to constant (observed experimentally to be less than 10), and each step requires $O(n)$ local minima to form the underestimator, a linear increase in number of local minimizations is expected. In addition, since every function evaluation requires $O(n^2)$ time while every gradient evaluation requires $O(n^3)$ time, due to the Lennard-Jones term, the underestimator method produces a final prediction in $O(n^4)$ time on average. The effect of increases in problem size on the total time for computing the global minimum can be seen in Figure 10.2. The time required to compute the global minimum as a function of the number of beads $n$ can be approximated (using a least squares fit) by the function $T(n) = 0.1311\, n^4 - 5.843\, n^3 + 94.82\, n^2 - 623.9\, n + 1377$ seconds, and therefore increases at an average rate of $O(n^4)$ as expected. In fact, the 22 bead problem took 36 minutes for the final prediction (Figure 10.3). This prediction was made with nine iterations of the global underestimator, and a total of 1950 local minimizations were performed. Further increases in problem size will face even larger increases in execution time: a 50 bead problem is roughly predicted to require 8 hours on the Cray Y-MP C90 / Cray T3D system (using all 64 processors). The global energy minimum conformations with corresponding global minimum function values, for $n = 16$ through $n = 22$, are given in Appendix B (Section 13).

As an alternate means for observing the time complexity of this algorithm, Figure 10.4 shows the solution time *per iteration* required to obtain the global energy minimum. That is, the total solution time is normalized by the number of iterations. Since the number of iterations has been observed to be less than 10 in all cases, the time complexity in this case is also $O(n^4)$. In this case, the function $\overline{T}(n) = 0.00880\, n^4 - 0.361\, n^3 + 5.66\, n^2 - 36.4\, n + 80.7$ seconds approximates this curve.

Finally, it must be noted that although this method is not *guaranteed* to find the global minimum conformation of the potential function $U$, it tends to perform reasonably well on

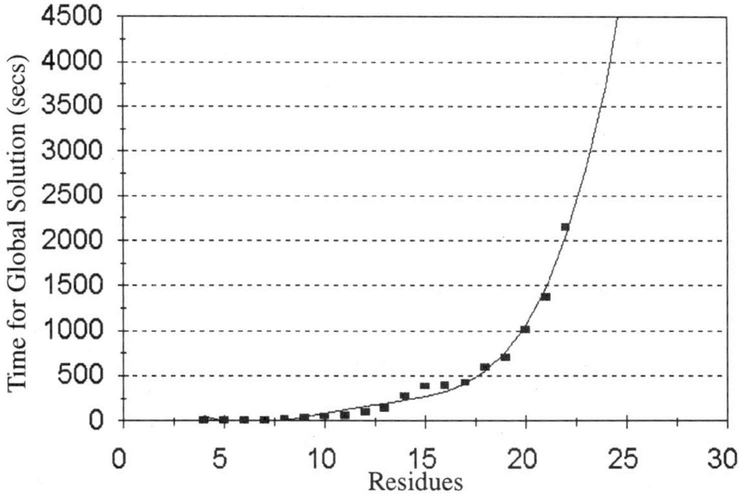


**Figure 10.2 Total CPU Time for Global Minimum Computation as a Function of the Number of Residues**

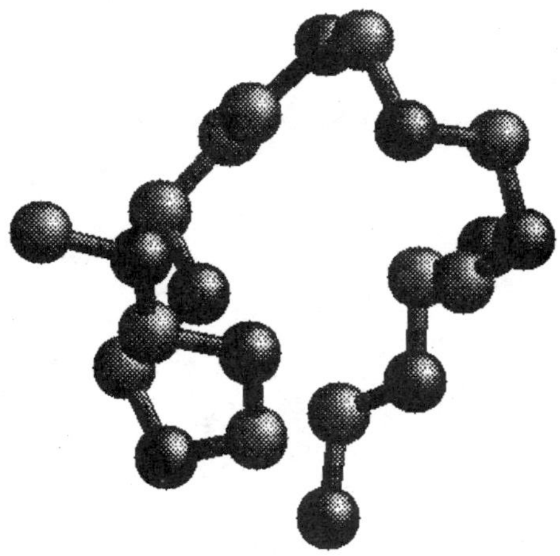

**Figure 10.3 Global Energy Minimum for 22 Bead Hydrocarbon**

problems of moderate size ($n \leq 30$). In fact, based on the computational results obtained for $n = 4$ though $n = 22$, the computed global minimum energy for the potential function $U$, as a function of the number of beads $n$, can be approximated (using a least squares fit) by the function $E(n) = 1.24 - 0.118\, n - 0.0202\, n^2$. Figure 10.5 shows this function superimposed on the computed global minima obtained by the algorithm of section 9 for all hydrocarbon chains of size $n \leq 22$ and also for $n = 25$ and $n = 30$. Notice that the function $E(n)$ does indeed correctly predict the global minimum energy for $n = 25$ and $n = 30$, the two largest problems. This important property permits estimation of the global minimum energy for larger molecular chains, and can also be used to accelerate the global minimization algorithm. Accordingly, the predicted global minimum energy for an $n = 50$ bead conformation is expected to obtain a global minimum energy of approximately -55.14 kcal. If

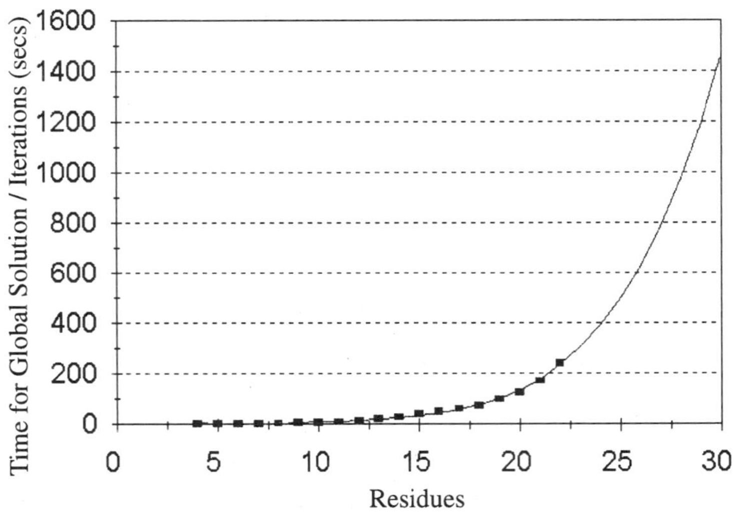


**Figure 10.4 Total CPU Time for Global Minimum Computation per Iteration as a Function of the Number of Residues**

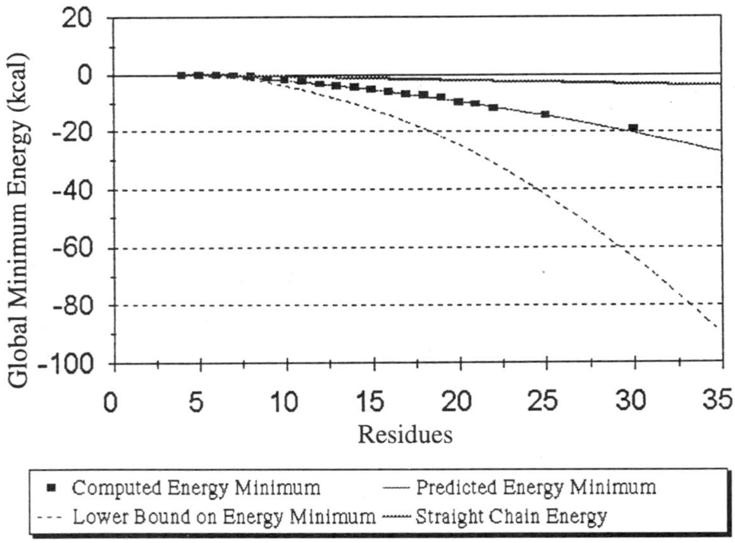


**Figure 10.5 Computed Global Energy Minima as a Function of the Number of Residues**

we define $f_0 = -\sum \varepsilon_{i,l}$, where the sum is computed over all pairs ($a_i$,$a_l$) of beads separated by more than two along the primary chain, then $f_0$ represents a lower bound on the potential function $U$ from (1). Figure 10.5 also shows $f_0$ (for $n$-chain hydrocarbons), for comparison with the computed energy minima, which is given by $f_0(n) = -1.05855 + 0.62445\, n - 0.0905\, n^2$, and the straight chain potential energy value (all torsions $\phi_i = 180°$) which can be regarded as a simple upper bound. Finally, it should be emphasized that while the minimum energy is a smooth function of the chain length, the corresponding minimum energy conformations (as illustrated in Appendix B) may differ substantailly as the chain length increases.

## 11. Conclusions

A convex quadratic global underestimator method was used to predict stable molecular conformations for $n$-chain homopolymers (hydrocarbons) of length 4 through 22, 25, and 30. For the convex global underestimator approach to realize its full potential, not only must massively parallel machines be used effectively to reduce the time required to compute the large number of local minima, but also, since the local minimization procedures rely on many function evaluations, improved methods for computing the potential function must be designed. Such improvements in the function evaluation time (including the conversions) would be of tremendous benefit, and perhaps will provide the key to reducing the local minimization times. Recent work in [6] has attempted to specifically address this problem.

## 12. Appendix A: Transformations and Analytic Derivatives

If the three dimensional cartesian position of bead $a_m$, $m=1,\ldots,n$, is represented by $(x_m, y_m, z_m) \in R^3$, then the following sequence of matrix products is sufficient to convert from internal molecular coordinates to the cartesian representation, and is based solely on the internal molecular coordinates of the previous $m$-1 beads in the chain:

$$\begin{bmatrix} x_m \\ y_m \\ z_m \\ 1 \end{bmatrix} = B_1 B_2 \ldots B_m \begin{bmatrix} 0 \\ 0 \\ 0 \\ 1 \end{bmatrix} \tag{9}$$

where the transformation matrices $B_i \in R^{4\text{x}4}$, $i=1,\ldots,m$, were first defined in [11] to be of the form

$$B_1 = \begin{bmatrix} 1 & 0 & 0 & 0 \\ 0 & 1 & 0 & 0 \\ 0 & 0 & 1 & 0 \\ 0 & 0 & 0 & 1 \end{bmatrix}, B_2 = \begin{bmatrix} -1 & 0 & 0 & -l_2 \\ 0 & 1 & 0 & 0 \\ 0 & 0 & -1 & 0 \\ 0 & 0 & 0 & 1 \end{bmatrix}$$

$$B_i = \begin{bmatrix} -\cos\theta_i & -\sin\theta_i & 0 & -l_i\cos\theta_i \\ \sin\theta_i\cos\varphi_i & -\cos\theta_i\cos\varphi_i & -\sin\varphi_i & l_i\sin\theta_i\cos\varphi_i \\ \sin\theta_i\cos\varphi_i & -\cos\theta_i\sin\varphi_i & \cos\varphi_i & l_i\sin\theta_i\sin\varphi_i \\ 0 & 0 & 0 & 1 \end{bmatrix} \tag{10}$$

for $i=3,\ldots,m$. Hence, in order to compute the cartesian positions of each of the $n$ beads $a_1$, $a_2$, ..., $a_n$, an O($n$) algorithm involving only 4x4 matrix products is sufficient.

For an $n$ bead molecular structure, the potential function $U$ depends on the $3n$ - 6 independent variables $l_k$, $k=2,\ldots,n$, $\theta_k$, $k=3,\ldots,n$, and $\varphi_k$, $k=4,\ldots,n$. Computing the gradient of the function $U$, with respect to the internal coordinates, will therefore require an analytic formulation for $\partial U/\partial\varphi_k$, $\partial U/\partial\theta_k$, and $\partial U/\partial l_k$. Each of these can be computed using a sequence of matrix products in a manner similar to the transformations above. To compute each, first notice from (1) that

$$\frac{\partial U}{\partial \varphi_k} = -\frac{\bar{n} V_{\bar{n}}}{2} \sin(\bar{n}\varphi_k + \delta) + \sum_{i,l} \frac{12\varepsilon_{i,l}}{r_{i,l}} \left( \left( \frac{\sigma_{i,l}}{r_{i,l}} \right)^{12} + \left( \frac{\sigma_{i,l}}{r_{i,l}} \right)^{6} \right) \frac{\partial r_{i,l}}{\partial \varphi_k} \tag{11}$$

$$\frac{\partial U}{\partial \theta_k} = k_\theta (\theta_k - \theta_0) + \sum_{i,l} \frac{12\varepsilon_{i,l}}{r_{i,l}} \left( \left( \frac{\sigma_{i,l}}{r_{i,l}} \right)^{12} + \left( \frac{\sigma_{i,l}}{r_{i,l}} \right)^{6} \right) \frac{\partial r_{i,l}}{\partial \theta_k} \tag{12}$$

$$\frac{\partial U}{\partial l_k} = k_l (l_k - l_0) + \sum_{i,l} \frac{12\varepsilon_{i,l}}{r_{i,l}} \left( \left( \frac{\sigma_{i,l}}{r_{i,l}} \right)^{12} + \left( \frac{\sigma_{i,l}}{r_{i,l}} \right)^{6} \right) \frac{\partial r_{i,l}}{\partial l_k}. \tag{13}$$

The dependence of $r_{i,l}$ on $\varphi_k$, $\theta_k$, and $l_k$ is given by the sequence of 4x4 matrix products in (9). To analytically compute the terms $\partial r_{i,l}/\partial \varphi_k$, $\partial r_{i,l}/\partial \theta_k$, and $\partial r_{i,l}/\partial l_k$, first define $w_m = (x_m, y_m, z_m, 1)^T$ and notice that $r_{i,l}^2 = (w_i - w_l)^T (w_i - w_l) = v_{i,l}^T v_{i,l}$ where

$$v_{i,l} = w_i - w_l = \left( \prod_{j=1}^{i} B_j \right) \left( I - \prod_{j=i+1}^{l} B_j \right) e_4 \tag{14}$$

with $e_4 = (0,0,0,1)^T$. Letting

$$B_{i,l} = \left( \prod_{j=1}^{i} B_j \right) \left( I - \prod_{j=i+1}^{l} B_j \right) \tag{15}$$

then $v_{i,l} = B_{i,l} e_4$ and $r_{i,l}^2 = (B_{i,l} e_4)^T (B_{i,l} e_4)$. Therefore,

$$\frac{\partial r_{i,l}}{\partial \varphi_k} = \frac{1}{2 r_{i,l}} \frac{\partial r_{i,l}^2}{\partial \varphi_k} = \frac{e_4^T B_{i,l}^T}{r_{i,l}} \left( \frac{\partial B_{i,l}}{\partial \varphi_k} \right) e_4$$

and similarly,

$$\frac{\partial r_{i,l}}{\partial \theta_k} = \frac{e_4^T B_{i,l}^T}{r_{i,l}} \left( \frac{\partial B_{i,l}}{\partial \theta_k} \right) e_4$$

$$\frac{\partial r_{i,l}}{\partial l_k} = \frac{e_4^T B_{i,l}^T}{r_{i,l}} \left( \frac{\partial B_{i,l}}{\partial l_k} \right) e_4 .$$

Finally, the dependence of $B_{i,l}$ on $\varphi_k$ ($k = 4,\ldots,n$), $\theta_k$ ($k = 3,\ldots,n$), and $l_k$ ($k = 2,\ldots,n$) can be expressed by the following three cases

$$\frac{\partial B_{i,l}}{\partial \alpha_k} = \begin{cases} \underline{0} & l < k \\ -\left( \prod_{j=1}^{k-1} B_j \right) \frac{\partial B_k}{\partial \alpha_k} \left( \prod_{j=k+1}^{l} B_j \right) & i < k \le l \\ \left( \prod_{j=1}^{k-1} B_j \right) \frac{\partial B_k}{\partial \alpha_k} \left( \prod_{j=k+1}^{i} B_j \right) \left( I - \prod_{j=i+1}^{l} B_j \right) & 1 \le k \le i \end{cases} \tag{16}$$

where $\alpha_k$ represents any one of the three parameters $\varphi_k$, $\theta_k$, or $l_k$, and $\underline{0}$ is the 4x4 zero matrix. Note that whereas computing the potential function $U$ requires $O(n^2)$ steps (due to the pairwise Lennard-Jones terms), the analytic evaluation of the gradient requires $O(n^4)$ steps.

## 13. Appendix B: Minimum Energy Conformations for $n$ = 16 - 22.

**Table 13.1**

| Global Minimum Conformation | | | | | | |
|---|---|---|---|---|---|---|
| $n = 16$ | $n = 17$ | $n = 18$ | $n = 19$ | $n = 20$ | $n = 21$ | $n = 22$ |
| 173° | 172° | 181° | 296° | 175° | 295° | 180° |
| 188° | 189° | 176° | 180° | 65° | 181° | 182° |
| 63° | 62° | 293° | 172° | 195° | 177° | 295° |
| 59° | 59° | 292° | 296° | 67° | 291° | 159° |
| 187° | 186° | 165° | 292° | 65° | 180° | 291° |
| 60° | 60° | 294° | 152° | 195° | 290° | 296° |
| 63° | 64° | 193° | 290° | 63° | 292° | 172° |
| 60° | 60° | 166° | 172° | 174° | 171° | 295° |
| 187° | 186° | 61° | 178° | 177° | 292° | 183° |
| 59° | 58° | 197° | 63° | 56° | 298° | 181° |
| 63° | 62° | 65° | 65° | 184° | 297° | 298° |
| 188° | 192° | 66° | 178° | 61° | 160° | 177° |
| 173° | 168° | 193° | 67° | 64° | 299° | 297° |
| | 185° | 67° | 67° | 186° | 181° | 288° |
| | | 181° | 173° | 62° | 183° | 152° |
| | | | 182° | 69° | 65° | 293° |
| | | | | 185° | 181° | 177° |
| | | | | | 182° | 175° |
| | | | | | | 59° |
| Global Minimum Potential Energy | | | | | | |
| **-5.783** | **-6.397** | **-7.634** | **-8.025** | **-9.766** | **-10.44** | **-11.81** |

## 14. References

1. R.A. Abagyan, *Towards Protein Folding by Global Energy Optimization*, Federation of European Biochemical Societies: Letters **325** (1993), 17-22.
2. S.A. Benner and D.L. Gerloff, *Predicting the Conformation of Proteins: Man versus Machine*, Federation of European Biochemical Societies: Letters **325** (1993), 29-33.
3. D.A. Case, *Computer Simulations of Protein Dynamics and Thermodynamics*, IEEE: Computer Science & Engineering, October (1993), 47-57.
4. K.A. Dill, *Dominant Forces in Protein Folding*, Biochemistry **29** (1990), 7133-7155.
5. D.M. Ferguson, A. Marsh, T. Metzger, D. Garret, and K. Kastella, *Conformational Searches for the Global Minimum of Protein Models*, Journal of Global Optimization **4** (1994), 209-227.
6. L. Greengard, *The Rapid Evaluation of Potential Fields in Particle Systems*, The MIT Press, Cambridge, 1988.

7. A.L. Lehninger, *Biochemistry: The Molecular Basis of Cell Structure and Function*, Worth Publishers, New York, 1970.
8. K. Merz and S. Le Grand, *The Protein Folding Problem and Tertiary Structure Prediction*, Birkhauser, Boston, 1994.
9. F.M. Richards, *The Protein Folding Problem*, Scientific American (1991), 54-63.
10. S. Sun, *Reduced representation model of protein structure prediction: statistical potential and genetic algorithms*, Protein Science **2** (1993), 762-785.
11. H.B. Thompson, *Calculation of Cartesian Coordinates and Their Derivatives from Internal Molecular Coordinates*, The Journal of Chemical Physics **47** (1967), 3407-3410.
12. V.H. Walke, *Computational Solutions to the Protein Folding Problem*, Trident Scholar Report No. 223, U.S. Naval Academy, Annapolis, Maryland (1994).

A.T. PHILLIPS, COMPUTER SCIENCE DEPARTMENT, UNITED STATES NAVAL ACADEMY, ANNAPOLIS, MD 21402

J.B. ROSEN, COMPUTER SCIENCE AND ENGINEERING DEPARTMENT, UNIVERSITY OF CALIFORNIA AT SAN DIEGO, SAN DIEGO, CA 92093

V.H. WALKE, COMPUTER SCIENCE DEPARTMENT, UNIVERISTY OF TEXAS AT AUSTIN, AUSTIN, TX 78712

DIMACS Series in Discrete Mathematics
and Theoretical Computer Science
Volume **23**, 1996

# Thermodynamics and kinetics of protein folding

Andrej Šali, Eugene Shakhnovich, and Martin Karplus

ABSTRACT. The number of all possible conformations of a polypeptide chain is too large to be sampled exhaustively. Nevertheless, protein sequences do fold into unique native states in seconds (Levinthal paradox). To determine how the Levinthal paradox is resolved, we use a lattice Monte Carlo model in which the global minimum (native state) is known. The necessary and sufficient condition for folding in this model is that the native state be a pronounced global minimum on the potential surface. This guarantees thermodynamic stability of the native state at a temperature where the chain does not get trapped in local minima. Folding starts by a rapid collapse from a random-coil state to a random semi-compact globule. It then proceeds by a slow, rate-determining search through the semi-compact states to find a transition state close to the native state from which the chain folds rapidly to the native state. The elements of the folding mechanism that lead to the resolution of the Levinthal paradox are the reduced number of conformations that need to be searched in the semi-compact globule ($\approx 10^{10}$ *versus* $\approx 10^{16}$ for the random coil) and the existence of many ($\approx 10^3$) transition states. The results have evolutionary implications and suggest principles for the folding of real proteins.

## 1. Introduction

The mechanism of protein folding is not understood, despite many studies devoted to this subject [1–4]. The essential question is how a polypeptide chain is able to fold rapidly, in ms to s, to the stable native state in spite of the very large number of possible conformations that exist for the chain (Levinthal paradox) [5]. The mechanisms of protein folding are proposed on the basis of conceptual discussions [6–9], simulations [10–14], and theory [15–17]. Theories of protein folding were recently reviewed in refs. [1, 3, 4, 18–27].

While the dynamics of the native state can be characterized relatively well by molecular dynamics simulations [28], much less is known about the potential surface

1991 *Mathematics Subject Classification.* Primary 65K10, 90C30; Secondary 92C05, 92E10.

*Key words and phrases.* Protein folding, mechanism, kinetics, thermodynamics, global minimization, energy landscape.

A.Š. was a Fellow of The Jane Coffin Childs Memorial Fund for Medical Research. This investigation has been aided by a grant from The Jane Coffin Childs Memorial Fund for Medical Research (A.Š.), by David and Lucille Packard Fellowship (E.S.), and by a grant from the National Science Foundation and a gift from Molecular Simulations Inc. (M.K.).

governing the non-native portion of conformation space that is involved in protein folding. It includes a wide range of structures which may differ by tens of angstroms. Concomitantly, instead of the time scale of picoseconds to nanoseconds that is required for exploring the neighborhood of the native state, the characteristic times corresponding to the motions in the full conformation space are in the nanosecond to second range. The existence of such a separation of time and length scales with fast local motions and slow large-scale motions makes it possible to introduce two simplifying concepts, which can serve as a basis for theoretical work on protein folding. The first simplification is an effective potential or potential of mean force and the second is a discretized description of the polypeptide chain. Both of these concepts are based on the idea of "preaveraging" the small-scale motions to obtain a "coarse grained" model, which can treat a molecule on the time and length scales at which protein folding occurs. This leads to simplified models of proteins that include only a subset of atoms [29]. and to discretized conformational space of various lattice models [30–32] that employ Monte Carlo (MC) dynamics to simulate the kinetics.

Recently, lattice models have been used to address a variety of aspects of the protein folding problem [9, 10, 12, 33–44].

In one particularly simple class of lattice models, the protein chain is represented as a string of beads on the 2D square lattice [32, 45] or 3D cubic lattice [16]. These models are generally not meant to simulate folding of any particular real amino acid sequence. However, the lattice models do capture the most essential features of proteins, including their heteropolymeric nature and the interactions between sequentially local and distant amino acid residues. Thus, the simple lattice models may be suitable for elucidating overall features of the folding process such as the main stages of folding. The advantages of simple lattice models are that simulations of many sequences are possible and that frequently the global energy minimum can be determined by enumeration.

As the model does not include side chains, the "native" state in the lattice model corresponds to a compact globule with the native fold of a real protein. Such structured globules may correspond to the experimentally observed molten globules, which (although expanded relative to the native state) preserve much of the backbone structure, and whose side chains are free to undergo dihedral angle transitions [46–49]. Molten globules appear to be a late stage in the folding of some proteins and their formation involves resolution of the Levinthal paradox [47, 49, 50]. For further discussion of the suitability of lattice models to describe real proteins and of MC to represent molecular dynamics see refs. [4, 37, 51, 52]. The purpose of this study is to examine some attributes of the kinetics of protein folding by the use of simple lattice models. For global optimization techniques, the reader is referred to a review paper by Pardalos and coworkers [53].

In this review, we summarize our results obtained by the use of MC simulations of 27-mer heteropolymers on a cubic lattice. We identified the features of the folding sequences that allow them to fold rapidly into a unique and stable native state [51] and the mechanism by which they achieve this state [12, 54].

## 2. Methods

We use a simplified model that consists of a 27-bead self-avoiding chain on a cubic lattice [10] (Figures. 1, 2). The native (lowest-energy) state can be determined exactly [10] and a survey of the folding behavior of many sequences is possible [51]. In addition, the full phase space density of the system (Figure 6) [12, 54] can be obtained and the thermodynamic properties can be calculated as a function of the folding "reaction coordinate" (Figure 7) [12, 54]. The model is sufficiently complex that the Levinthal paradox exists: *i.e.*, some sequences find the native state in only $\approx 10^7$ MC (MC) steps even though there are $\approx 10^{16}$ possible conformations (Figure 6b).

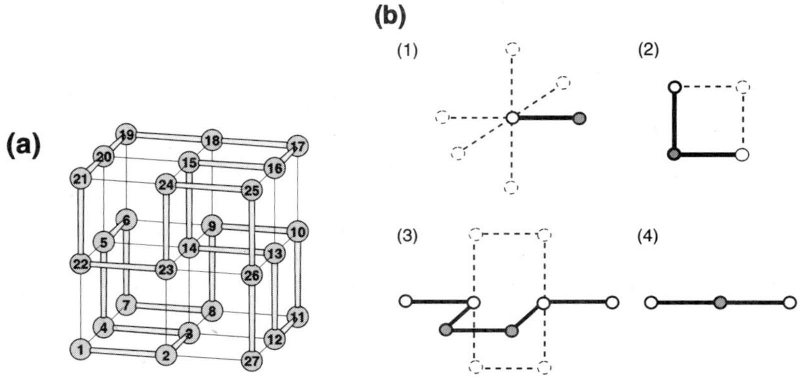


FIGURE 1. Lattice model of protein folding. *a*, An example of a compact self-avoiding structure of a chain of 27 monomers (filled numbered circles) with 28 contacts (thin lines). The structure shown is the native state of the folding random sequence 43 used as an example in this paper. The total energy of a conformation is the sum of contact energies: $E = \sum_{i<j} \Delta(r_i, r_j)B_{ij}$, where $r_i$ are the positions of monomers $i$, $B_{ij}$ are the contact energies for pairs of monomers $i, j$, and $\Delta(r_i, r_j)$ is 1 if monomers $i$ and $j$ are in contact and is 0 otherwise; two monomers are in contact if they are not successive in sequence and are at unit distance from each other. The values of the $B_{ij}$ are obtained from a Gaussian distribution with a mean $B_o$ and standard deviation $\sigma_B$. This particular model for $B_{ij}$ corresponds to a heteropolymer with a random sequence of monomers of many different types whose heterogeneity is measured by $\sigma_B$ [15]. The parameter $B_o$ is an overall attractive term that emulates the hydrophobic effect observed in globular proteins. A particular *sequence* is defined by the matrix of contact energies $B_{ij}$. In terms of their magnitude and standard deviation, the $B_{ij}$'s correspond to the contact energies in real proteins [12], such as those described by Miyazawa and Jernigan [55]. The *native* conformation is the compact self-avoiding chain with the lowest energy. *b*, The three types of possible MC moves (1–3). Situation 4 shows a conformation where no move of the central monomer is possible. The current conformation is shown in thick lines. Possible new conformations are shown in dashed lines. A move is possible if all new positions are unoccupied. The monomers that are being moved are shown in dark gray.

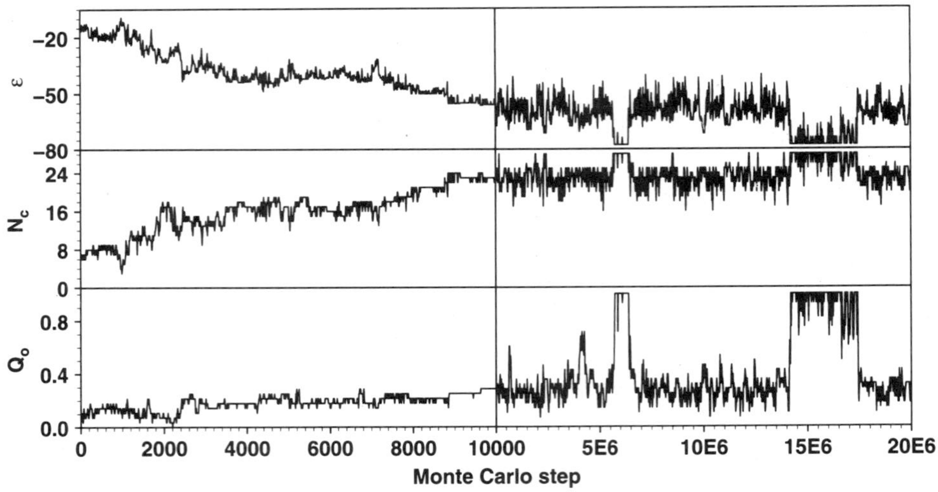


FIGURE 2. Typical trajectory for a folding random sequence ($T = 1.3$). Energy, $\epsilon$ (in units of $k_BT$); the number of contacts, $N_c$; fraction of the number of contacts in common with the native state, $Q_o$. The instantaneous values of these quantities are plotted every 10 MC steps in the first part of the trajectory ($\leq$ 10000 MC steps) and every 20000 steps in the subsequent part. The folding trajectory starts with a random-coil conformation and consists of local MC moves described in Figure 1b [51]; one MC step corresponds to a move and a test of its acceptance with the Metropolis criterion [56]. In the first part of the trajectory, $\approx$ 50% of the MC moves are accepted, while only 5–10% of the moves are successful in the subsequent part of the simulation.

## 3. Results and Discussion

**3.1. A pronounced energy minimum is necessary and sufficient for rapid folding to a stable native state.** In the first part of the analysis, 200 sequences with random interactions were generated and subjected to MC folding simulations (Figure 3) [51]. Of these, 30 chains found the known native state in a short time. These chains correspond to actual protein sequences in the present model; the remaining sequences, which do not fold, do not correspond to protein sequences and serve as controls. The 30 folding sequences were analyzed and compared with the non-folding sequences. Several suggested mechanisms for resolving the Levinthal paradox do not apply to the present model; *i.e.*, the features assumed to be responsible for rapid folding in these mechanisms are found to be the same for the folding and non-folding sequences. These include a high number of short *versus* long range contacts in the native state [8], a high content of secondary structure in the native state [7], a strong correlation between the native contact map and the interaction parameters [30], and the existence of a high number of low energy states with near-native conformations [10]. Moreover, there is no repetitive trapping of the non-folding sequences in the same local minimum, so that the native state cannot be a metastable state [57]. The only significant difference between the folding and non-folding sequences is that the native state is at a pronounced energy minimum (Figure 3). This is the necessary and sufficient condition for a sequence to fold rapidly in the present model (Figure 4). It is necessary because no sequences without a pronounced energy minimum fold to the native state and it is sufficient because all sequences with such a minimum do fold.

An essential element of the study was to examine a large number of sequences and separate those that fold from those that do not. Consequently, a temperature at the neighbourhood of $T_m$, the mid-point of the folding transition, was used to speed up the reaction and save computer time (Figure 3). Optimization of the folding rate was important to verify that certain sequences did not fold. If folding to the native state were always possible under the simulation conditions, there would not have been any nonfolding sequences and our computer experiment would have failed. Figure 5 shows the temperature dependence of the folding rate over a range where 0 to 50% of the polymers are in the native state at equilibrium. Linear folding kinetics is observed throughout the temperature range; this justifies the use of a relatively high temperature for the folding experiments (Figure 3).

The reason for the correlation between folding and stability is that significant portions of the potential energy surface of the model system are "rugged". In particular, the random collapsed state that is sampled in the three-stage random search (3SRS) mechanism (see below) is a multiminimum surface on which the search for the native state requires surmounting many intervening barriers. This can be done on a reasonable time scale only if the folding temperature is sufficiently high for there to be a significant probability of overcoming such barriers. However, at a high temperature, the majority of the random sequences have a ground state that is not stable; in other words, the Boltzmann probability of being in any of the excited states is too large unless a sizeable energy gap separates the native state from the excited states. As the temperature at which the folding simulations are done is near the midpoint of the thermodynamic transition temperature between the native and denatured states, the simulation temperature is high enough to overcome the barriers only for the strongly folding sequences with a particularly stable ground state . The transition temperature for the nonfolding sequences is so low that the 27-mer gets trapped in a metastable well. This qualitative argument only explains why the pronounced energy minimum is necessary for folding. The explanation for why it is also sufficient is provided by the 3SRS mechanism discussed below. It is likely that the necessity of the energy gap condition is general for reasonable lattice models and for real proteins, while its sufficiency may be of more limited applicability.

The importance of the temperature in the protein folding reaction and the relation between an energy gap and folding, which are clearly demonstrated by the 27-mer simulations [12, 51], have been discussed previously. Based on statistical mechanical arguments and spin glass theory, Bryngelson and Wolynes [17, 58] suggested that there are two temperatures that need to be considered in determining the folding properties of a sequence. One is the folding temperature and the other is the glass transition temperature. They suggested that the glass transition temperature corresponds to the temperature below which the chain is frozen into a random low energy conformation because it does not have enough energy to overcome the barriers separating such conformations. On this basis, they concluded that the temperature at which the sequence folds must be higher than the glass transition temperature. Further, it was shown that random sequences do not satisfy this condition and so would be likely to be trapped in metastable states [17, 58]. Bryngelson and Wolynes introduced specific biases toward the native state to make

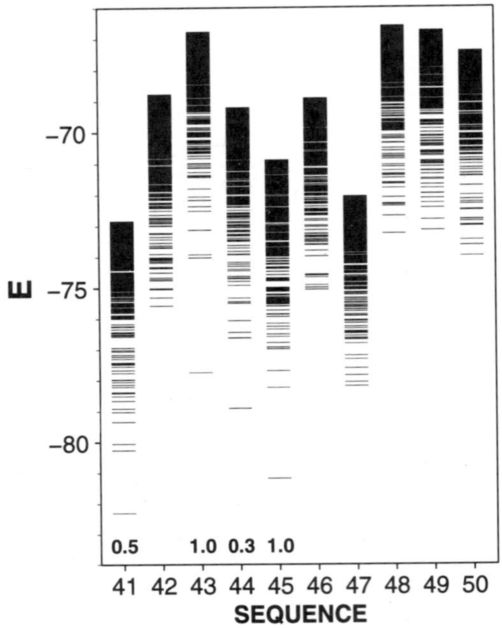


FIGURE 3. Energy spectra for 10 folding and non-folding random sequences. The energies of the 400 lowest compact self-avoiding conformations are shown. The native state corresponds to the bottom bar. The numbers below the spectra show the foldicities of the corresponding sequences; if no number is given, foldicity is 0. A sequence *folds* in a given MC simulation if it finds the native conformation within $50 \cdot 10^6$ MC steps. *Foldicity* of a given sequence is defined as the fraction of 10 MC runs that started with a random conformation and reached the native conformation under a given set of conditions. A sequence is a *folding sequence* if the native conformation is structurally unique and foldicity is high ($\geq 0.4$) under conditions where the native structure is thermodynamically stable. A sequence is a *non-folding sequence* if the foldicity is 0.0. There are 24 intermediate sequences that we do not consider here. Optimal values for parameters $B_o$ ($-2$) and $\sigma_B$ (1) were determined by exhaustive sampling of foldicity as a function of these two parameters [51]. Each sequence is studied at a temperature, $T_x$, slightly above the midpoint of the folding transition [51]. An order parameter that describes a transition of a chain from a degenerate state with many backbone conformations to a state with few, possibly only one, backbone conformation is $X(T) = 1 - \sum_i^M p_i^2$, where $p_i = \exp(-\frac{\epsilon_i}{k_B T})/Z$ and $Z = \sum_i^M \exp(-\frac{\epsilon_i}{k_B T})$. $k_B$ is the Boltzmann constant set to 1 in this study, $p_i$ is the Boltzmann probability for a system to be in state $i$, $M$ is the number of all compact self-avoiding states, and $Z$ is the configuration partition function of the chain. $T_x$ is defined such that $X(T_x) = 0.8$ where the native state has a weight almost invariably larger than 0.2 relative to other compact-self avoiding conformations.

folding possible. The existence of such biases on the entire potential surface corresponds to the principle of "minimum frustration" [17,58], which is closely related to

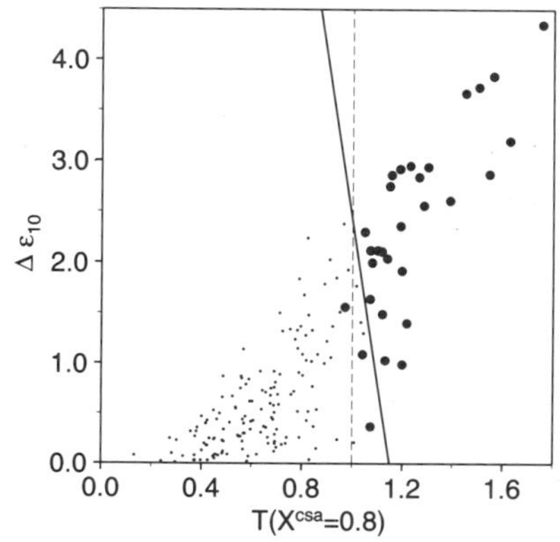


FIGURE 4. Discrimination between strongly folding (circles) and non-folding sequences (dots). $\Delta\epsilon_{10}$ is the energy gap between the native state and the second most stable compact self-avoiding conformation. $T_x$ is the temperature at which the MC simulations are performed. The larger are $\Delta\epsilon_{10}$ and $T_x$, the more pronounced the global energy minimum is. The continuous line separates the two groups of sequences by minimizing the number of the folding and non-folding sequences in the non-folding and folding parts of the plot, respectively. The dashed line indicates the critical temperature, defined as the temperature below which almost no folding is observed [51]. $T_x$ is determined for each sequence separately such that the native state had a high probability to be reasonably stable (see Figure 3) [51].

the "consistency" or "harmony" principle proposed by Gõ and Abe [59]. One way of introducing the bias is by the use of associative-memory Hamiltonians [60–64] which have been employed successfully in a variety of applications; e.g., it was shown that the ratio between the folding and glass transition temperatures, whose maximization was assumed to lead to faster folding, is proportional to the ratio of the energetic separation of the native state from the denatured states and the range of energies corresponding to the denatured states [60].

Simulations using other simple lattice models confirmed that the pronounced global energy minimum is associated with rapid folding [37, 39, 65]. An additional confirmation came from the success of designing stable folding sequences by minimizing the energy of a given native state [34, 65–67].

**3.2. Random sequences fold by the three-stage random search mechanism.** The insights concerning the role of the temperature and the energy gap provided by the 27-mer computer experiment [51] and theoretical analyses [17] do not, in themselves, provide the specific mechanism by which a model resolves the Levinthal paradox. For this purpose, further examination of the results for the strongly folding sequences in the 27-mer was required [12, 54].

To explore the mechanism of folding, the density of states was determined as a function of the energy and the reaction coordinate (the number of native contacts) (Figure 6); a related reaction coordinate, the number of residues in the native conformation, has been used previously [17]. If there were a nucleus of a small number of native contacts that would lead rapidly and with high probability

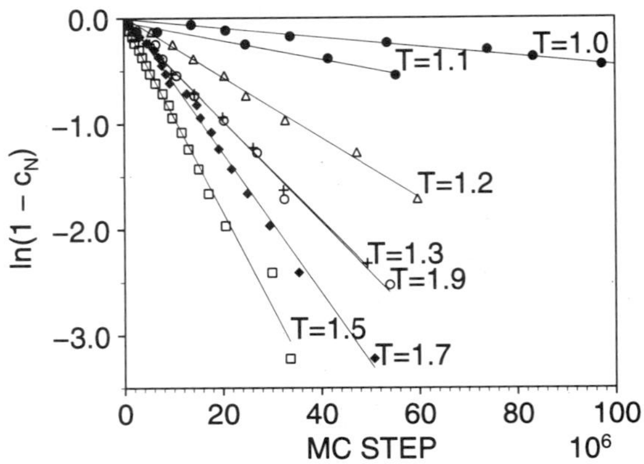


FIGURE 5. Kinetics of conversion of the denatured state into the native state for sequence 43 as a function of temperature. Kinetics of conversion of the denatured state into the native state for sequence 43 as a function of temperature. For each temperature, the probability, $C_N$, that the native state is reached in a given number of MC steps is estimated from 100 folding trials. The lines are linear least-squares fits to the points determined at the temperature as indicated on the plot. It can be seen that the results satisfy the simple unimolecular rate equation $dC_N/\,dt = kC_D$, where $k$ is the rate constant, $C_D$ is the probability of being in the denatured state. The line at $T = 1.0$ is linear in the whole time range explored (up to $400 \cdot 10^6$ MC steps where $C_N \approx 1$).

to the native state, the fraction of native contacts would not be a good reaction coordinate; instead, the fraction of the contacts present in the nucleus should be used. However, a folding nucleus is not present in the model, confirming that the fraction of native contacts is a suitable reaction coordinate. From the density of states, the mean energy, entropy, and free energy for a given temperature were calculated (Figure 7). Above the critical temperature ($T_c \approx 1$), the energy and entropy decrease smoothly as the chain approaches the native state. The free energy has a maximum corresponding to the transition region that separates the denatured and native states. This barrier, which makes the reaction a cooperative transition, is dominated by the entropic contribution, as can be seen from the temperature dependence of the free energy. Below the critical temperature, the reaction profile corresponding to the free energy becomes rugged [15, 17] and folding is much slower because the chain is likely to be trapped in one of the many local minima.

Analyses of individual trajectories for both strongly folding and non-folding sequences, as well as the calculated density of states and reaction profiles [12, 54], demonstrated that the sequences in the present model fold according to the three-stage random search (3SRS) mechanism at temperatures above 1.0 (Figure 8). The time history of the folding process (Figure 1b) shows that there is a rapid collapse in $\approx 10^4$ MC steps to a semi-compact random globule; *i.e.*, the number of contacts increases, the energy decreases, while the fraction of native contacts remains below 0.3. In this way, the total of $\approx 10^{16}$ random-coil conformations is effectively reduced to $\approx 10^{10}$ random semi-compact globule states. The fast collapse results from a large energy gradient (Figure 1b) and the presence of many empty lattice points.

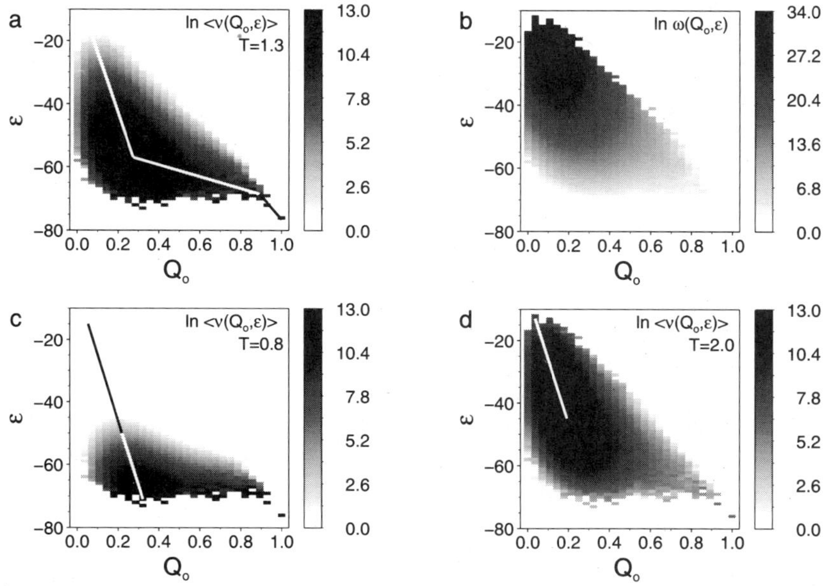


FIGURE 6. Density of states for a folding random sequence, as obtained from a long MC simulation. *a*, The logarithm of the average occupancy of the $(Q_o, \epsilon)$ bins, $\langle \nu(Q_o, \epsilon) \rangle$, at $T = 1.3$. Each bin spans $1/28$ $Q_o$ units and 1 energy unit. The occupancy of a bin is the number of MC steps that remain or result in any of the conformations corresponding to the bin. The average is calculated from 20 independent MC sampling simulations of $50 \cdot 10^6$ steps each [12, 54]. The broken line indicates a typical folding pathway. *b*, The logarithm of the density of states, $\omega(Q_o, \epsilon)$. The density of states is calculated from $\langle \nu(Q_o, \epsilon) \rangle$ by the use of $\frac{\langle \nu(Q_o,\epsilon) \rangle}{\langle \nu(Q_o=1,\epsilon=\epsilon_o) \rangle} = \frac{\omega(Q_o,\epsilon)}{\omega(Q_o=1,\epsilon=\epsilon_o)} \frac{\exp[-\epsilon/(k_B T)]}{\exp[-\epsilon_o/(k_B T)]}$ where $\epsilon_o$ is the energy of the native state and $k_B$ is the Boltzmann constant set to 1; because $\omega(Q_o = 1, \epsilon = \epsilon_o) = 1$, the density of states is $\omega(Q_o, \epsilon) = \frac{\langle \nu(Q_o,\epsilon) \rangle}{\langle \nu(1,\epsilon_o) \rangle} \exp(-\frac{\epsilon_o - \epsilon}{k_B T})$. The calculation is accurate because thermodynamic equilibrium at the present $Q_o, \epsilon$ resolution is achieved, as demonstrated by a small fractional error in $\langle \nu(Q_o, \epsilon) \rangle$ (less than 10% in the populated parts of the phase space). The summation of $\omega(Q_o, \epsilon)$ over all $Q_o$ and $\epsilon$ results in $\approx 1.1 \cdot 10^{16}$ self-avoiding chains; this is in good agreement with the extrapolation of the exact enumerations for shorter chains: $\nu^{L-1}/12 = 2.2 \cdot 10^{16}$ where $L$ is the number of monomers, $\nu = 4.68$ is the average number of monomer states, and division by 12 corrects for symmetry [68]. Semi-compact states are defined as the states with energy less than $-45 k_B T$; summation over the appropriate region of $\omega(Q_o, \epsilon)$ yields $\approx 10^{10}$ such states. The transition states correspond to all the states with $0.8 \leq Q_o < 1$; there are $\approx 10^3$ such states. *c*, The average occupancy of the bins at a low $T$ ($T = 0.8$) as calculated from $\omega(Q_o, \epsilon)$. The line shows the rapid collapse of a random coil chain into a random semi-compact frozen conformation. *d*, The calculated average occupancy of the bins at a high $T$ ($T = 2.0$). The line shows a rapid collapse of the random coil chain into a region consisting of many interconverting semi-compact random states; the ground state is no longer stable so the chain spends most of its time in the entropically favored denatured states. In contrast, at $T = 1.0$, the native state occurs for $\approx 40\%$ of the time while still being kinetically accessible.

In the second stage, which is rate limiting, the chain searches for one of the $\approx 10^3$ transition states. The transition region consists of all states from which the chain folds rapidly to the native state. The transition states are structurally similar to the native state, with 23–26 (80–95%) of the native contacts. Generally, the mean first passage time, $\tau$, for finding any of the $n$ states among the total of $N$ states by a random search which explores $r$ states per unit of time is $N/(n \times r)$. For the present model, $\tau \approx 10^{10}/(10^3 \times 1) = 10^7$, identical to the observed time scale. This indicates that the rate limiting stage in folding consists of a random search for a transition state in the semi-compact part of the phase space. In the third stage, the chain rapidly (within $\approx 10^5$ MC steps) attains the native conformation from any one of the transition states. The relationship of each of the three stages in the 3SRS-mechanism to other mechanisms is discussed in ref. [4].

The second, rate-limiting stage is "random" in the sense that, over many trajectories, the microscopic states are occupied according to their Boltzmann probabilities and that there are many different microscopic states with comparable Boltzmann probabilities at the random-coil, random globule, and transition-state stages of folding. There is only a small difference between the folding and random non-folding sequences in the way they explore the phase space at the same absolute temperature. That is, above the glass transition temperature, the folding sequences tend to find their native states no more than an order of magnitude faster than the non-folding sequences [39, 51, 54]. However, in the neighbourhood of $T_m$, the midpoint of the folding transition, the non-folding sequences fold much slower. This is clearly different from the funnel hypothesis of protein folding which assumes that folding sequences fold because they have a single large folding funnel leading to the native conformation and that non-folding sequences do not fold because they have multiple pathways leading to several conformations [11].

The pronounced energy minimum is the necessary condition for the folding of a 27-mer on a lattice because it guarantees that the native state is stable above the critical temperature, where the rearrangements required in the rate limiting stage are energetically possible. It is also a sufficient condition because a random search of compact globules with random structures can rapidly find a transition state that folds to the stable native state in a short time. While a non-folding sequence may also fold slowly to its native state above the critical temperature, it would not be stable and, therefore, could not correspond to a real protein.

Surfaces can be constructed for which resolution of the Levinthal paradox is trivial (*e.g.*, a smooth descent to an energy minimum [59], or only local interactions stabilizing the native state [71], as in the helix-coil transition). However, this does not obviate the fact that the large size of the configuration space is a necessary condition for a paradox to exist. The 27-mer model satisfies this condition with $10^{16}$ configurations, while short oligomers that have been extensively studied on a two-dimensional square lattice [14, 37, 38, 45] may not; for example, in ref. [45], simulations involving more than $10^5$ MC steps were used to fold a 13-mer on a square lattice that has only $\approx 4 \cdot 10^4$ conformers. As was pointed out previously, the shape of the configuration space, as well as the number of conformers, is important [11, 51, 71]. The landscape of the 27-mer is clearly sufficiently complex to have a Levinthal paradox; otherwise all sequences would have folded rapidly to the stable native state.

The size of the search for the native state is greatly reduced when the chain is semi-compact as it is in real proteins [69]. Nevertheless, in the 27-mer model

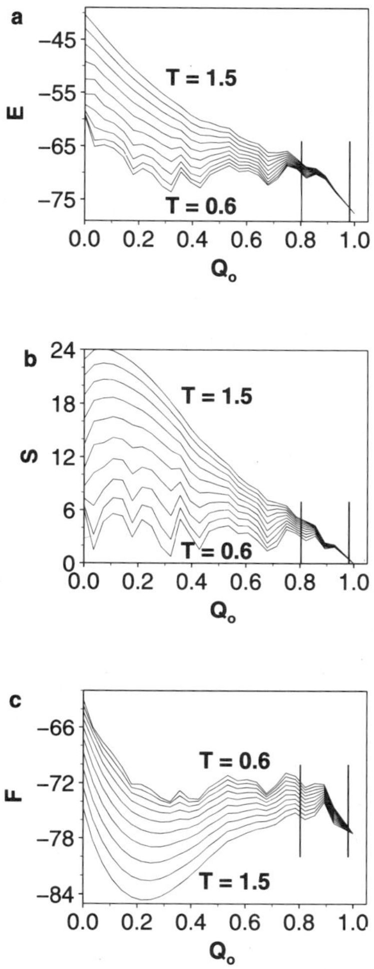


FIGURE 7. Reaction profiles as a function of the reaction coordinate for the folding random sequence at $0.6 \leq T \leq 1.5$. *a*, Energy, $E$. *b*, Entropy, $S$. *c*, Free energy, $F$. The transition state lies between the two vertical lines at $0.83 \leq Q_o \leq 0.96$. Profiles are calculated in temperature intervals of 0.1, using the partition function $Z(Q_o, T) = \sum_\epsilon \omega(Q_o, \epsilon) \exp(-\epsilon/kT)$. $E(Q_o, T) = \sum_\epsilon \omega(Q_o, \epsilon)\ p_\epsilon\ \epsilon$, $p_\epsilon = \exp(-\epsilon/k_B T)/Z(Q_o, T)$, $F(Q_o, T) = -k_B T \ln Z(Q_o, T)$, and $S(Q_o, T) = [E(Q_o, T) - F(Q_o, T)]/T$.

as in real proteins, a random search of a collapsed globule cannot find the native state in the observed time. It is the existence of a transition region, consisting of a large number of states, that reduces the search time to realistic values when combined with the search of the random compact globule. Extrapolation of the folding time to real proteins suggests that the 3SRS-mechanism could be effective for small proteins (Figure 8). However, the mechanism breaks down for long chains because the folding time is expected to increase exponentially with chain length; *i.e.*, the number of semi-compact states increases faster than the number of the

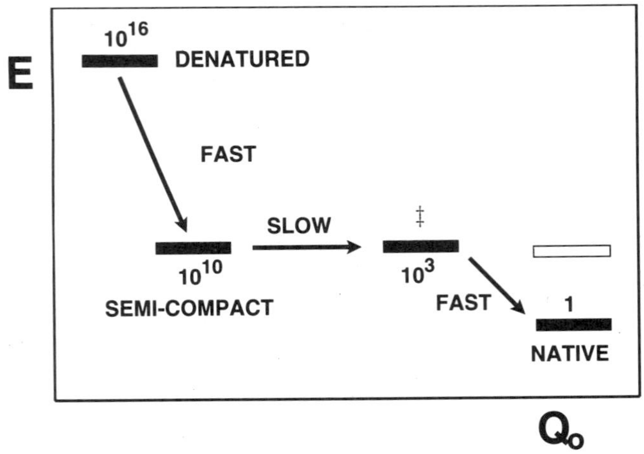


FIGURE 8. Three-stage random search mechanism of folding for the present model (see text). The numbers of the states were obtained from Figure 6. The relative rates are obtained from 100 folding simulations (Figure 2). The empty rectangle indicates the energy of the native state of a non-folding sequence; in the non-folding sequence, the ground state would not be populated at a temperature sufficiently high to avoid being trapped as in Figure 6c. The other states of a non-folding sequence appear at essentially the same positions as the corresponding states of a folding sequence. It can be estimated that a real protein of 80 residues has $2.57^{79} \approx 10^{32}$ possible mainchain conformations of which $1.7^{79} \approx 10^{18}$ are semi-compact [69]. The number of the transition states can be extrapolated from a 27-mer as follows. In a 27-mer, $10^3$ out of the $10^{10}$ semi-compact states are transition states; thus, in an 80-residue protein, approximately $10^6$ out of the $10^{18}$ semi-compact states are likely to be transition states. According to a molecular dynamics simulation of native hydrated myoglobin at 300 K [70], there are 52 mainchain dihedral angle changes per 153 residues per 100 ps or 3 transitions per residue in 1 ns; conformational transitions in the semi-compact state are likely to be faster. Use of these numbers for an 80 residue protein that folds by the three-stage mechanism yields a rate determining step of $10^{12}$ transitions which would correspond to a folding time to a molten globule state of about 3 sec. This is close to the time scale observed in real proteins.

transition states. Thus, a modification of the present mechanism is required for larger proteins.

It is likely that proteins existing early in evolution were small enough to fold according to the 3SRS-mechanism [72]. Since the pre-biotic and early biotic environment was hot, unusually thermostable proteins were required, such as those found in the most primitive bacteria that live at temperatures as high as 105° [73]. If so, the stability condition required a native state that is a very low energy minimum for which the folding problem was solved simultaneously. As evolution progressed, longer proteins evolved. These proteins had to fold on the same time scale. One way of achieving this is by evolving proteins with sequences that have a larger difference between the native and non-native contact energies than the random folding sequences of the present model. Such sequences would have an even more pronounced global minimum in their potential surfaces, in line with the "consistency principle" [59] and the "principle of minimal frustration" [58]. It is also in accord with the existence of a nucleus for folding [8], or the early appearance of secondary structural elements [7, 74], neither of which are found in the folding sequences of the present model. As a result, the collapse would not result in random semi-compact globules and the very favorable native contacts would lead more directly to a native-like molten globule state; *i.e.*, the folding pathway in Figure 6a would approximate a straight line from the random-coil to the "native state". Such a mechanism has been observed in folding simulations of long chains with highly stable native states [13, 33, 75]. Actual proteins could use an intermediate mechanism that might vary with the external conditions.

The results of this study may have implications for the prediction of the structure of a protein from its amino acid sequence. The success of the 3SRS-mechanism in finding the pronounced global minimum on a potential surface suggests, at least for small proteins, that the bottleneck in structure prediction may be the derivation of a suitable potential function rather than the design of folding algorithms.

## Acknowledgments

We are grateful to Aaron Dinner, Daša Šali, Oleg Ptitsyn, Alexander Gutin, Georgios Archontis, Amedeo Caflisch, Oren Becker, and Lloyd Dimitrius for discussions concerning the protein folding problem. The computations were done on IBM RS/6000, Silicon Graphics Iris 4D, SUN Sparcstation, DEC Decstation, DEC Alphastation, and NeXT workstations. The paper is based on references [12, 51] and is essentially the same as ref. [76].

## References

[1] T. Creighton, editor, *Protein Folding*, W.H. Freeman and Company, New York, 1992.

[2] K. M. Jr. and S. L. Grand, editors, *The Protein Folding Problem and Tertiary Structure Prediction*, Birkhäuser, Boston, 1994.

[3] M. Karplus and E. Shakhnovich, Protein folding: Theoretical studies of thermodynamics and dynamics, in *Protein Folding*, edited by T. Creighton, pages 127–196, W.H. Freeman and Company, New York, 1992.

[4] M. Karplus and A. Šali, Curr. Opin. Struct. Biol. **5**, 58 (1995).

[5] C. Levinthal, How to fold graciously, in *Mossbauer Spectroscopy in Biological Systems, Proceedings of a Meeting held at Allerton House, Monticello, IL*, edited by P. Debrunner, J. Tsibris, and E. Münck, pages 22–24, University of Illinois Press, Urbana, 1969.

[6] M. Karplus and D. Weaver, Prot. Science **3**, 650 (1994).

[7] P. Kim and R. Baldwin, Ann. Rev. Biochem. **59**, 631 (1990).

[8] D. Wetlaufer, Proc. Natl. Acad. Sci. USA **70**, 697 (1973).

[9] K. Fiebig and K. Dill, J. Chem. Phys. **98**, 3475 (1993).

[10] E. Shakhnovich, G. Farztdinov, A. Gutin, and M. Karplus, Phys. Rev. Lett. **67**, 1665 (1991).

[11] P. Leopold, M. Montal, and J. Onuchic, Proc. Natl. Acad. Sci. USA **89**, 8721 (1992).

[12] A. Šali, E. Shakhnovich, and M. Karplus, Nature **369**, 248 (1994).

[13] V. Abkevich, A. Gutin, and E. Shakhnovich, Biochem. **33**, 10026 (1994).

[14] C. Camacho and D. Thirumalai, Proc.Natl.Acad.Sci. USA **90**, 6369 (1993).

[15] E. Shakhnovich and A. Gutin, Biophys. Chem. **34**, 187 (1989).

[16] E. Shakhnovich and A. Gutin, Nature **346**, 773 (1990).

[17] J. Bryngelson and P. Wolynes, J. Phys. Chem. **93**, 6902 (1989).

[18] A. Caflisch and M. Karplus, Molecular dynamics studies of protein and peptide folding and unfolding, in *The protein folding problem and tertiary structure prediction*, edited by K. M. Jr. and S. L. Grand, pages 193–230, Birkhäuser, Boston, 1994.

[19] H. Frauenfelder and P. Wolynes, Physics Today **47**, 58 (1994).

[20] V. Daggett and M. Levitt, Curr. Opin. Struct. Biol. **4**, 291 (1994).
[21] K. Dill, Curr. Opin. Struct. Biol. **3**, 99 (1993).
[22] H. Chan and K. Dill, Physics Today **46**, 24 (1993).
[23] G. Rose and T. Creamer, Proteins **19**, 1 (1994).
[24] G. Rose and R. Wolfenden, Annu. Rev. Biophys. Biomol. Struct. **22**, 381 (1993).
[25] M. Levitt, Curr. Opin. Struct. Biol. **1**, 224 (1991).
[26] R. Abagyan, FEBS Lett. **325**, 17 (1993).
[27] J. Skolnick, A. Kolinski, and A. Godzik, Proc. Natl. Acad. Sci. USA **90**, 2099 (1993).
[28] C. B. III, M. Karplus, and B. Pettit, *Proteins: A Theoretical Perspective of Dynamics, Structure and Thermodynamics*, John Wiley & Sons, New York, 1988.
[29] M. Levitt and A. Warshel, Nature **253**, 694 (1975).
[30] N. Gõ and H. Abe, Biopolymers **20**, 1013 (1981).
[31] J. Skolnick and A. Kolinski, Science **250**, 1121 (1990).
[32] K. Lau and K. Dill, Macromolecules **22**, 3986 (1989).
[33] V. Abkevich, A. Gutin, and E. Shakhnovich, J. Chem. Phys. **101**, 6052 (1994).
[34] S. Ramanathan and E. Shakhnovich, Physical Review E **50**, 1303 (1994).
[35] A. Dinner, A. Šali, M. Karplus, and E. Shakhnovich, J. Chem. Phys. **101**, 1444 (1994).
[36] J. Ngo, J. Marks, and M. Karplus, Computational complexity, protein structure prediction, and the levinthal paradox, in *The Protein Folding Problem and Tertiary Structure Prediction*, edited by K. M. Jr. and S. L. Grand, pages 433–506, Birkhäuser, Boston, 1994.
[37] H. Chan and K. Dill, J. Chem. Phys. **100**, 9238 (1994).
[38] C. Camacho and D. Thirumalai, Phys. Rev. Lett. **71**, 2505 (1993).
[39] N. Socci and J. Onuchic, J. Chem. Phys. **101**, 1519 (1994).
[40] E. O'Toole, J. Chem. Phys. **98**, 3185 (1993).
[41] A. Kolinski and J. Skolnick, Proteins **18**, 338 (1994).
[42] D. Hinds and M. Levitt, J. Mol. Biol. **243**, 668 (1994).
[43] D. Covell, J. Mol. Biol. **235**, 1032 (1994).
[44] M.-H. Hao and H. Scheraga, J. Phys. Chem. **98**, 4940 (1994).
[45] R. Miller et al., J. Chem. Phys. **96**, 768 (1992).
[46] O. Ptitsyn, J. Prot. Chem. **6**, 273 (1987).
[47] M. Harding, D. Williams, and D. Woolfson, Biochemistry **30**, 3120 (1991).
[48] E. Shakhnovich and A. Finkelstein, Biopolymers **28**, 1667 (1989).
[49] Z.-Y. Peng and P. Kim, Biochemistry **33**, 2136 (1994).
[50] D. Dolgikh et al., FEBS Lett. **136**, 311 (1981).
[51] A. Šali, E. Shakhnovich, and M. Karplus, J. Mol. Biol. **235**, 1614 (1994).
[52] J. Skolnick and A. Kolinski, J. Mol. Biol. **221**, 499 (1991).
[53] P. Pardalos, G. Xue, and D. Shalloway, J. Glob. Opt. **4**, 117 (1994).
[54] A. Šali, E. Shakhnovich, and M. Karplus, In preparation .
[55] S. Miyazawa and R. Jernigan, Macromolecules **18**, 534 (1985).
[56] N. Metropolis, A. Rosenbluth, M. Rosenbluth, A. Teller, and E. Teller, J. Chem. Phys. **21**, 1087 (1953).
[57] J. Honeycutt and D. Thirumalai, Biopolymers **32**, 695 (1992).
[58] J. Bryngelson and P. Wolynes, Proc. Natl. Acad. Sci. USA **84**, 7524 (1987).
[59] N. Gõ and H. Abe, Adv. Biophysics **18**, 149 (1984).
[60] R. Goldstein, Z. Luthey-Schulten, and P. Wolynes, Proc. Natl. Acad. Sci. USA **89**, 4918 (1992).
[61] M. Friedrichs, R. Goldstein, and P. Wolynes, J. Mol. Biol. **222**, 1013 (1991).
[62] M. Friedrichs and P. Wolynes, Science **246**, 371 (1989).
[63] M. Sasai and P. Wolynes, Phys. Rev. Lett. **65**, 2740 (1990).
[64] A. Caliri, H. Bohr, P. Wolynes, M. Sasai, and P. Wolynes, Phys. Rev. A **46**, 7979 (1992).
[65] E. Shakhnovich, Phys. Rev. Lett. **72**, 3907 (1994).
[66] E. Shakhnovich and A. Gutin, Proc. Natl. Acad. Sci. USA **90**, 7195 (1993).
[67] E. Shakhnovich and A. Gutin, Prot. Eng. **6**, 793 (1993).
[68] M. Sykes, J. Chem. Phys. **39**, 410 (1963).
[69] K. Dill, Biochemistry **24**, 1501 (1985).
[70] R. Loncharich and B. Brooks, J. Mol. Biol. **215**, 439 (1990).
[71] R. Zwanzig, A. Szabo, and B. Bagchi, Proc. Natl. Acad. Sci. USA **89**, 20 (1992).
[72] E. D. Iorio et al., Proc. Natl. Acad. Sci. USA **90**, 2025 (1993).

[73] K. Stetter, Life at the upper temperature border, in *Frontiers of Life*, edited by J. T. T. Vân, J. Mounolou, J. Schneider, and C. McKay, pages 195–212, Editions Frontières, Gif-sur-Yvette, France, 1992.
[74] M. Karplus and D. Weaver, Nature **260**, 404 (1976).
[75] A. Dinner, A. Šali, and M. Karplus, In preparation .
[76] A. Šali, E. Shakhnovich, and M. Karplus, Protein folding studied by Monte Carlo simulations, in *Distance-Based Approaches to Protein Structure Determination II*, edited by H. Bohr and S. Brunak, CRC Press, Florida, USA., In press.

*E-mail address*: `sali@rockvax.rockefeller.edu`

DEPT. OF CHEMISTRY, 12 OXFORD ST., HARVARD UNIVERSITY, CAMBRIDGE, MA 02138, USA
*Current address*: The Rockefeller University, 1230 York Ave., New York, NY 10021, USA

*E-mail address*: `eugene@diamond.harvard.edu`

DEPT. OF CHEMISTRY, 12 OXFORD ST., HARVARD UNIVERSITY, CAMBRIDGE, MA 02138, USA

*E-mail address*: `marci@tammy.harvard.edu`

DEPT. OF CHEMISTRY, 12 OXFORD ST., HARVARD UNIVERSITY, CAMBRIDGE, MA 02138, USA

DIMACS Series in Discrete Mathematics
and Theoretical Computer Science
Volume **23**, 1996

# Beyond Optimization: Simulating the Dynamics of Supercoiled DNA by a Macroscopic Model

Gomathi Ramachandran and Tamar Schlick

ABSTRACT. The development and application of macroscopic dynamic models for large biomolecules is as much of an art as it is a science. Here we discuss scaling and parameterization issues for a macroscopic B-spline model of supercoiled DNA in the context of the Langevin framework. We show how scaling the masses and the damping constant, $\gamma$, is computationally advantageous (for enhanced sampling) and how, by calibration, the corresponding physical timestep and damping constant are obtained. Our timestep estimate for a mass scaled model, roughly 20 ps, is consistent with a rigorous calibration for the dynamic B-spline model performed earlier. An interesting finding here is that the B-spline model, though different in details from a bead model, is essentially equivalent in the Langevin framework; thus, its computational simplicity can be advantageous. Another finding is that the optimal $\gamma$ for sampling is smaller (by two orders of magnitude) than the value needed to reproduce the experimental translational diffusion constant, $D_t$. This has been noted previously for proteins and appears to be an important aspect of modeling macromolecules in view of the sampling problem, as well as numerical integration. Using the smaller $\gamma$ has the effect, in the diffusive regime, of scaling the times computed in a Langevin trajectory by a corresponding factor. This leads to an effective timestep in our simulation around 10 ns. Performance of the implicit integrator used here is also discussed, demonstrating that numerical damping effects are negligible for this macroscopic model at the $\gamma$ used for maximal sampling. Furthermore, computational gain can be achieved over explicit integration. Finally, an estimate on the effects of hydrodynamics on the translational friction constant, $f_T$, is presented, on the basis of the Kirkwood-Riseman equation. We find that $f_T$ is reduced by approximately a factor of two for DNA of 1000 base pairs, making the free-draining limit (i.e., no hydrodynamics) quite reasonable. The results presented here – scaling, parameterization, and integration in the context of a dynamic macroscopic model – have general applicability to simplified models of biopolymers. They also add the needed quantitative information so that slow dynamic processes involving supercoiled DNA, such as slithering, can be interpreted in terms of physical timescales from such macroscopic dynamic simulations.

1991 *Mathematics Subject Classification.* Primary 92C05; Secondary 92-08, 60H10, 82C31.

*Key words and phrases.* Molecular Dynamics, Langevin Dynamics, Configurational Sampling, DNA Supercoiling, Macroscopic Model, Elastic-Rod Model, Mass Scaling, Translational Diffusion Constants, Hydrodynamics.

This work is supported by the National Institutes of Health (Research Resource RR08102) and an Alfred P. Sloan fellowship. T.S. is an investigator of the Howard Hughes Medical Institute.

The modeling of large biomolecular systems such as proteins and nucleic acids, as a complement to experiment, poses many computational challenges to a wide range of biological and physical scientists. Biomolecular structures are governed by a complex energy landscape and therefore exhibit functional activities over an enormous range of spatial and temporal scales. Developing appropriate mathematical models and simulation methodologies to investigate these systems forms an interdisciplinary task of great importance.

The study of DNA supercoiling, of interest to us, is particularly fascinating biologically and computationally challenging. Supercoiling is the higher order twisting and folding of the global double-helical axis itself. It condenses the DNA significantly, facilitating DNA's packaging within the cell and also mediating many fundamental processes that involve template-directed synthesis such as replication and transcription [**10, 8, 1**]. Unlike proteins, the very large topologically constrained DNA systems associated with supercoiling – thousands of base pairs (bp) – are inherently floppy in the aqueous environment and therefore adopt many, thermally accessible configurational states. Thus, the challenge in developing and applying macroscopic (i.e., not all-atom) models of supercoiled DNA is proper consideration of the wide range of possible conformations. Energy minimization provides an important preliminary description of these biomolecules, but additional statistical mechanical descriptions, such as available by Monte Carlo (MC) and molecular dynamics (MD), are essential for understanding the biological activities of supercoiled DNA.

Several researchers have examined supercoiled DNA properties analytically on a macroscopic level on the basis of the elastic-rod framework [**19, 4, 15, 14, 33, 16**, for example]. These analytical treatments naturally led to numerical models for simulation (see below). More recently, simple numerical models of proteins have been developed, based on lattice models, bead models, and others, and applied to protein folding [**5, 9**, and references cited therein]. In the case of proteins, most macroscopic models focus on obtaining energy minima of the system. For DNA, numerical optimization provides guiding equilibria information [**12, 11, 31, 32, 37, 2, 35**, for example], but MC and dynamic approaches provide more global views such as configurational distributions and dynamic pathways (recently reviewed in [**27**]). For example, several studies have revealed behavior as a function of the internal superhelical stress and the external ionic strength [**35, 30, 3**].

In general, dynamic simulations in the context of macroscopic models have been far fewer than energy minimization for both proteins and nucleic acids. The difficulty lies in extracting quantitative or semi-quantitative dynamic information from these macroscopic models. While the link between macroscopic model and detailed structure is often nontrivial to establish for static properties (the position of the global minimum for example; see the article by Scheraga and co-workers on minimization approaches in this collection [**17**]), the link to dynamic properties is more challenging. This difficulty arises because parameters such as mass, timestep, velocity, etc., are typically associated with large groups of the polymer (e.g., 4 bp, 2 residues) and also often scaled for computational reasons. Thus, the estimation of rates and timescales of dynamic processes from a macroscopic model requires the generation of experimentally comparable data in order to perform a rough 'calibration'. Naturally, there is also the question of the overall effect of the approximations made on the results obtained. In this respect, the development and application of dynamic macroscopic models is as much of an art as it is a science.

A recent macroscopic DNA model by Schlick and Olson for dynamic simulations is based on a B-spline representation, an elastic plus electrostatic energy function, and Langevin dynamics [**31, 30**]. The influence of damping was recently incorporated by parameterizing the viscosity variable (the damping constant $\gamma$) in the Langevin formalism [**26**]. In this paper, we discuss the scaling and parameterization issues in this model in the context of the Langevin framework. These concepts are of general interest to modelers.

In particular, there are two types of scaling that we address: mass scaling (mathematical and model specific) and physical scaling ($\gamma$ dependent and experimentally based). We show how scaling the masses and $\gamma$ is computationally advantageous (for enhanced sampling) and how, by calibration, the corresponding physical timestep and damping constant are obtained. Our timestep estimate for this mass scaled model, roughly 20 ps, agrees excellently with a rigorous theoretical analysis and numerical calibration for the dynamic B-spline model performed earlier [**29**]. That calibration involved equating the expected kinetic energy of the system to the actual value obtained numerically; it also considered the effects of numerical damping due to the implicit-Euler integrator used.

The second scaling arises from the use of the optimal sampling $\gamma$ rather than the experimentally determined $\gamma$. It leads to an effective timestep around 10 ns, a value that agrees well with our previous estimate in the nanosecond range [**26**] based on comparison with experimental translational diffusion constants, $D_t$.

An interesting finding here is that the B-spline model, though different in details from a bead model, is essentially equivalent in the Langevin framework; thus, its computational simplicity can be advantageous (fewer control points than beads). Another finding is that the optimal $\gamma$ for enhanced sampling is smaller (by two orders of magnitude) than the value needed to reproduce the experimental $D_t$. Indeed, this is consistent with work on polypeptide models by Pastor and co-workers [**20**] and highlights the complex issues involved in dynamic macroscopic treatments.

Numerical performance of the LI integrator used here is also discussed, demonstrating that numerical damping effects are negligible for this macroscopic model at the $\gamma$ used for maximal sampling. Furthermore, computational gain can be achieved over explicit integration.

Finally, we estimate the effect of hydrodynamic interactions on the translational friction constant, $f_T$, by using the Kirkwood-Riseman equation. We find that neglect of hydrodynamic interactions for this macroscopic model is acceptable (a factor of two for the 1000 bp system examined) and thus the free-draining limit is quite reasonable.

The results presented here – scaling, parameterization, and integration in the context of a dynamic macroscopic model – have general applicability to macroscopic models of biopolymers. They also add the needed quantitative information so that interesting dynamic processes involving supercoiled DNA, such as slithering, can be interpreted in terms of physical timescales from such macroscopic dynamic simulations.

## 2. Numerical Model and Simulation Details

**2.1. Energy Model.** We use the B-spline model to represent the DNA duplex curve in terms of a small number, $N_b$, of control points [**31**] (e.g., 14 per 1000 bp). From these control points we generate curve points (e.g., 8 or 16 per spline segment), which are used for energy evaluation. For the 1000 base pair DNA system, we assign to each spline point an arbitrary mass of 1 amu. We normalize masses accordingly for other plasmid lengths, i.e., 28 amu total for 2000 bp.

The potential energy $E$ consists of elastic and electrostatic components. The elastic terms include homogeneous bending ($E_B$) and twisting ($E_T$) potentials. The Coulombic term ($E_C$) is in Debye-Hückel form. A simple harmonic term ($E_L$) is also included to ensure that the total chain length, $L$, remains very close to its target value, $L_0$. The energy components are given as follows:

$$E_B = \frac{A}{2} \oint \kappa^2(s)\, \mathrm{d}s \,, \tag{1}$$

$$E_T = \frac{2\pi^2 C}{L_0} (\Delta Lk - Wr)^2 \,, \tag{2}$$

$$E_L = K(L - L_0)^2 \,, \tag{3}$$

$$E_C = B(\beta) \int\int \frac{e^{-\beta r_{ij}}}{r_{ij}}\, \mathrm{d}l_i\, \mathrm{d}l_j . \tag{4}$$

In these expressions, $s$ denotes arc length; $\kappa$ denotes curvature; and the integrals are evaluated over the total length of the curve. The bending and torsional-rigidity constants, $A$ and $C$, are taken from experimental measurements [**27, 25**], and the parameter $K$ is a stiffness constant chosen for computational reasons (i.e., not physical stretching, which might be a future addition). The writhing number, $Wr$, is a geometric variable describing the DNA shape [**36**]; $\Delta Lk$ is the linking number deficit, defined as $Lk - Lk_0$, where $Lk$, the linking number, is a topological invariant describing the intertwining of the DNA strands about one another, and $Lk_0$ is the corresponding value for relaxed DNA [**36**]. In the electrostatic potential, $\beta$ is the Debye-Hückel screening parameter ($0.33\sqrt{c_s}$ at room temperature, where $c_s$ is the molar salt concentration) [**30**]; $B(\beta)$ is a salt-dependent coefficient [**30**]; $r_{ij}$ is the distance between points $i$ and $j$ along the DNA chain; and $l_i$ denotes the length of each charged segment along the chain. Full details of the electrostatic treatment are given in [**30**], and details of the B-spline model and the elastic-energy functions are given in [**31**].

**2.2. Systems and Parameters.** We consider here closed circular DNA systems of 1000, 2000, and 3000 bp. We use $A = 1.28 \times 10^{-19}$ erg cm for the elastic bending constant, corresponding to the nonelectrostatic contributions to the bending persistence length [**21**]. For the torsional-rigidity constant, we use $A/C = 1.5$, to yield $C = 0.85 \times 10^{-19}$ erg cm, corresponding to a root-mean-square (rms) twist angle of 7.3° [**30**]. We have recently suggested that a higher value ($C = 1.8 \times 10^{-19}$ erg cm, rms twist angle of 5.0°) might better correspond to experimental observations [**27, 25**] but use here a previous value consistent with

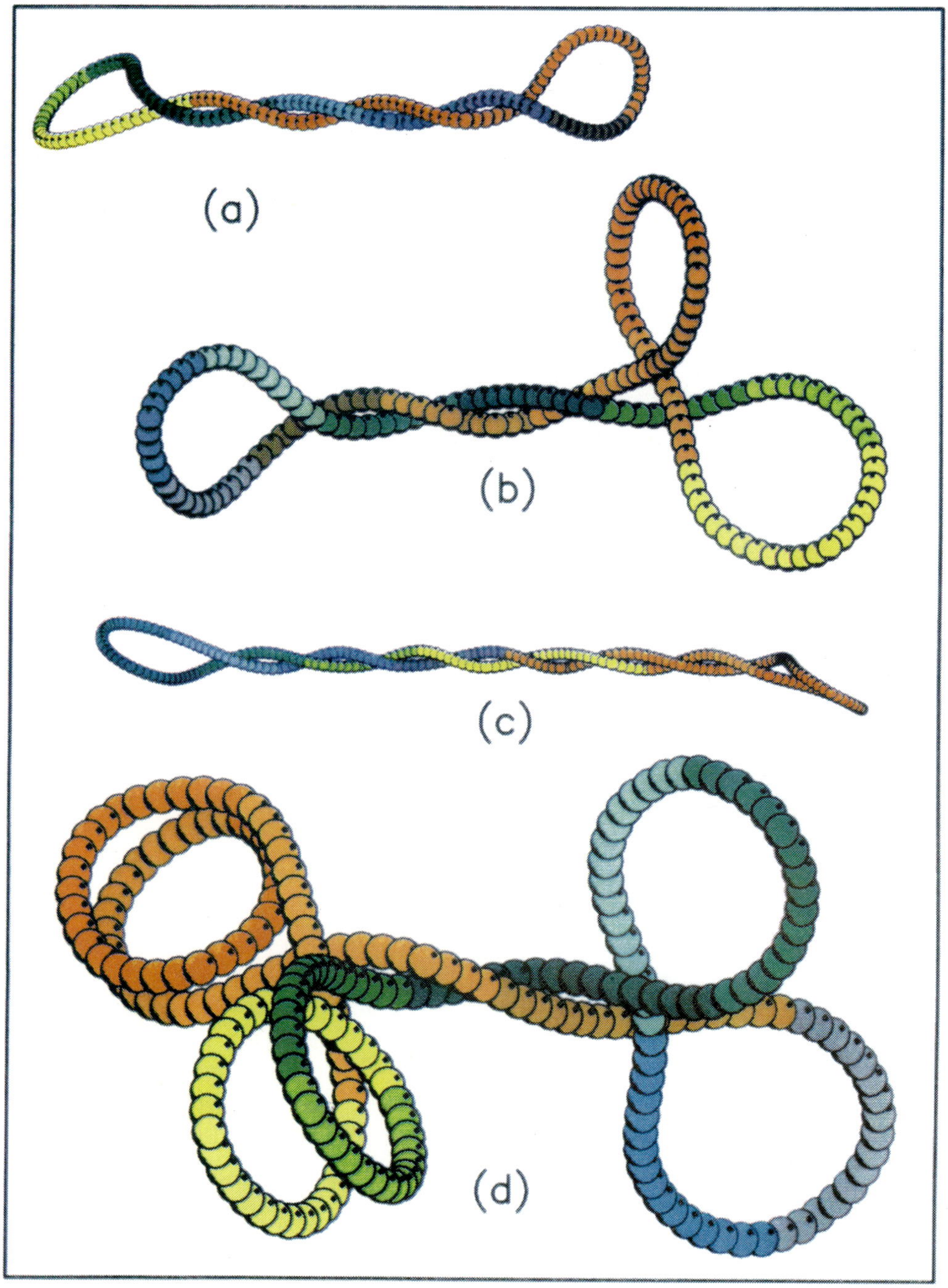


FIGURE 1. *Representative equilibria for supercoiled DNA systems of 2000 and 3000 bp obtained by energy minimization.* Sizes, total energies (in kcal/mol), and writhing numbers for the configurations shown are: (a) 2000 bp, $E = 145.3$, $Wr = -5.02$; (b) 2000 bp, $E = 148.8$, $Wr = -4.49$; (c) 3000 bp, $E = 212.2$, $Wr = -8.33$; and (d) 3000 bp, $E = 227.2$, $Wr = -6.64$. The dynamic simulations described in this work are started from the configuration of lowest energy (straight interwound).

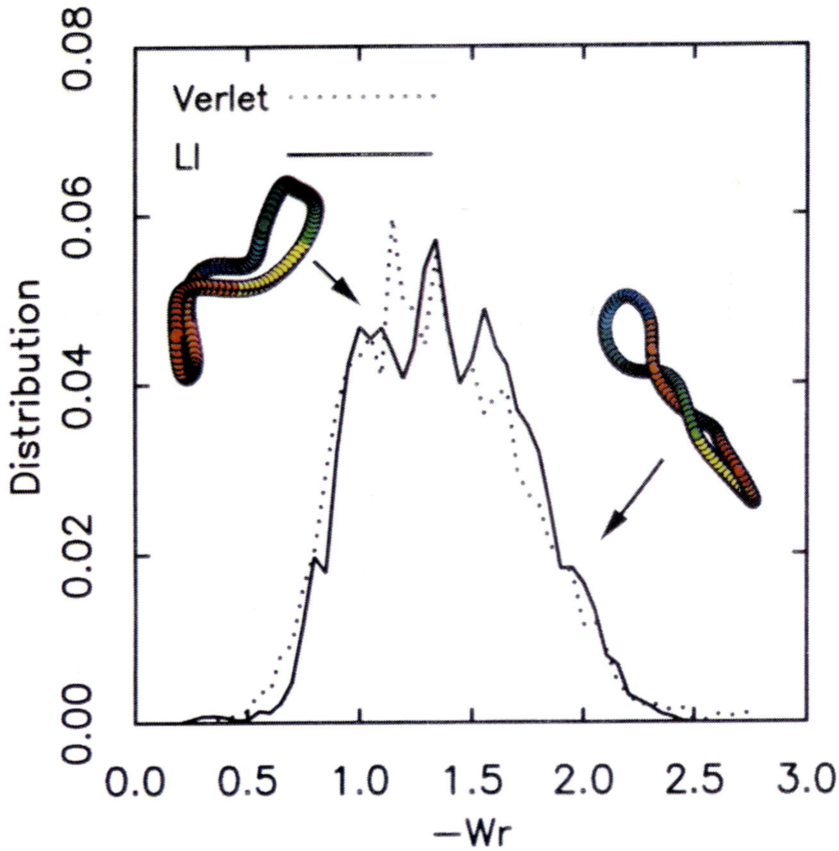


FIGURE 2. *Writhe distributions for a DNA system of 1000 bp: Verlet vs. LI.* The distribution for each scheme corresponds to the set of three trajectories, as described in Table 1. Each trajectory is started from the energy minimized structure with $Wr = -2.05$, shown on the right part of this figure. Note that the writhe distributions shift to smaller magnitudes of $Wr$ (more open structures), demonstrating the entropic lowering of the writhe [**26**]. A typical dynamically-generated configuration that is less compact ($Wr = -1.09$) is also shown, on the left.

| Scheme | $\langle Wr \rangle$ | Standard deviation of $Wr$ |
|---|---|---|
| LI | $-1.37$ | 0.39 |
| LI | $-1.34$ | 0.32 |
| LI | $-1.42$ | 0.35 |
| Verlet | $-1.37$ | 0.38 |
| Verlet | $-1.28$ | 0.32 |
| Verlet | $-1.39$ | 0.42 |

TABLE 1. **Comparison of the LI and Verlet Schemes: Dynamic Averages for DNA of 1000 bp.** Trajectories for the Verlet scheme are computed with a program timestep of $\simeq 5$ fs for 600,000 iterations, while those for the LI scheme are computed with a program timestep of $\simeq 100$ fs for 30,000 iterations (same total time). The value of $\gamma = 9.77 \times 10^{10}$ s$^{-1}$ (program units). See text for details.

our earlier paper which presented $D_t$ values [**26**]. There is, unfortunately, great uncertainty in the appropriate $C$ value for superhelical DNA [**25**]. The results regarding $D_t$, though, are not expected to change much in this range of $C$.

Throughout this work, we use the superhelical density $\sigma = -0.0525$ to match typical physiological superhelical densities of DNA. The superhelical density is defined as $\sigma = \Delta Lk/Lk_0$, where $Lk_0 = N/h$ with $N$ denoting the number of bp and $h$ denoting the helical repeat (10.5 bp for B-DNA). A salt concentration of 0.1 M sodium is used in all these studies, and the temperature, $T$, is set to 300 K. The dynamic simulations are started at the lowest energy interwound equilibrium configuration found by energy minimization, as shown in Figs. 1 and 2. Branched (as opposed to straight) configurations are also seen in Fig. 1 and often detected in electron-microscopy photographs of long DNA.

**2.3. Algorithms.** For the dynamic trajectories, we use a Langevin dynamics protocol with $\gamma = 9.77 \times 10^{10}$ s$^{-1}$ (in program units) as calibrated in [**26**] to give an approximate representation of solvent damping as well as maximal sampling. The Langevin equation of motion is integrated numerically by the Langevin/implicit-Euler ("LI") algorithm with a timestep of $\tau \simeq 100$ fs in program units (see next section for further details). The scaling discussed in the next section will determine what $\tau$ corresponds to in physical time for the macroscopic model. Full algorithmic details are given in other works [**28, 24, 38**].

The LI trajectories computed with this value of $\gamma$ are in good agreement with explicit (Verlet) trajectories. This is because numerical damping issues [**38**] are not relevant for this macroscopic model at moderate $\gamma$. (They would be significant at low $\gamma$). See Table 1 for compared dynamic averages of the writhing number, $Wr$, from the two simulations. We also find that other statistical properties, such as the distribution of writhe (Fig. 2), do not differ substantially. The significant width of the distributions in both cases reflects thermal deviations from the enthalpic minimum [**26**].

The variation in time of the writhing number and energy components (bending, twisting, Coulombic), shown in Fig. 3, as well as the total kinetic and potential energies (Fig. 4), are also similar for Verlet and LI. The larger fluctuations in $E_C$ for Verlet most likely result from larger fluctuations in total contour length (and hence $E_L$), a high-frequency term that is effectively damped in LI. Of course, differ-

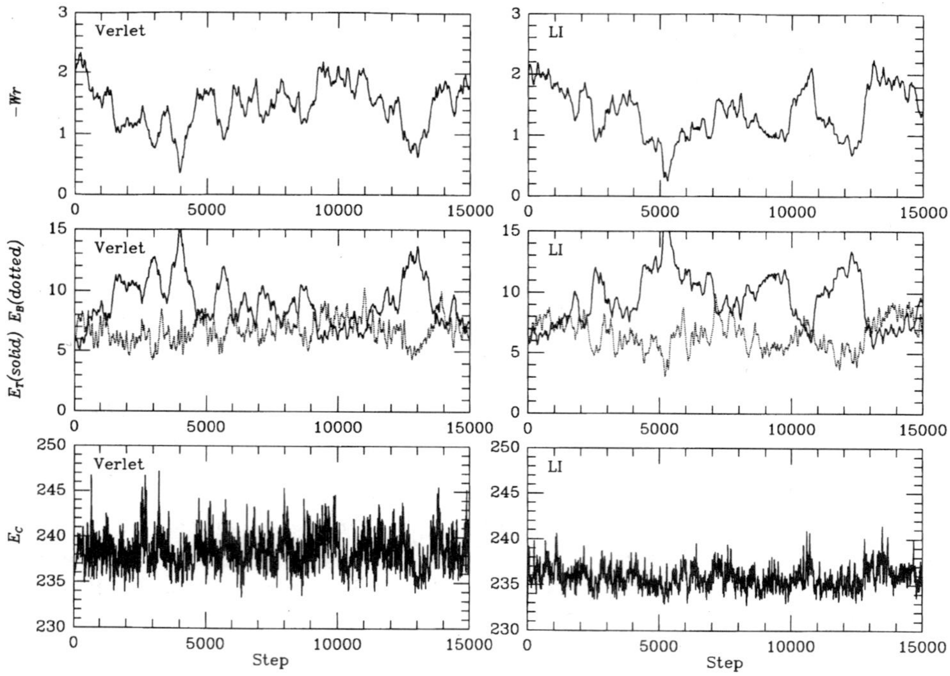


FIGURE 3. *The time evolution of the writhe and energy components (bending, twisting, and Coulombic) for a DNA system of 1000 bp: Verlet vs. LI integration.* Data correspond to the first trajectory for each set of Table 1. For clarity only half of the total trajectory is shown and the 'step' refers to the LI timestep. Panels on the left correspond to Verlet, and panels on the right correspond to LI.

ences are expected due to the stochastic nature of the Langevin formulation. That is, there is no exact trajectory in a stochastic simulation. LI, however, offers a significant computational advantage: it is almost three times faster than the Verlet trajectory for the same total time. This difference makes LI useful for computing converged properties for larger DNA systems, such as translational diffusion constants.

**2.4. Computational Performance.** All computations are performed on a Silicon Graphics Power Challenge computer with four 75 MHZ IP21 processors. An MD simulation of 30,000 iterations for a DNA system of 1000 bp requires approximately 21 CPU hours with the LI scheme (program timestep of $\simeq$ 100 fs) in comparison to 58 CPU hours for a Verlet trajectory covering the same total time with a program timestep of $\simeq$ 5 fs (600,000 iterations). CPU times scale similarly for longer DNA.

## 3. Results and Discussion

In this section, we discuss in turn the scaling of masses, timestep, and damping constant in the Langevin equation. We show consistency of translational diffusion constants in the diffusive limit between Langevin theory and experiment,

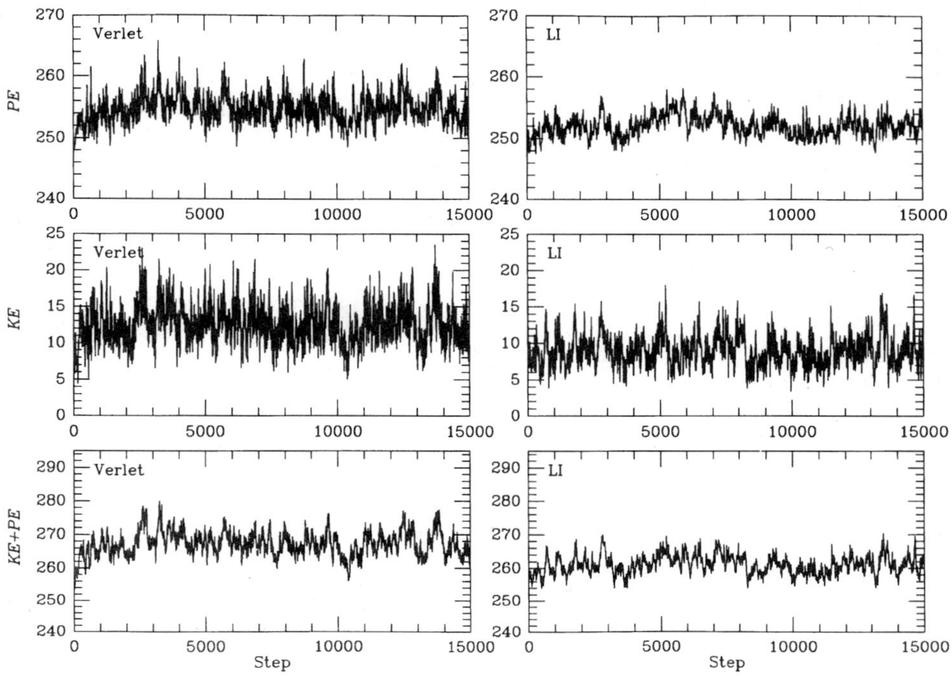


FIGURE 4. *The time evolution of the energies for a DNA system of 1000 bp: Verlet vs. LI integration.* Shown from top to bottom are the potential, kinetic, and total energies. See Fig. 3 caption.

and we calibrate the physical damping constant for supercoiled DNA. We also discuss the effect of using a computationally advantageous $\gamma$ (in the optimal sampling regime) and show that, in the diffusive limit, this represents an additional scaling in time for correspondence with physical processes.

**3.1. Mass Scaling and Associated Timestep.** We rescale the masses of our independent variables (the spline control points) for computational reasons, namely to speed up the dynamics so that interesting phenomena can be captured more rapidly. As we will show, even though the energy depends on the curve points, which are constructed from the control points, the dynamics propagated is essentially equivalent to that of a bead model.

The effect of rescaling the masses is to rescale the time of the motion. This can be seen by considering the Langevin equation of motion in its simplest form:

$$M\frac{d^2\mathbf{X}(t)}{dt^2} = -\mathbf{g}_E(\mathbf{X}(t)) - \gamma M\frac{d\mathbf{X}(t)}{dt} + \mathbf{R}(t), \tag{5}$$

where $M$ is the diagonal mass matrix; $\gamma$ is the collision frequency (or damping constant); $\mathbf{X}$ and $\frac{d\mathbf{X}(t)}{dt}$ are the collective coordinate and velocity vectors of the molecular system, respectively; $\mathbf{g}_E(\mathbf{X})$ is the gradient vector of the potential energy $E$; and $\mathbf{R}(t)$ is a random force vector, whose mean, $\langle\mathbf{R}(t)\rangle$, is zero and whose covariance matrix is given by:

$$\langle\mathbf{R}(t_1)\mathbf{R}(t_2)^T\rangle = 2\gamma k_B T M\delta(t_1 - t_2). \tag{6}$$

Above, $k_B$ is Boltzmann's constant and $\delta$ is the usual Dirac delta function.

Now, each particle has a real mass $m$ from association with the physical system. If the masses are identical, as in our model, then all entries in the diagonal mass matrix $M$ are equal. In the program, we use rescaled masses, $m'$ (instead of $m$), which define the diagonal mass matrix $M'$. Consider the transformation in time $t \to \alpha t'$, i.e.,

$$t' = t/\alpha, \tag{7}$$

where

$$\alpha = \sqrt{m/m'}. \tag{8}$$

For example, in simulating 1000 bp, we use 14 control points as the independent variables and assign a mass of 1 amu to each. Assuming each base pair has a mass of approximately 615 amu, $\alpha = \sqrt{(6.15 \times 10^5/14)/1} = 210$. In general, scaling to lighter particles produces $\alpha > 1$, a larger physical time unit than the program value. Under this transformation, the Langevin equation becomes, in the new time frame $t'$:

$$M' \frac{d^2\mathbf{X}(t')}{dt'^2} = -\mathbf{g}_E(\mathbf{X}(t')) - (\Gamma\gamma)M' \frac{d\mathbf{X}(t')}{dt'} + \mathbf{R}(t'), \tag{9}$$

where the matrix $\Gamma$ is diagonal with each entry equal to $\alpha$. If we set

$$\gamma' = \alpha\gamma, \tag{10}$$

we recover the form of Eq. 5, where $1/\gamma'$ is now in units $t'$. Note that the units of the two pairs, $\{\gamma, t\}$ and $\{\gamma', t'\}$ must be consistent. Our previously reported $\gamma$ is actually in inverse units of $t/\alpha$ rather than $t$ (which is in seconds) [**26**].

In our program, since $\alpha = 210$, we obtain a mass scaled timestep $\tau \simeq 20$ ps corresponding to the integration timestep of $\tau' \simeq 100$ fs. This rescaled timestep coincides with the rigorous lower bound to the timestep estimated previously [**29**] by a theoretical and numerical treatment. This estimate was based on the expected kinetic energy for the system. The effects of numerical damping due to the implicit integrator (at low $\gamma$) were also considered by approximating the effective temperature in the simulation, leading to an upper bound of 150 ps [**29**]. However, numerical damping is negligible in our current simulations as discussed above, where $\gamma$ is not as small as in the inertial regime [**26**]. Indeed, the larger value of $\gamma = 4.65 \times 10^8 \mathrm{s}^{-1}$ chosen in this work for enhanced sampling considerations [**26**] leads to an effective temperature very close to the input temperature of 300 K (see Table 2). Thus, only the lower bound of [**29**] is relevant and the timestep $\tau$ is the value of 20 ps derived above by mass scaling.

**3.2. Translational Diffusion Constants — Theory vs. Simulation.** The Langevin formalism for modeling solvent-solute coupling is a very simple, though useful one [**22**]. In its simplest form, it certainly ignores the effects of hydrodynamic interactions, the finite time of solvent relaxation, and solute memory. However, this approach is often useful in diffusive regimes where the above effects can be considered as small corrections to the overall motion. In particular, one can choose $\gamma$ to reproduce exactly the experimentally known translational diffusion constant, $D_t$. This is because the numerical value for $D_t$ in the diffusive limit of the Langevin framework corresponds to

$$D_t = \frac{k_B T}{\Sigma m\gamma}, \tag{11}$$

| Scheme | $\langle T \rangle$ (K) | standard deviation of $T$ |
|---|---|---|
| LI | 293.4 | 248.7 |
| LI | 287.6 | 235.2 |
| LI | 277.7 | 226.8 |
| Verlet | 312.7 | 257.5 |
| Verlet | 292.7 | 236.4 |
| Verlet | 284.8 | 223.5 |

TABLE 2. **Comparison of the LI and Verlet Schemes: Kinetic Temperatures.** See legend to Table 1. The mean temperatures are reported from the kinetic energy of the center of mass, $E_k$, according to the formula: $T = 2E_k/3k_B$. The large standard deviation in the center-of-mass temperature is expected in this regime of stochastic dynamics.

where $\Sigma m$ denotes the total mass of the particles (beads, atoms, etc.). Note that $D_t$ is configuration independent in the simple Langevin equation; configuration dependence arises with the incorporation of hydrodynamic effects in the form of a friction kernel [**22**].

From Eq. 11, we can estimate what $D_t$ should be for our simulations in the high $\gamma$ regime and compare this to computed values. From a numerical simulation $D_t$ can be calculated by using the Einstein-Stokes equation, which relates the mean square fluctuations of the system to $D_t$ according to:

$$(12) \qquad 6tD_t(t_2) = \langle | \mathbf{X}_{cm}(t_1) - \mathbf{X}_{cm}(t_1 + t_2) |^2 \rangle,$$

where $\mathbf{X}_{cm}$ is the center-of-mass position vector of the DNA supercoil. For example, for our DNA system of 1000 bp, working in the lighter-mass scaled time units, denoted as $t'$, and taking $k_BT = 4.14 \times 10^{-21}$ J, $\Sigma m = 14$ amu $\times 1.6605 \times 10^{-27}$ kg/amu, and $\gamma' = 9.77 \times 10^{10}$ $(t')^{-1}$ results in $D_t = 1.82 \times 10^{-6}$ m$^2/t'$. Converting this to mass scaled time (dividing $D_t$ by $\alpha$) gives $D_t = 8.67 \times 10^{-9}$ m$^2$/s. Indeed, we find that our model agrees well with this theoretical result as the diffusive limit is approached (Table 3). For moderate $\gamma$, our computed $D_t = 3.11 \times 10^{-6}$ m$^2/t'$ or $14.8 \times 10^{-9}$ m$^2$/s (Table 3, first row of data) most likely contains contributions of the faster motions, coming from global bending and twisting [**23, 32**]. The trajectory is also relatively short compared to the velocity relaxation time [**26**]. Indeed, at a much higher $\gamma$ ($\simeq 10^{13}$ $(t')^{-1}$ or $5 \times 10^{10}$ s$^{-1}$), the results are in excellent quantitative agreement with the expected $D_t$ (Table 3, second row of data) due to the faster relaxation time [**26**]. The trends for 2000 bp are identical (third and fourth data rows of Table 3).

We also find that these $D_t$ results are independent of the starting configuration of the DNA. In essence, it is only the motion of the center-of-mass that is relevant to the diffusion of the molecule and all effects of varying configurational shape are absorbed into $\gamma$. Therefore, $\gamma$ is an adjustable parameter that can be chosen on the basis of Eq. 11 to reproduce the experimental values of $D_t$ in this simple Langevin framework. The close agreement between the theoretical and numerical values validates the procedure and integrator used and also shows the essential equivalence of the B-spline and bead model in this regard. Of course, incorporation

| Base pairs | $\gamma'$ ($(t')^{-1}$) | $\gamma$ ($s^{-1}$) | $D_t$ ($m^2\ s^{-1}$), simulation | $k_BT/\Sigma m\gamma$ ($m^2\ s^{-1}$) |
|---|---|---|---|---|
| 1000 | $9.77 \times 10^{10}$ | $4.65 \times 10^{8}$ | $14.8 \times 10^{-9}$ | $8.67 \times 10^{-9}$ |
| 1000 | $9.77 \times 10^{12}$ | $4.65 \times 10^{10}$ | $8.44 \times 10^{-11}$ | $8.67 \times 10^{-11}$ |
| 2000 | $9.77 \times 10^{10}$ | $4.65 \times 10^{8}$ | $6.53 \times 10^{-9}$ | $4.33 \times 10^{-9}$ |
| 2000 | $9.77 \times 10^{12}$ | $4.65 \times 10^{10}$ | $4.41 \times 10^{-11}$ | $4.33 \times 10^{-11}$ |

TABLE 3. **Translational Diffusion Constants, $D_t$: Theory vs. Simulation.** Shown in the fourth column are computed $D_t$ values as estimated from the mean square fluctuations (Eq. 12). The last column contains values of $D_t$ expected from Eq. 11. All reported values for $D_t$ are given in real time units. Values of $\gamma$ are given in both time units for direct comparison with [**26**]. Results for DNA of 1000 bp at $\gamma = 4.65 \times 10^{10}$ s$^{-1}$ ($\gamma' = 9.77 \times 10^{12}$ $(t')^{-1}$) are taken from [**26**]. A very good agreement is noted in the diffusive limit ($\gamma = 4.65 \times 10^{10}$ s$^{-1}$, second and fourth data rows) for both DNA systems of 1000 and 2000 bp.

| Base pairs | $D_t$ ($m^2\ s^{-1}$), experimental [**18**] | $\gamma$ ($s^{-1}$) | $\gamma$ ($(t')^{-1}$) program units |
|---|---|---|---|
| 1000 | $> 9.5 \times 10^{-12}$ | $< 4.27 \times 10^{11}$ | $< 9.00 \times 10^{13}$ |
| 2000 | $6.0 \times 10^{-12}$ | $3.34 \times 10^{11}$ | $7.09 \times 10^{13}$ |
| 3000 | $4.6 \times 10^{-12}$ | $2.94 \times 10^{11}$ | $6.16 \times 10^{13}$ |

TABLE 4. **Estimation of $\gamma$ from Experimental $D_t$.** Shown are the values of the collision frequency, $\gamma$, corresponding to the experimentally derived $D_t$ for 1000, 2000, and 3000 bp, as computed using Eq. 11 in the text. The experimental values of $D_t$ [**18**] are subject to uncertainty. The values for 2000 and 3000 bp are derived from the straight line in Fig. 3 of [**18**], while the value for 1000 bp is estimated [**7**] to be larger than the value extrapolated from that straight line [**18**].

of hydrodynamic effects is required for interpreting biological processes associated with supercoiling on a finer level.

**3.3. Translational Diffusion Constants — Simulation vs. Experiment.** Up to now, we have seen that our model and timestep agree exactly with results expected from a bead model of DNA in the Langevin framework on the basis of the $D_t$ formula (Eq. 11). To determine the physically relevant $\gamma$, we now examine experimental $D_t$ values (Table 4). These $D_t$ values for closed-circular DNA systems of 1000, 2000, and 3000 bp [**18**] are given along with the values of $\gamma$ needed to reproduce them. The experimental values themselves are prone to uncertainty [**23, 34**]; hence, as more reliable estimates for $D_t$ are obtained, improved parameterization can be made. Note that the rough dependence of $D_t$ on size, estimated from experiment, namely $D_t \sim N^{-0.6}$ [**18**], departs from the Langevin dependence (Eq. 11), which is $N^{-1}$ to a first approximation if a constant $\gamma$ is used (also seen from the last column of Table 3).

We find that the corresponding value of $\gamma$ ($\approx 3 \times 10^{11}$ s$^{-1}$) is roughly 700 times our chosen value for optimal sampling. Actually, we suggested a broad range of $\gamma$ for optimal sampling, between $4.65 \times 10^8$ and $4.19 \times 10^9$ s$^{-1}$ [**26**]; therefore, the physical $\gamma$ for reproducing $D_t$ is between 70 to 700 times the $\gamma$ for optimal sampling. Indeed, this difference between the physical $\gamma$ to reproduce $D_t$ in the Langevin framework and the computational optimal $\gamma$ for sampling is well appreciated. In a work of Pastor and co-workers it was shown that the maximal isomerization rate for peptide models is obtained at $\gamma = 2$ ps$^{-1}$, while typical collision frequencies of protein atoms exposed to solvent having a viscosity of 1 cp are $\gamma$ = 50–60 ps$^{-1}$ [**20**].

Since algorithms in the diffusive regime depend on the ratio $\gamma/\Delta t$, scaling of time-dependent quantities can be done in correspondence with scaling of $\gamma$. This is because the friction, and hence the relaxation time, is proportional to the viscosity. Using a smaller $\gamma$ is also computationally advantageous since the product of $\gamma$ and the timestep affects numerical behavior. A smaller $\gamma$ can permit a larger timestep for the same accuracy.

In our case, choosing $\gamma = 4.65 \times 10^8$ s$^{-1}$ (smaller than the physical $\gamma$ by a factor of 700) yields an effective timestep of order 10 ns ($20 \times 700$ ps) according to this direct scaling. This value of $\tau$ agrees well with our previous estimate of the timestep in the nanosecond range, based on comparison to experiment [**26**].

In the non-diffusive case, this direct scaling is not necessarily true. This is because the various modes in the system are coupled differently to the solvent in a complex way. However, the $\gamma$ we use here ($4.65 \times 10^8$ s$^{-1}$) is close to the diffusive regime, as evidenced by the rapid decay of the velocity autocorrelation function, shown in Fig. 12(d) of [**26**]. Thus the scaling proposed here, leading to an effective physical timestep in the nanosecond range, should be a good approximation.

In sum, the optimal computational $\gamma$ allows simulations to capture more easily global processes that would require longer simulations for adequate statistics. Since the scaling is understood in the diffusive regime, choosing the optimal $\gamma$ for a simulation is advantageous.

**3.4. Hydrodynamic Effects.** In general, hydrodynamic interactions lower the total friction for a large polymer. This is because the solvent is immobilized to a certain extent within the polymer domain and moves with the biomolecule in solution. To obtain a rough estimate of how the translational friction constant, $f_T$, would change in the presence of hydrodynamic interactions for our macroscopic DNA model, we now apply the Kirkwood-Riseman (KR) approximation for bead hydrodynamics [**22**]. In the absence of hydrodynamic interactions (the "free-draining" limit) $f_T$ is the sum of the friction constants of the individual beads, i.e., $f_T = N_b \zeta$, where $N_b$ is the number of beads and $\zeta = \gamma m$.

The KR equation for $f_T$, which approximately incorporates hydrodynamic interactions, can be written as:

$$f_T = \frac{N_b \zeta}{1 + \frac{\zeta}{6\pi\eta N_b} \sum_{\substack{i,j \\ i \neq j}} \langle \frac{1}{r_{ij}} \rangle}, \tag{13}$$

where $\eta$ is the viscosity of the solvent and $\langle \frac{1}{r_{ij}} \rangle$ is the mean inverse distance between beads $i$ and $j$ averaged over an ensemble of configurations. Thus, the KR approximation is based on a preaveraged Oseen configuration tensor [**22**]; the true hydrodynamic interaction matrix requires inversion of all configuration-dependent tensors in the ensemble and is not practical for large floppy systems. Now, from Stokes' equation (assuming stick boundary conditions) one can estimate $\zeta$ to be:

$$\zeta = 6\pi\eta r_e, \tag{14}$$

where $r_e$ is the effective hydrodynamic radius of each bead. Substituting this expression for $\zeta$ into the KR expression (Eq. 13) yields:

$$f_T = \frac{N_b\zeta}{1 + \frac{r_e}{N_b}\sum_{\substack{i,j \\ i \neq j}} \langle \frac{1}{r_{ij}} \rangle}. \tag{15}$$

For our 1000 bp DNA model, $N_b = 14$ and we take $r_e = 30$ Å at 0.1 M salt, since the effective DNA diameter at this environment is 60 Å. The quantity $\sum_{i,j} \langle \frac{1}{r_{ij}} \rangle$ (where the $i \neq j$ in the sum is hereafter assumed) must be calculated numerically over an ensemble of possible configurations. As a quick upper bound, consider that the largest distance between two beads corresponds to the largest distance in a closed-circular chain; interwound configurations are more compact (Figs 1, 2). This largest distance is $L/\pi$ where $L$ is the contour length of the DNA (3400 Å for 1000 bp DNA) and corresponds to 1082 Å. Thus, the smallest $\sum_{i,j} \langle \frac{1}{r_{ij}} \rangle$ is $N_b(N_b-1)\pi/L$, which leads to an upper bound for $f_T$ of $10.3\zeta$. This value reflects the deviation from $14\zeta$, the free-draining limit.

From our dynamics simulation (with $\gamma$ in the optimal sampling regime), we have analyzed the entire configurational ensemble to find that $\sum_{i,j} \langle \frac{1}{r_{ij}} \rangle \simeq 0.4036$ Å$^{-1}$. This is based on a mean value for $\langle 1/r_{ij} \rangle$ for each distinct $\{i, j\}$ pair over 12,000 configurations from the first LI trajectory in Table 1. (If we convert this to a single $\langle r_{ij} \rangle$ value, we get 451 Å). Using the above value for the sum we find that $f_T \simeq 7.5\zeta$. A 10% underestimate of $f_T$ is also expected [**22**]. Thus, the total translational friction constant might be reduced by roughly a factor of two when hydrodynamic interactions are incorporated. This implies that we are not neglecting significant interactions that will affect behavior by orders of magnitude. We are, therefore, in the limit where the Langevin formalism is appropriate and the free-draining limit reasonable. In particular, the relative ordering of relaxation times for the different modes of the DNA [**32**] is likely to be correct.

## 4. Summary

From the theoretical and numerical analysis described here, we found the following.

1. **Implicit integration.**
The implicit LI algorithm exhibits negligible numerical damping for the range of $\gamma$ chosen for optimal sampling in our macroscopic model. A computational gain of a factor of three in comparison to a standard explicit integrator is found. Thus, as used here, LI is appropriate for the macroscopic model of DNA where high-frequency modes are essentially absent. At low $\gamma$, however, numerical damping will be evident, even for a macroscopic model [**26**]. Thus LI must be applied with care. More generally, LI can be useful in the context of other macroscopic models or formulations that include a counteracting mechanism for numerical damping. Scheraga and co-workers [**13**] have used LI in this way to suggest a folding pathway for the protein BPTI.

2. **Mass and time scaling in the Langevin formalism.**
Scaling the masses in a macroscopic dynamic model based on the Langevin equation is a convenient way to achieve computational efficiency. Rescaling to lighter masses accelerates the timescales. The physical time $t$, corresponding to the simulated time $t'$ is given by $t' = t/\alpha$, where $\alpha = \sqrt{m/m'}$, the square root of the ratio of the actual mass to the program mass used for the independent variables (assumed to be all equivalent). Note that $\gamma$ must also be scaled to the new time unit, i.e., $\gamma' = \alpha\gamma$, for consistency (Eq. 9 vs. 5). In our DNA model, we have shown that this mass scaling results in an associated integration timestep of 20 ps corresponding to a program timestep of 100 fs. This calibrated value of 20 ps is consistent with our previous rigorous theoretical and numerical analysis [**29**]. Moreover, it completes the previous work because it shows that the upper bound of [**29**] is not relevant here.

3. **Langevin $D_t$ – theory vs. simulation.**
Our B-spline model reproduces excellently the theoretically expected value of the translational diffusion constant, $D_t = k_B T/\Sigma m\gamma$, in the diffusive limit (Table 3). This validates our simulation protocol, both model and numerical integration algorithm.

4. **Optimal computational $\gamma$ vs. experimentally based $\gamma$.**
Comparison with the experimental values of $D_t$ produces an estimate of the $\gamma$ to reproduce experimental $D_t$ values (Table 4): this $\gamma$ (of the order of $10^{13}$ s$^{-1}$) is roughly two orders of magnitude larger than the range of $\gamma$ used for optimal configurational sampling. This is consistent with the trends noted for proteins and shows that a smaller $\gamma$ is numerically advantageous for focusing on global collective motions and thereby improving statistics. Since algorithms in the diffusive regime depend on the ratio $\gamma/\Delta t$, scaling of time-dependent quantities can be done for a simulation at smaller $\gamma$. In the high $\gamma$ regime, numerical integration is more of a problem, since the product of $\gamma$ and the timestep affects numerical behavior. As a side remark, note that the experimental results suggest a size scaling in $D_t$ of $N^{-0.6}$ (also the limit at high molecular weight for flexible chains [**6**]), which departs from the Langevin and rigid-rod limit [**6**] of $N^{-1}$, where $N$ is the number of bp; this implies that to match the experimental value of $D_t$ under the Langevin formulation, $\gamma$ should be size dependent.

5. **Physical timestep for DNA.**
The calibrated value of 20 ps for the mass scaled formulation is associated with our $\gamma$ for maximal sampling (point 4 above). The effective physical timestep must take into consideration the physical $\gamma$ to reproduce the experimental $D_t$. This physical scaling, valid in the diffusive regime, yields an effective timestep in the nanosecond range ($\sim$ 10 ns) corresponding to our mass scaled integration timestep of 20 ps. This value is in good agreement with our previous estimate based on comparison to physical measurements [**26**].

6. **Hydrodynamic effects.**
By using the KR approximation based on a preaveraged configuration tensor, we find that hydrodynamic interactions might reduce the total friction constant from $14\zeta$ to $7.5\zeta$ for a 1000 bp DNA system. Thus, neglect of these

interactions in the moderate viscosity regime is quite reasonable and should not lead to dramatic differences of orders of magnitude. In particular, the relative ordering of relaxation times for the different modes for the DNA [**32**] is likely to be correct. Hydrodynamic interactions, however, are likely to increase in importance as DNA size increases.

7. **The B-spline and bead models.**
From the $D_t$ analysis discussed in this work, we find that the B-spline model is equivalent to a bead model for quantitative measurements. That is, the $D_t$ values expected from Eq. 11 are obtained exactly in the diffusive limit (Table 3). The B-spline representation is nonetheless more compact and results in computational efficiency.

We hope that the results presented here will be useful for the development and application of other macroscopic models for proteins and nucleic acids.

## Acknowledgements

We thank David Shalloway and Panos Pardalos for organizing a stimulating meeting at DIMACS and for inviting this contribution. We are indebted to Rich Pastor for his invaluable insights and comments regarding Langevin dynamics.

## References

1. A. D. Bates and A. Maxwell, *DNA topology*, Oxford Univ. Press, New York, 1993.
2. W. R. Bauer, R. A. Lund, and J. H. White, *Twist and writhe of a DNA loop containing intrinsic bends*, Proc. Natl. Acad. Sci. USA **90** (1993), 833–837.
3. J. Bednar, P. Furrer, A. Stasiak, J. Dubochet, E. H. Engelman, and A. D. Bates, *The twist, writhe and overall shape of supercoiled DNA change during counterion-induced transition from loosely to a tightly interwound superhelix. Possible implications for DNA structure* in vivo, J. Mol. Biol. **235** (1994), 825–847.
4. C. J. Benham, *Onset of writhing in circular elastic polymers*, Phys. Rev. A **39** (1989), 2852–2856.
5. J. D. Bryngelson, J. N. Onuchic, N. D. Socci, and P. G. Wolynes, *Funnels, pathways, and the energy landscape of protein folding: a synthesis*, Proteins: Struc. Func. Gen. **21** (1995), 167–195.
6. C. R. Cantor and P. R. Schimmel, *Biophysical chemistry*, W. H. Freeman and Company, San Francisco, 1980.
7. G. Chirico and J. Langowski, *Kinetics of DNA supercoiling studied by Brownian dynamics simulation*, Biopolymers **34** (1994), 415–433.
8. N. R. Cozzarelli and J. C. Wang (eds.), *DNA topology and its biological effects*, Cold Spring Harbor Laboratory Press, Cold Spring Harbor, New York, 1990.
9. K. A. Dill, S. Bromberg, K. Yue, K. M. Fiebig, D. P. Yee, P. D. Thomas, and H. S. Chan, *Principles of protein folding — a perspective from simple exact models*, Protein Science **4** (1995), 561–602.
10. M. D. Frank-Kamenetskii, *Unravelling DNA*, VCH Publishers Inc., New York, NY, 1993.
11. M.-H. Hao and W. K. Olson, *Global equilibrium configurations of supercoiled DNA*, Macromolecules **22** (1989), 3292–3383.
12. ———, *Modeling DNA supercoils and knots with B-spline functions*, Biopolymers **28** (1989), 873–900.
13. M.-H. Hao, M. R. Pincus, S. Rackovsky, and H. A. Scheraga, *Unfolding and refolding of the native structure of bovine pancreatic trypson inhibitor studied by computer simulations*, Biochemistry **32** (1993), 9614–9631.
14. J. E. Hearst and N. G. Hunt, *Statistical mechanical theory for the plectonemic DNA supercoil*, J. Chem. Phys. **95** (1991), 9322–9327.
15. N. G. Hunt and J. E. Hearst, *Elastic model of DNA supercoiling in the infinite-length limit*, J. Chem. Phys. **95** (1991), 9329–9336.
16. F. Jülicher, *Supercoiling transitions of closed DNA*, Phys. Rev. E **49** (1994), 2429–2435.
17. J. Kostrowicki and H. A. Scheraga, *Some approaches to the multiple-minima problem in protein folding*, 1995, in these proceedings.
18. J. Langowski and U. Geisen, *Configurational and dynamic properties of different length superhelical DNAs measures by dynamic light scattering*, Biophysical Chemistry **34** (1989), 9–18.
19. M. Le Bret, *Catastrophic variation of twist and writhing of circular DNAs with constraint?*, Biopolymers **18** (1979), 1709–1725.

20. R. J. Loncharich, B. R. Brooks, and R. W. Pastor, *Langevin dynamics of peptides: The frictional dependence of isomerization rates of N-acetylalanyl-N'-methylamide*, Biopolymers **32** (1992), 523–535.
21. G. S. Manning, *A procedure for extracting persistence lengths from light-scattering data on intermediate molecular weight DNA*, Biopolymers **20** (1981), 1751–1755.
22. R. W. Pastor, *Techniques and applications of Langevin dynamics simulations*, The Molecular Dynamics of Liquid Crystals (G. R. Luckhurst and C. A. Veracini, eds.), Kluwer Academic Press, Dordrecht, The Netherlands, 1994, pp. 85–138.
23. R. Pecora, *DNA: A model compound for solution studies of macromolecules*, Science **251** (1991), 893–898.
24. C. S. Peskin and T. Schlick, *Molecular dynamics by the backward Euler method*, Comm. Pure App. Math. **42** (1989), 1001–1031.
25. G. Ramachandran and T. Schlick, *On buckling transitions in superhelical DNA: Dependence on elastic constants and DNA size*, preprint, 1995.
26. ———, *Solvent effects on the dynamics of supercoiled DNA explored by Langevin dynamics simulations*, Phys. Rev. E **51** (1995), 6188–6203.
27. T. Schlick, *Modeling superhelical DNA: Recent analytical and dynamic approaches*, Curr. Opin. Struc. Biol. **5** (1995), 245–262.
28. T. Schlick, B. E. Hingerty, C. S. Peskin, M. L. Overton, and S. Broyde, *Search strategies, minimization algorithms and molecular dynamics simulations for exploring conformational spaces of nucleic acids*, Theoretical Biochemistry and Molecular Biophysics (D. L. Beveridge and R. Lavery, eds.), vol. 1, Adenine Press, Schenectady, New York, 1991, pp. 39–58.
29. T. Schlick, B. Li, and M.-H. Hao, *Calibration of the timestep for molecular dynamics of supercoiled DNA modeled by B-splines*, Structural Biology: State of the Art 1993, Proceedings of the Eighth Conversations, Volume I (Schenectady, New York) (R.H. Sarma and M.H. Sarma, eds.), Adenine Press, 1994, pp. 157–174.
30. T. Schlick, B. Li, and W. K. Olson, *The influence of salt on the structure and energetics of supercoiled DNA*, Biophysical Journal **67** (1994), 2146–2166.
31. T. Schlick and W. K. Olson, *Supercoiled DNA energetics and dynamics by computer simulation*, J. Mol. Biol. **223** (1992), 1089–1119.
32. T. Schlick, W. K. Olson, T. Westcott, and J. P. Greenberg, *On higher buckling transitions in supercoiled DNA*, Biopolymers **34** (1994), 565–597.
33. Y. Shi and J. E. Hearst, *The Kirchoff elastic rod, the nonlinear Schrödinger equation, and DNA supercoiling*, J. Chem. Phys. **101** (1994), 5186–5200.
34. S. S. Sorlie and R. Pecora, *A dynamic light scattering study of a 2311 base pair DNA restriction fragment*, Macromolecules **21** (1988), 1437–1449.
35. A. V. Vologodskii and N. R. Cozzarelli, *Conformational and thermodynamic properties of supercoiled DNA*, Ann. Rev. Biophys. Biomol. Struc. **23** (1994), 609–643.
36. J. H. White, *An introduction to the geometry and topology of DNA structures*, Mathematical Methods for DNA Sequences (M. S. Waterman, ed.), CRC Press, Boca Raton, Florida, 1989.
37. Y. Yang, I. Tobias, and W. K. Olson, *Finite element analysis of DNA supercoiling*, J. Chem. Phys. **98** (1993), 1673–1686.
38. G. Zhang and T. Schlick, *LIN: A new algorithm combining implicit integration and normal mode techniques for molecular dynamics*, J. Comp. Chem. **14** (1993), 1212–1233.

The Howard Hughes Medical Institute and New York University, Chemistry Department and Courant Institute of Mathematical Sciences, 251 Mercer Street, New York, New York 10012, U.S.A.

*E-mail address*: schlick@nyu.edu, ramachan@franklin.biomath.nyu.edu

DIMACS Series in Discrete Mathematics
and Theoretical Computer Science
Volume **23**, 1996

# A hierarchical approach to the prediction of the quaternary structure of GCN4 and its mutants

Michal Vieth, Andrzej Kolinski, C.L. Brooks, III, and Jeffrey Skolnick*

ABSTRACT. A hierarchical approach to protein folding is employed to examine the folding pathway and predict the quaternary structure of the GCN4 leucine zipper. Structures comparable in quality to experiment have been predicted. In addition, the equilibrium between dimers, trimers and tetramers of a number of GCN4 mutants has been examined. In five out of eight cases, the simulation results are in accordance with the experimental studies of Harbury, et al [**1**].

## 1. Introduction

An outstanding unsolved problem in molecular biology is the prediction of protein structure from amino acid sequence. As a step in this direction, we have examined the folding pathway and the native structure GCN4 leucine zipper. In addition, the equilibrium between various GCN4 mutants has been explored. The basic idea is to employ a reduced model to stimulate the early stages of folding and to obtain the native state topology. Restrained molecular dynamics are then used to produce full atom models [**2**]. We have also developed a general formalism to estimate the equilibrium between various multimeric species. An overview of the procedure to predict the native conformation of a protein from sequence is presented in Figure 1 [**3, 4**]. Folding of the model chains commences on a very high coordination lattice, from a pair of random unfolded chains. Following dimer assembly to a parallel, left handed, in register coiled coil, the resulting structures are refined on the lattice to produce a family of native structures whose rms from native ranges from 2.3 to 3.7 Å. Subsequently, full atom models are built, the structures are solvated and refined, using molecular dynamics with the CHARMM potential [**5**]. The resulting family of structures is indistinguishable from the native one when subjected to the molecular dyanmics refinement protocol. The average structure from this family has a backbone heavy atom rms 0.8 Å, all heavy atoms in the dimerization interface differ by 1.31 Å rms from native, and all heavy atoms differ from the crystal structure by 2.29 Å rms. These studies were the first time that protein quaternary structures of this quality have been obtained from random

1991 *Mathematics Subject Classification.* Primary 65K10, 90C30; Secondary 92C05, 92E10.
*Key words and phrases.* Protein folding, GCN4 Leucine zipper, Computer simulations, Lattice protein models.

TABLE 1. Comparison of simulation results with experiment on GCN4 mutants. *indicates those results which apparently disagree with experiment. In the VI case, the degree of association could not be assigned experimentally.

| | Experiment | Simulation |
|---|---|---|
| Residues at positions a d | Dominant species | Dominant species |
| wild type | 2 | 2 |
| I L | 2 | 2 |
| I I | 3 | 3 |
| L I | 4 | 3,4! |
| V I | ? | 3! |
| L V | 3 | 3 |
| V L | (2,3) | 2! |
| L L | 3 | 3 |

unfolded conformations. In a recent paper in Science by Harbury et. al. [**1**], they examined the shift in equilibrium between parallel, coiled coil dimers, trimers and tetramers associated with various residue identities of the two hydrophobic residues in the coiled heptad repeat.

To further test the validity of the model, we prepared a series of dimers, trimers, and tetramers for each of the eight species that have been experimentally studied. To compare with the experiment, we have to calculate the equilibrium constants associated with the dimer, $D$, trimer, $T$, and tetramer, $R$, species

$$3D \leftrightarrow 2T \tag{1}$$

$$2D \leftrightarrow R \tag{2}$$

with $(L)$, the number of chains in species $L = D, T, R$. The equilibrium constants are

$$K_{DT} = \frac{(T)^2}{(D)^3} = \frac{VZ_T^2}{Z_D^3} \tag{3}$$

$$K_{DR} = \frac{(R)}{(D)^2} = \frac{VZ_R}{Z_D^2} \tag{4}$$

with $V$ the total accessible volume of the system. The key to the calculation of the partition function is to use a generalization of a rotational isomeric approach originally applied to 24 neighbor lattice by Skolnick and Kolinski [**6**] to the present lattice. With the neglect of excluded volume between residues that are more than 4 neighbors down the chain (a fairly minor approximation in the denatured state of proteins), we can calculate the configurational partition function using a transfer matrix approach that is logically equivalent to Zimm-Bragg helix coil theory [**7**]. For the native molecule, we approximate the partition function by the Boltzmann factor of the energy of the most probable conformation times the product of lattice accessible lattice states. The latter are decomposed into the product of the probabilities of accessible three bond states. (Sampling problems preclude the inclusion of higher order correlations.) The comparison with experiment is shown in Table 1.

## 2. Conclusions

These simulations on the folding of coiled coils demonstrate that it is possible to predict quaternary structure of simple folding motifs from amino acid sequence alone. Furthermore, the ability to predict the dominant species provides some confidence that the phenomenological potentials have captured aspects of the experimental situation. Future work will entail extension to larger coiled coils and extension of the model to predict the quarternary structure of some small globular proteins.

## References

[1] P. Harbury, T. Zhang, P. Kim S., T. Alber, *Science*, **262**, 1401–1407 (1993).
[2] M. Vieth, A. Kolinski, C.L. Brooks III, J. Skolnick, *J. Mol. Biol.*, (submitted), (1995).
[3] A. Kolinski, J. Skolnick, *Proteins*, **18**, 338–352 (1994).
[4] M. Vieth, A. Kolinski, C.L. Brooks III, J. Skolnick, *J. Mol. Biol.*, **237**, 361–367 (1994).
[5] B.R. Brooks et al., *J. Comp. Chem.*, **4**, 187–217 (1983).
[6] J. Skolnick, A. Kolinski, *J. Mol. Biol.*, **221**, 499–531 (1991).
[7] D. Poland, H.A. Scheraga, *Theory of helix–coil transitions in biopolymers*, (Academic Press, New York, 1970).

THE SCRIPPS RESEARCH INSTITUTE, 10666 NORTH TORREY PINES ROAD, LA JOLLA, CA 92037

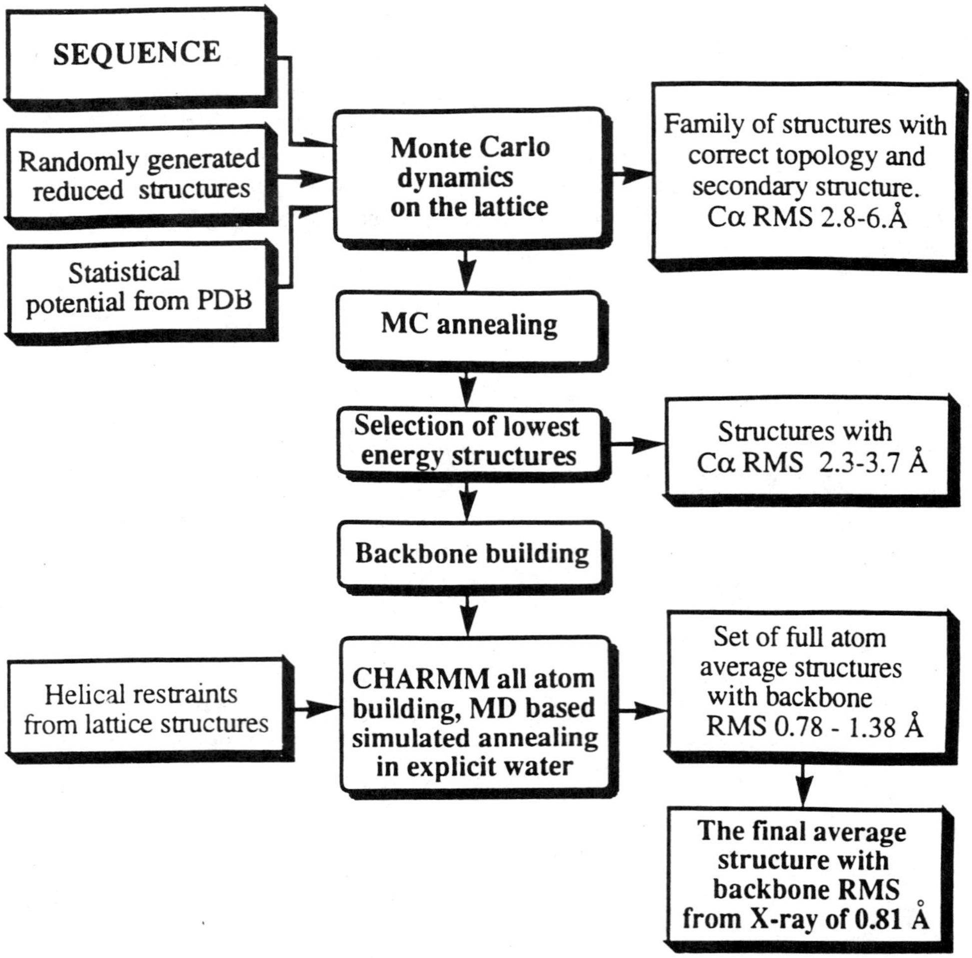


FIGURE 1. An overview of the hierarchical approach. The left hand column displays the input information, the central column displays the protocol, and the right-hand column shows the resulting output.

DIMACS Series in Discrete Mathematics
and Theoretical Computer Science
Volume **23**, 1996

# Rapid Evaluation of Potential Energy Functions in Molecular And Protein Conformations

G.L. Xue, A.J Zall, and P.M. Pardalos

ABSTRACT. One of the biggest obstacles in computer simulation of large scale molecular and protein conformations is the evaluation of potential energy functions and its gradient vector. Given a system involving $n$ particles, conventional evaluation of the potential energy function requires $O(n^2)$ operations, making the simulation of large systems prohibitive. We observe that in most of the potential energy functions used, only the term involving pair-wise potential energy requires $O(n^2)$ operations. In this paper, we study algorithms for the rapid evaluation of pair-wise potential energy function and its gradient vector. In particular, we have implemented a variant of the Appel's algorithm. This algorithm requires $O(n \log n)$ operations. Our computational result shows that the Appel algorithm becomes faster than the conventional algorithm when the problem size $n$ becomes 3000 or larger.

## 1. Introduction

One of the grand challenge problems in molecular biophysics is the `protein folding problem`: computing the native 3-dimensional conformation (folded state) of a globular protein given its amino acid sequence. A related problem is the `molecular conformation problem`: computing the native 3-dimensional conformation of a cluster of atoms given a potential energy function that involves two-body interatomic pair potentials. The success of computational methods to solve these problems hinges on two factors: First, a suitable potential energy function must be defined to reliably predict the native states of the system as the global minimizer of the potential energy function. Second, minimization methods must be available to efficiently locate the global minimizer of the potential energy function. Since these potential functions usually have many local minimizers and the evaluation of both the function and gradient vector is expensive, computing global

---

1991 *Mathematics Subject Classification.* Primary: 03G10, 06B35, 90B40, 90C27, 92B05; Secondary: 49M07, 51D20, 65K05 70F10, 92C05.

*Key words and phrases.* Algorithms, divide and conquer, nonconvex potential energy functions, grand challenge problems, protein folding, molecular conformation.

This research was supported in part by National Science Foundation grant No. NSF ASC-9409285.

minimizers of such potential functions is very expensive for small- and medium-sized problems and prohibitive for large-sized problems.

In this paper, we single out the most expensive part of the potential energy functions and study algorithms for the rapid evaluation of this function and its gradient. The rest of this paper is organized as follows. In section 2, we identify the computational task by single out the most expensive part used in most potential energy functions. In section 3, we present the conventional algorithm for the evaluation of the function and gradient. In section 4, we discuss a divide and conquer algorithm for the rapid evaluation of the potential energy function and its gradient. Computational results are presented in section 6.

## 2. The Computational Task

A simplified protein model consists of a linear chain of $n$ beads centered at $x_1, x_2, \cdots, x_n$, in a 3-dimensional space. For every pair of consecutive beads $x_{i-1}$ and $x_i$, there is a bond length $||x_i - x_{i-1}||_2$ which is the distance between them. For every three consecutive beads $x_{i-2}, x_{i-1}$, and $x_i$, let $\theta_i$ be the bond angle corresponding to the relative position of the third bead with respect to the line containing the previous two. Likewise, for every four consecutive beads $x_{i-3}, x_{i-2}, x_{i-1}$, and $x_i$, let $\phi_i$ represent the torsion angle corresponding to the relative position of the fourth bead with respect to the plane containing the previous three. In addition to the above, there is a Lennard-Jones potential which characterizes the 2-body interactions between every pair of beads. Let the potentials corresponding to bond length, bond angle, torsion angle, and Lennard-Jones be defined as

$$V_1(x) = \sum_{j=2}^{n} K_{bond} \times (||x_j - x_{j-1}||_2 - bond_0)^2, \tag{1}$$

$$V_2(x) = \sum_{j=3}^{n} K_\theta \times (\theta_j - \theta_0)^2, \tag{2}$$

$$V_3(x) = \sum_{j=4}^{n} K_\phi \times (1 + \cos(3 \times \phi_j)), \tag{3}$$

$$V_4(x) = \sum_{j=2}^{n} \sum_{i=1}^{j-1} K_{LJ} \times v_{LJ}(||x_j - x_i||), \tag{4}$$

where $K_{bond}$ and $K_\theta$ are the bond stretching and angle bending force constants, $K_\phi$ is the torsional constant, $K_{LJ}$ is the Lennard-Jones coefficients, and $bond_0$ and $\theta_0$ are idea values for the bond length and bond angle, respectively. The **protein conformation problem** is to find an optimal configuration for the chain (positions of the $n$ beads) such that the total potential energy function

$$V(x) = V_1(x) + V_2(x) + V_3(x) + V_4(x) \tag{5}$$

is globally minimized. Detailed description of this simplified protein model can be found in [**4**].

We note that in the above formulation of the potential energy function, $V_1, V_2$ and $V_3$ can all be computed in $O(n)$ operations while the computation of $V_4$ requires $O(n^2)$ operations. Therefore we will concentrate on the rapid evaluation of $V_4$ which happens to be the potential energy function used in most molecular conformation problems.

The molecular conformation problem is to find a global minimizer of a predefined potential energy function of a cluster of atoms. Of the many cases of interest, the simplest potential energy function is that of two-body central forces between the component atoms. Although this potential energy function does not actually model the real system in many cases–ignoring angle dependent conditions, bond constraints, and dihedral angle conditions–it provides a useful prototype that enormously simplifies the design of computer programs. This simplest potential energy function is also very important because it has been the most expensive part of the more general potential energy functions.

Given a cluster of $n$ atoms in 3–dimensional space, the potential energy function of the cluster is defined as the summation (over all of the pairs) of the two-body interatomic pair potentials. Let the center of the $n$ atoms be $x_1, \cdots, x_n$. The potential energy function of the cluster is defined as

$$V_n(x_1, \cdots, x_n) = \sum_{1 \le i < j \le n} v_{pair}(||x_i - x_j||), \tag{6}$$

where $||x_i - x_j||$ is the Euclidean distance between $x_i$ and $x_j$ and $v_{pair}(r)$ is an interatomic pair potential. In this proposal, $n$ will be used to represent the number of atoms or beads in the given system if not stated otherwise.

Although many types of function $v_{pair}(r)$ may be used in physical models, it is necessary to apply some restrictions if we want $v_{pair}(r)$ to have satisfactory behavior. In general, we would like $v_{pair}(r)$ to be continuous at least down to a hard core $r_{min}$ and to possess derivatives up to the second order over the interval $(r_{min}, \infty)$. More specifically, we are only interested in cases where $v_{pair}(r)$ is a `well potential` satisfying the following conditions:

(C_1) $v_{pair}(r) \to 0^-$ as $r \to \infty$;
(C_2) $v_{pair}(r) \to \infty$ as $r \to r_{min}$ and $r_{min} \ge 0$;
(C_3) $v'_{pair}(r_0) = 0$ for a unique $r_0$ with $r_{min} < r_0 < \infty$;
(C_4) $v''_{pair}(r_0) > 0$ and $v(r_0) < 0$.

There are a number of pair potential energy functions which satisfy the above conditions [**6**]. A well-known example is the Lennard-Jones pair potential function

$$v_{LJ}(r) = (\frac{\sigma}{r})^{12} - 2(\frac{\sigma}{r})^6, \tag{7}$$

where $\sigma$ is a positive constant. For the Lennard-Jones potential function, the global minimum of -1 is achieved at a unique point $r = \sigma$. A plot of this function with $\sigma = 1$ is given in Figure 1.

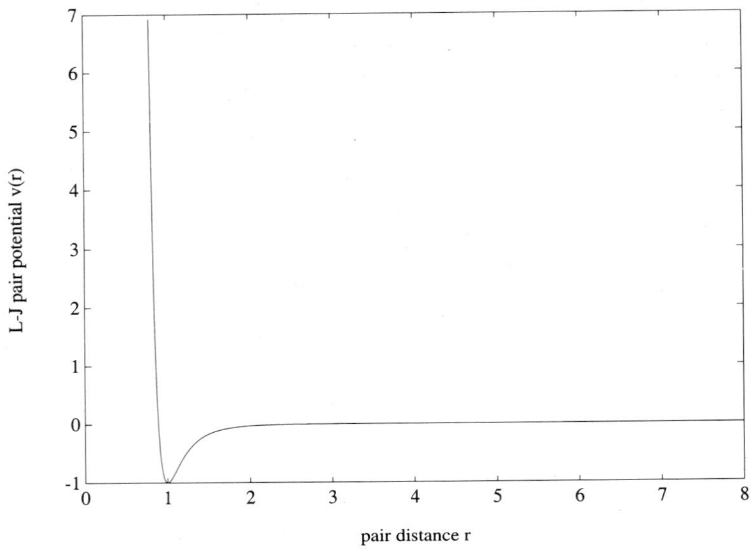


FIGURE 1. Plot of the Lennard-Jones Pair Potential Function

It is clear from the above discussions that the most expensive part in the evaluation of potential energy functions is the evaluation of the 2-body potential energy functions. Therefore we identify our computational task as the following. Given a cluster of $n$ particles centered at $x_1, x_2, \cdots, x_n$ in 3-dimensional space, compute

$$f(x_1, x_2, \cdots, x_n) = \sum_{1 \le i < j \le n} v_{LJ}(||x_j - x_i||) \tag{8}$$

and its gradient vector, where $v_{LJ}$ is the Lennard-Jones pair potential energy function defined in (7).

## 3. The Conventional Algorithm

The easiest way to code up the computation of $f(x)$ and its gradient vector is to follow the definitions in (8) and (7) and use a $O(n^2)$ operation method. A `FORTRAN` 77 code is presented in Figure 2 to serve as the basis of the comparison of our new algorithm.

The first author of the present paper have used this subroutine in many simulations [**9, 10**]. The code represent a highly optimized implementation of the $O(n^2)$

```
      subroutine lj3d(n3, x, f, g)
      integer    n3
      double precision x(n3), f, g(n3)
      double precision xx, yy, zz, rij, temp
c
      n = n3 / 3
      f = 0.0
      do 00010, i = 1, n3
        g(i) = 0.0
00010 continue
c
      do 00030, j = 2, n
        do 00020, i = 1, j-1
          ix = (i-1) * 3
          jx = (j-1) * 3
          xx = x(jx+1) - x(ix+1)
          yy = x(jx+2) - x(ix+2)
          zz = x(jx+3) - x(ix+3)
          rij = xx * xx + yy * yy + zz * zz
          temp = 1.0 / rij ** 3
          f = f + temp * (temp - 2.0)
          temp = 12 * temp * (temp - 1.0) / rij
          g(jx+1) = g(jx+1) - xx * temp
          g(jx+2) = g(jx+2) - yy * temp
          g(jx+3) = g(jx+3) - zz * temp
          g(ix+1) = g(ix+1) + xx * temp
          g(ix+2) = g(ix+2) + yy * temp
          g(ix+3) = g(ix+3) + zz * temp
00020   continue
00030 continue
      return
      end
```

FIGURE 2. Conventional algorithm for the evaluation of the Lennard-Jones pair potential function and its gradient vector (FORTRAN 77).

operation algorithm. Note that the gradient vector is also computed in $O(n^2)$ operations.

## 4. A Divide and Conquer Algorithm

In 1985, Appel [**1**] proposed a divide and conquer algorithm for many-body simulations. His algorithm has a proved complexity of $O(n \log n)$. Appel's algorithm is based on the following idea. Given two clusters of particles each consisting of $m$ and $n$ particles with unit mass. If the two clusters are *well separated*, they can treated as two particles with masses $m$ and $n$ and located at the centers of gravity of each cluster. Instead of spending $O(m \times n)$ time to compute the interactions between pairs of particles (one from each cluster), we can spend constant time computing the single pair interaction between the two *big particles.* The Appel algorithm can be summarized by the following two pieces of C codes.

```
void TwoNode(CLUSTER A, CLUSTER B)
{
  float d;
  integer i, j;

  d = distance(A,B);
  if(((radius(A)/d) <= lambda) && ((radius(B)/d) <= lambda))
    newton(A,B);
  else
    for(i = 0; i < 8; i++)
      for(j = 0; j < 8; j++)
        TwoNode(A.child[i], B.child[j]);
}


void ComputeAccel(CLUSTER A)
{
  integer i, j;

  if(!leaf(A)){
    for(i = 0; i < 8; i++)
      ComputeAccel(A.child[i]);
    for(i = 0; i < 8; i++)
      for(j = i + 1; j < 8; j++)
        TwoNode(A.child[i], A.child[j]);
  }
}
```

FIGURE 3. Appel's divide and conquer algorithm (C).

At the beginning, a *computational box* is computed which contains all the particles of interest. During the execution of the algorithm, the computational box is subdivided into smaller and smaller computational boxes until the box contains no more than one particle. The partition of computational box for 500 particles in 2-dimensional space is illustrated in Figure 4.

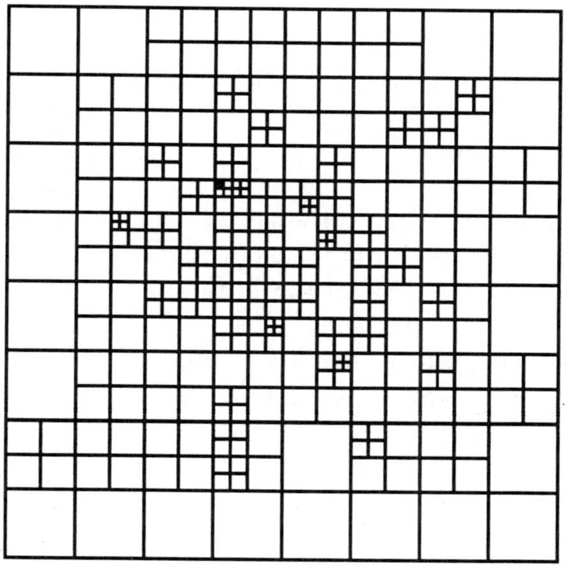

FIGURE 4. Subdivision of space for 500 particles in 2-dimensional space.

The Appel algorithm requires an Oct-Tree data structure to work in three dimensional space. Each internal node in the tree represents a three-dimensional cube having eight children to represent the sub-cubes which are created by subdividing the computational box along each dimension. Each leaf node represents an actual particle in space. Every node contains a weight, a potential, an acceleration, and the dimensions of the computational box represented by each sub-cluster. After the tree is populated the division of the computational box is similar to that in Figure 4.

Procedure ComputeAccel in Figure 3 is the procedure called from the driver routine. It does a depth first tree traversal calling TwoNode for each pair of children of every internal node. Assuming that there are $8^m$ particles uniformly distributed, then there are $m+1$ levels in the tree making the total number of clusters

$$\sum_{j=0}^{m} 8^j = \frac{1}{7}(8^{m+1} - 1) \tag{9}$$

and since ComputeAccel is called for every cluster in our computational box, ComputeAccel requires $O(n)$ operations.

Function TwoNode, also in Figure 3, requires two clusters as inputs and calculates the potential between them and the acceleration caused by the attraction. TwoNode is a recursive function which examines the size and distance between the two clusters and if they are *well separated* calculates the potential between the two clusters. If the two clusters are not *well separated* then TwoNode is called recursively for each pair $p_i$ and $p_j$ where $p_i$ is a child of cluster 1 and $p_j$ is a child of cluster 2.

Our implementation consists of essentially three phases: building the tree, compute the potential and partial movement vectors at internal nodes, propagating down the movement vectors.

To build the tree, we begin with a cubical cell large enough to contain all the particles. We load the particles into the root cell one at a time. If at any time two particles fall into the same cell we subdivide the cell along each of the three dimensions into eight sub-cells. After subdividing any cell we place the existing cluster into its appropriate sub-cell and then try to insert the particle into its appropriate sub-cell. This subdivision continues until the particle is the lone particle in its cell.

Each node in the tree requires the following: a weight, a gravitational center, a potential, and a movement vector for the corresponding cluster. The internal node in particular requires pointers for each of the eight children and a pointer to the parent. The leaf nodes require a pointer to the actual particle and a pointer to the parent.

When all $n$ particles have been stored, the system will have been partitioned into a number of cubical cells of different sizes with at most one particle per cell. These particle bearing cells are combined to make larger cells containing a cluster of particles which are then combined to make even larger cells until we reach the root which contains the whole system.

Every time the procedure TwoNode is called, we check whether the two clusters are well separated. If they are, TwoNode will compute a portion of pair potential by

$$v = m_i m_j (r_{ij}^{-12} - 2r_{ij}^{-6}) \ , \tag{10}$$

where $m_i$ and $m_j$ are the weights of the two clusters and $r_{ij}$ is the distance between the gravity centers of the two cluster. It will also compute the portion of the pair potential within each cluster. Two compute a portion of the gradient vector, we take partial derivatives of the function in equation (*) with respect to the gravity center of (for example) the first cluster and store its value (divided by the weight of this cluster) at the node corresponding to this cluster. This value will be pushed down to the leaves in the subtree. This is exactly the way we compute derivatives using the chain-rule.

## 5. Computational Results

We have implemented the divide and conquer algorithm and performed test runs on SUN4 workstations for uniformly distributed particles for $n = 1000$ to

10000 with a step-size of 1000. The new algorithm was compared with the conventional algorithm [**11**].

For each problem size, we randomly generate 5 sets of particles and the average time is recorded. We pick a specific run for the comparison of function values. The timing results as well as computed function values are reported in Table 1. Although the error bound of the Appel algorithm is not very good (only of the second order), the computational results obtained show that we actually have a very good accuracy as indicated in that lower part of the table.

| | | | T(Appel) | |
|---|---|---|---|---|
| Particles | T(Naive) | p = 0.01 | p = 0.1 | p = 0.15 |
| 1000 | 5.2900 | 26.570 | 14.920 | 9.250 |
| 2000 | 18.800 | 103.040 | 45.770 | 23.660 |
| 3000 | 43.030 | 242.050 | 79.580 | 41.350 |
| 4000 | 78.120 | 406.930 | 116.350 | 62.630 |
| 5000 | 126.280 | 631.290 | 154.100 | 73.980 |
| 6000 | 187.580 | 901.390 | 221.880 | 91.870 |
| 7000 | 261.740 | 1241.060 | 265.020 | 113.350 |
| 8000 | 312.540 | 1670.490 | 295.640 | 138.260 |
| 9000 | 395.140 | 2059.510 | 368.790 | 165.260 |
| 10000 | 484.270 | 2466.400 | 387.840 | 175.020 |
| | | | F(Appel) | |
| Particles | F(Naive) | p = 0.01 | p = 0.1 | p = 0.15 |
| 1000 | 19.5244859021 | 19.5244859021 | 19.5244862087 | 19.5244884056 |
| 2000 | -0.8464628227 | -0.8464628227 | -0.8464626571 | -0.8464615933 |
| 3000 | -1.7483789036 | -1.7483789036 | -1.7483788191 | -1.7483782079 |
| 4000 | -0.8618296862 | -0.8618296862 | -0.8618295974 | -0.8618288447 |
| 5000 | -2.0655934265 | -2.0655934265 | -2.0655933546 | -2.0655927142 |
| 6000 | -1.8339295511 | -1.8339295511 | -1.8339292791 | -1.8339279456 |
| 7000 | -2.3551348738 | -2.3551348739 | -2.3551347667 | -2.3551340396 |
| 8000 | -0.9035939420 | -0.9035939420 | -0.9035938946 | -0.9035935316 |
| 9000 | -1.8952912780 | -1.8952912782 | -1.8952912351 | -1.8952907884 |
| 10000 | -0.2710981434 | -0.2710981434 | -0.2710981077 | -0.2710751688 |

TABLE 1. Computed potential and average time comparison.

The crossover point for the computation of potential energy function is illustrated in Figure 5. We are please to find that the crossover occurs at 3000 particles.

The timing results for the computation of both potential energy function and the gradient vector is illustrated in Figure 6 and 7. Here the crossover point is around 7000.

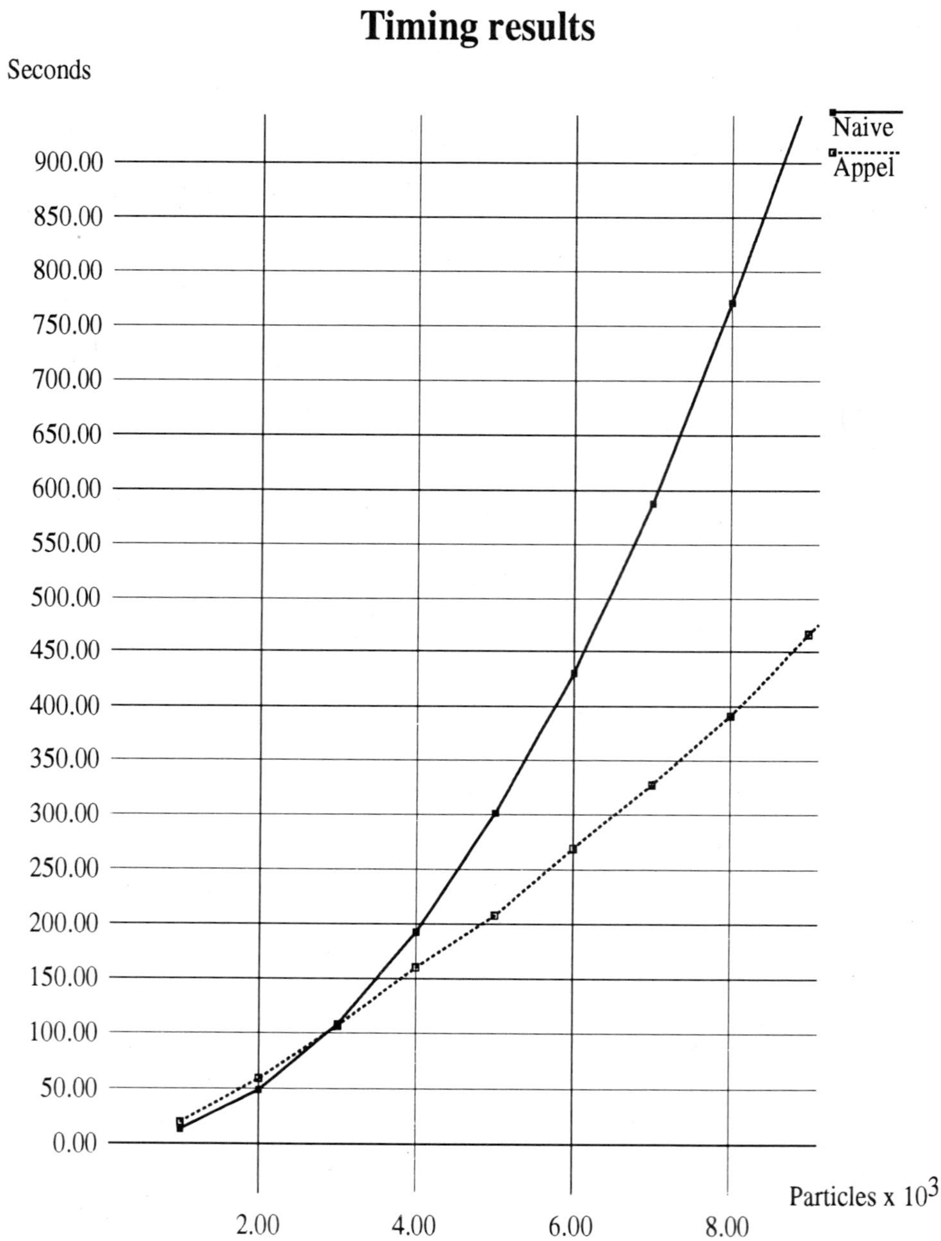


FIGURE 5. The Crossover Point for the Calculation of the Potential Field of a System of Uniformly Distributed Particles Using Appel's and the Naive Method.

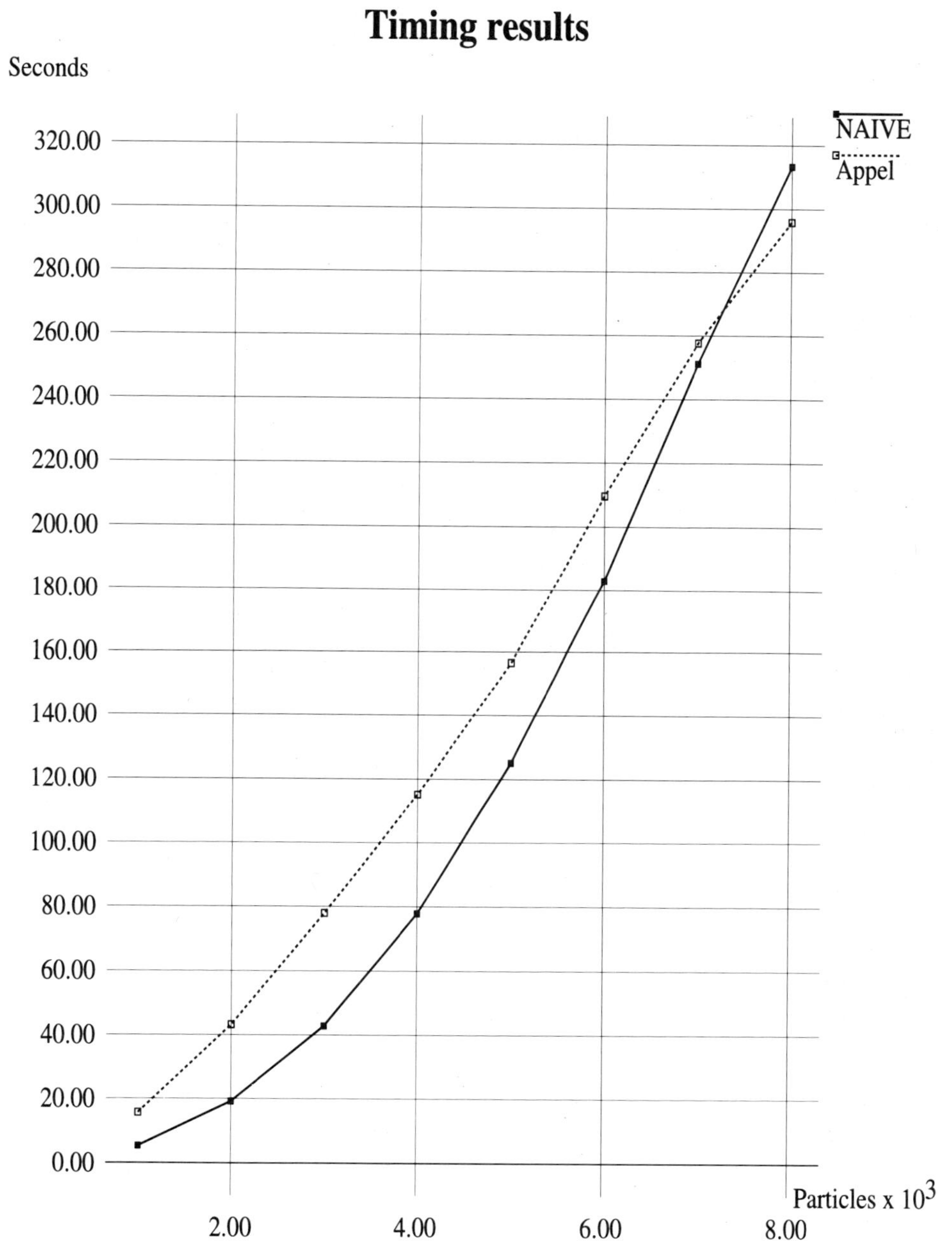


FIGURE 6. The Crossover Point for the Calculation of the Potential Field and Movement Vectors of a System of Uniformly Distributed Particles Using Appel's and the Naive Method.

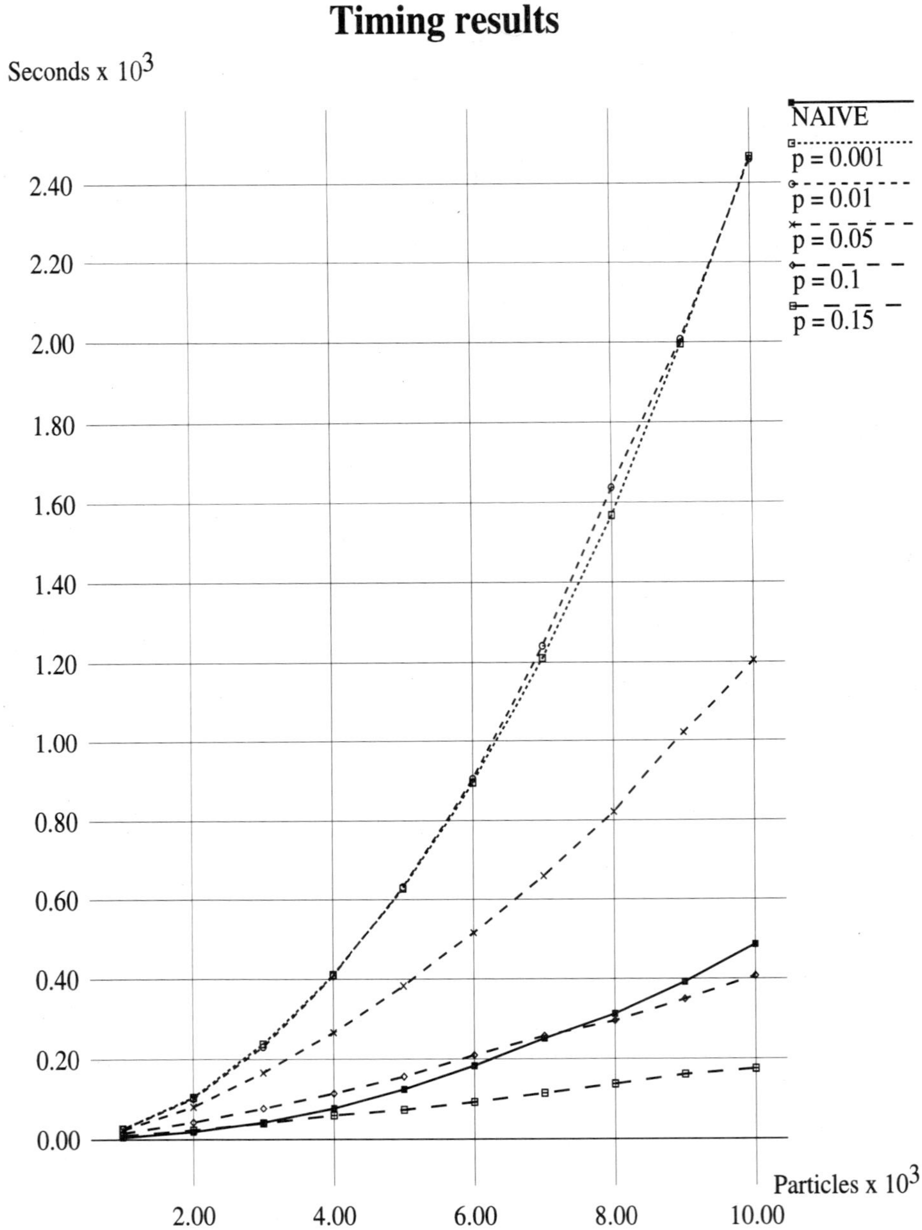


FIGURE 7. Evaluation of the Potential Field and Movement Vectors of a System of Uniformly Distributed Particles Using Different Values of $p$ for Appel's Algorithm.

## 6. Conclusions and Research in Progress

We have described a variant of the Appel algorithm and its computational results on the Lennard-Jones potential function. These results show that the Appel algorithm has very good speed and accuracy in practice. Upon completion of this paper, we have finished an implementation of an improved version of this algorithm which runs in $O(n)$ (linear) time. This will be reported in a forth coming paper.

## References


[1] A. W. Appel, An efficient program for many-body simulation, *SIAM Journal on Scientific and Statistical Computing* Vol. 6(1985), pp. 85-103.

[2] J. E. Barnes and P. Hut, A hierarchical $O(n \log n)$ force-calculation algorithm, *Nature* 324(1986), pp. 446–449.

[3] K. Esselink, The order of Appel's algorithm, *Information Processing Letters* Vol. 41(1992), pp. 141–147.

[4] D.M. Ferguson, A. Marsh, T. Metzger, D.G. Garrett, and K. Kastella. Conformational searches for the global minimum of protein models. *Journal of Global Optimization*, 4:209–228, 1994.

[5] L. Greengard and V. I. Rokhlin, A fast algorithm for particle simulations, *Journal of Computational Physics* 73 (1987): 325–348.

[6] P.M. Pardalos, David Shalloway and G.L. Xue, Optimization Methods for Computing Global Minima of Nonconvex Potential Energy Functions, *Journal of Global Optimization*, Vol. 4(1994), pp. 117-136.

[7] M. R. Hoare, Structure and Dynamics of Simple Microclusters, *Advances in Chemical Physics* Vol. 40(1979), pp. 49–135.

[8] K. E. Schmidt and M. A. Lee, Implementing the Fast Multipole Method in Three Dimensions, *Journal of Statistical Physics* Vol. 63(1991), pp. 1223–1235.

[9] G.L. Xue, Molecular Conformation on the CM-5 by Parallel Two-Level Simulated Annealing, *Journal of Global Optimization*, Vol. 4(1994), pp. 187-208.

[10] G.L. Xue, Improvement on the Northby Algorithm for Molecular Conformation: Better Solutions, *Journal of Global Optimization*, Vol. 4(1994), pp. 425-440.

[11] Alan Zall, *The rapid evaluation of Lennard-Jones functions for molecular and protein conformations*, M.S. Thesis, Department of Computer Science and Electrical Engineering, University of Vermont, 1995.



(G.L. Xue) Department of Computer Science and Electrical Engineering, College of Engineering and Mathematics, The University of Vermont, Burlington, VT 05405 USA

*E-mail address*, G.L. Xue: `xue@emba.uvm.edu`

(A.J. Zall) Department of Computer Science and Electrical Engineering, College of Engineering and Mathematics, The University of Vermont, Burlington, VT 05405 USA

*E-mail address*, A.J. Zall: `zall@emba.uvm.edu`

(P.M. Pardalos) Department of Industrial and Systems Engineering, University of Florida, Gainesville, FL 32611 USA

*E-mail address*, P.M. Pardalos: `pardalos@ufl.edu`

DIMACS Series in Discrete Mathematics
and Theoretical Computer Science
Volume **23**, 1996

# SIMULATED ANNEALING CALCULATIONS FOR OPTIMIZATION OF NANOCLUSTERS: THE ROLES OF QUENCHING, NUCLEATION, AND ISOMERIZATION IN CLUSTER MORPHOLOGY

MARKUS M. ZACHARIAS AND DIONISIOS G. VLACHOS

ABSTRACT. The Monte Carlo simulated annealing method is employed to determine structures of small Lennard-Jones clusters. Typical parameters of the quenching schedule are examined to study the important characteristics of simulated annealing in determining global and local minimum energy structures of nanoclusters. We have found that the probability of obtaining the global minimum energy decreases sharply as the quenching rate increases. When cluster quenching is fast, the isomer obtained depends significantly on the initial atom coordinates, whereas when quenching is sufficiently slow, the isomer obtained is the global minimum energy. The optimum fraction of successful atom displacements in exploring the potential hypersurface is about 0.3. The roles of cluster size, initial and final annealing temperature, and nature of interatomic potential in the probability of formation of various isomers are studied. Situations where lack of ergodicity of the Monte Carlo algorithm is possible, are also discussed.

## CONTENTS



1991 *Mathematics Subject Classification.* Primary 49D30, 60J27.

*Key words and phrases.* Optimization, simulated annealing, molecular structures, ergodicity.

This work was supported in part by NSF under grant number CTS-9410994 and the Engineering Computing Services of the University of Massachusetts.

This paper is in final form and no version of it will be submitted for publication elsewhere.

## 1. Introduction

Clusters of atomic dimensions (nanophase materials) are formed by homogeneous and heterogeneous nucleation in many disciplines including solidification processes, atmospheric aerosols (soot), catalysts, microelectronics, ceramics, pharmaceutical and food systems (proteins), etc. Despite the recent progress in novel experimental techniques such as scanning tunneling microscopy and atomic force microscopy, direct observation of structures of systems containing a few atoms is still beyond the capabilities of present experimental techniques. Additional complications due to defects and impurities in real systems and interactions between systems make understanding of small clusters a formidable task. Mathematical modeling can provide substantial understanding on nanophase materials and complement current experiments.

Key factors affecting the properties of small clusters are their geometric and electronic structures. The most stable structure corresponds to the global minimum free energy F

$$F = E - TS, \tag{1.1}$$

where E is the cohesive energy, T is the temperature, and S is the entropy of the cluster. The most stable isomer is expected under equilibrium conditions. Local minima in free energy, called metastable isomers, are typically formed when clusters are produced in far from equilibrium conditions, e.g., in fast growth of materials, rapid quenching of liquids, and irreversible chemical reactions.

Small clusters of inert gases such as argon are simpler to study and can be used as prototypes for understanding the properties of small molecular systems. Cluster structures (isomers) have been determined using several optimization techniques. The first extensive studies on Lennard-Jones (LJ) cluster structures were done by Hoare and Pal by growing seeds of certain symmetry at 0 K (Hoare and Pal, 1972; Hoare and Pal, 1971). Two methods, which are widely used to determine local minima of an objective function are the steepest descents method (Stillinger and Weber, 1982; Stillinger and Weber, 1983; Press et al., 1986) and the conjugate gradient method (Press et al., 1986; Polak, 1971). A parallel algorithm has also been suggested for minimizing molecular energy potential surfaces (Xue et al., 1991; Maier et al., 1992). It combines discrete, lattice-based search (Northby, 1987) with continuous optimization techniques. The influence of the range of the pair potential on optimization has been discussed in several studies (Stillinger and Weber, 1988;

Stillinger and Stillinger, 1990; Braier et al., 1990; Chang and Berry, 1992) along with the number of distinct minima and transition states. Recently, Maranas and Floudas proposed a global energy minimization technique (Maranas and Floudas, 1992) which proves the global structure within some tolerance but is at present limited in computational efficiency. A survey paper with many optimization references related to the topic has been recently published (Pardalos et al., 1994).

One major difficulty in determining structures of clusters is that the entropy S is not a closed form, known function of the cluster size, the temperature, and the material. Thus, most minimization methods are applicable only to 0 K where minimization of the potential energy E alone is carried out. The cohesive energy is a function of the $3n$ coordinates of n atoms

$$E(\mathbf{r}) = E(x_1, y_1, z_1, \ldots, x_n, y_n, z_n). \tag{1.2}$$

Due to translational and rotational invariance, $E(\mathbf{r})$ is in fact a function of $3n - 6$ coordinates (parameters). Six redundant parameters are often eliminated by setting, for example,

$$x_1 = y_1 = z_1 = y_2 = z_2 = z_3 = 0. \tag{1.3}$$

Typically, an initial set of coordinates $\mathbf{r}^o = [x_1^o, y_1^o, z_1^o, \ldots, x_n^o, y_n^o, z_n^o]^T$ is provided as a starting point in parameter space ($^T$ denotes transpose) for optimization at 0 K (ground state).

The simulated annealing technique (Kirkpatrick et al., 1983; Wille, 1987) in conjunction with Monte Carlo (MC) or molecular dynamics simulations is an alternative method which has been introduced to solve N-incomplete minimization problems. Minimization using molecular simulations considers high temperature effects and allows isomerization from local minima to the global minimum energy. In this way, structures can be examined at finite temperatures, and the role of materials processing (non equilibrium conditions) in cluster structures can be studied.

We have determined cluster structures for $n < 35$ atoms using simulated annealing for both the LJ potential and many body potentials (Vlachos et al., 1992). We have found that the structures of the most stable LJ clusters change from noncrystalline for $n < 31$ atoms to quasicrystalline for $n > 30$ atoms. At present, the roles of the parameters involved in simulated annealing calculations for cluster optimization are not well understood. As an example, the acceptance ratio (fraction of accepted MC trials) which controls the speed of exploration of the potential hypersurface is usually taken to be 0.5 (Allen and Tildesley, 1989). Even though it is generally known that sufficiently far from equilibrium conditions result in metastable and amorphous clusters (Rose and Berry, 1993), the effect of irreversibility of a process on cluster structures is not well known. This issue is central in tailoring materials with specific structures and properties. Here we examine how different parameters of the quenching schedule affect the formation probabilities of cluster isomers.

## 2. Interatomic potentials

The LJ 6-12 potential is employed to describe cohesion of materials. The energy of a cluster of n atoms is determined by

$$E = 4\epsilon \sum_{i<j=1}^{n} [(\sigma/r_{ij})^{12} - (\sigma/r_{ij})^{6}], \tag{2.1}$$

where $r_{ij}$ is the distance between atoms i and j, and the parameters $\epsilon$ and $\sigma$ are chosen to fit the cohesive energy and interatomic distance of the bulk crystal (Kittel, 1986). For bulk Ni-crystal considered here at 0 K, $\epsilon = 0.5186$ eV and $\sigma = 0.2283$ nm.

Embedded-atom many body potentials (Daw and Baskes, 1984) describe more accurately the cohesion of metals than pairwise additive potentials. The potential energy of a solid of n atoms is given by

$$E = \sum_{i=1}^{n} F_i(\rho_i) + \frac{1}{2} \sum_{i=1}^{n} \sum_{j=1(\neq i)}^{n} \phi_{ij}(r_{ij}), \tag{2.2}$$

where $\phi_{ij}(r_{ij})$ is a two-body central potential between atoms i and j which accounts for the electrostatic core-core repulsive interaction

$$\phi_{ij}(r_{ij}) = Z(r_{ij})^2 / r_{ij}. \tag{2.3}$$

Here, $Z(r_{ij})$ is the effective charge of each neutral, screened atom, and $F_i(\rho_i)$ is the energy to embed atom i into the background electron density $\rho_i$. The electron density at the location of the ith atom is approximated by a superposition of atomic electron densities

$$\rho_i = \sum_{j=1(\neq i)}^{n} \rho_j^a(r_{ij}), \tag{2.4}$$

where $\rho_j^a$ is the contribution by the jth atom. The atomic densities are evaluated from Hartree-Fock calculations (Clementi and Roetti, 1974). The EA is a semiempirical method having its roots in density functional theory. Here we use the parameters for Ni reported by Daw and Baskes (1984).

## 3. Monte Carlo simulated annealing calculations

Annealing is the physical process of heating up a solid at high temperatures and cooling it down. Depending on the rate of the cooling process, materials with different degree of crystallinity can be obtained. When cooling is sufficiently slow, the free energy of the solid may converge to the global minimum energy.

The MC simulated annealing method consists of a biased random walk that samples the objective function in the space of independent variables. In this method, a cluster is heated up at an initial temperature $T_i$, and the temperature is decreased according to a quenching schedule which will be described below. The annealing is combined with the Metropolis walk algorithm (Metropolis et al., 1953), a Monte Carlo technique that compares random numbers with probabilities derived from statistical mechanics. This algorithm is illustrated in Figure 1 and is described next.

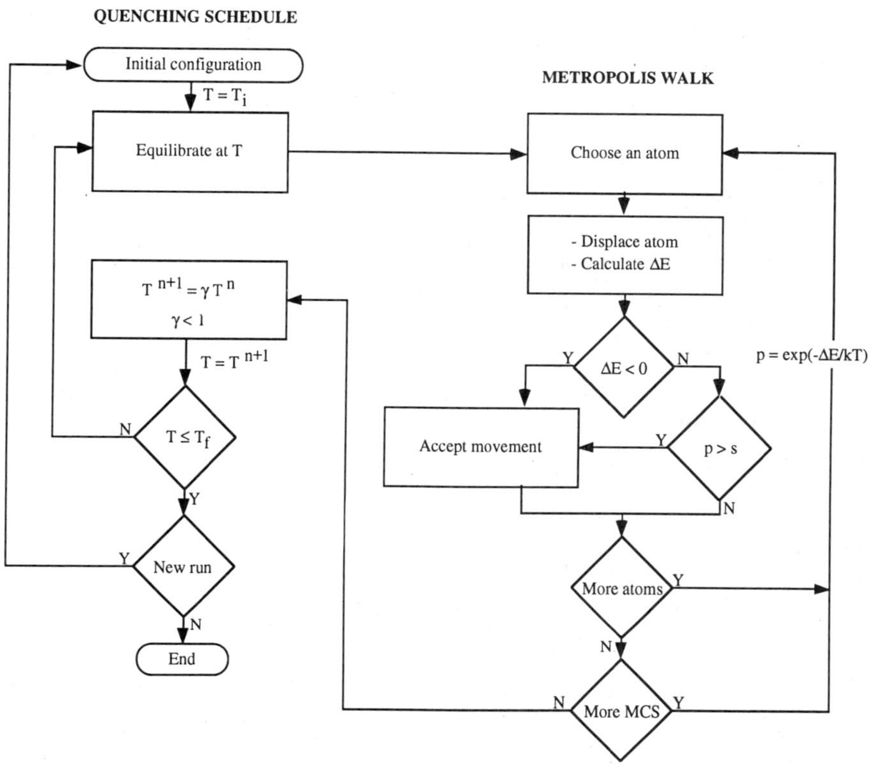


FIGURE 1. Flow chart illustrating the Monte Carlo simulated annealing algorithm.

**3.1. Metropolis walk.** At each temperature the Metropolis walk algorithm is applied. The time spent for equilibration at each temperature determines the degree of isomerization of a cluster, and the ratio of isomerization rate to quenching rate is a key factor in controlling the morphology of clusters. This algorithm involves the following steps:

(i) Choice of an atom (randomly or sequentially);

(ii) Displacement of this atom in three dimensions at random, based on a maximum displacement $\delta r_{max}$ (Allen and Tildesley, 1989);

(iii) Calculation of the energy change $\Delta E$ associated with this displacement;

(iv) Acceptance of the movement if the energy of the system is lowered $\Delta E < 0$; if the energy of the system increases, $\Delta E > 0$, the probability of acceptance is calculated using a Boltzmann factor $exp(-\Delta E/kT)$, i.e., the acceptance probability is

$$p = \begin{cases} 1, & \Delta E < 0, \\ exp(-\Delta E/kT), & \Delta E > 0. \end{cases}$$

For $\Delta E > 0$, p is compared with a random number s in the interval $(0, 1)$, and the movement is accepted if $p > s$. The Metropolis walk enables the cluster to move uphill in energy, and transitions (isomerizations) from a local to the global minimum energy can occur.

(v) Repetition of steps i-iv for a number of MC steps (MCS). One MCS is defined as $n$ MC attempts, where $n$ is the number of atoms of a cluster.

The maximum displacement of atoms determines the rate of moving in phase space. Low $\delta r_{max}$ implies that most trials are accepted and results in slow movement in phase space. When $\delta r_{max}$ is large, nearly all trial moves are rejected, and there is again slow movement in phase space. The speed of exploring the potential hypersurface is controlled through the acceptance ratio $\alpha$ which is the fraction of successful MC trials (Allen and Tildesley, 1989). The maximum displacement of atoms is dynamically adjusted by increasing or decreasing $\delta r_{max}$ by 5% every twenty MCS to keep the ratio of accepted MC trials equal to $\alpha$.

Figure 2 shows $\delta r_{max}$ versus temperature for various values of $\alpha$. Data points are ensemble averages from simulations with up to $10^7$ MCS at each temperature after appropriate equilibration, and the curves just connect the points. Large (small) atom displacements are allowed when $\alpha$ is small (large). The maximum atom displacement varies considerably with temperature as shown in Figure 2. When quenching is slow, the algorithm is capable of adjusting $\delta r_{max}$ to the appropriate value. In contrast, at high quenching rates the algorithm has no time to adjust $\delta r_{max}$, except when the number of MCS at each temperature is high. For these conditions, the temperature dependence of $\delta r_{max}$ is set in the program for every acceptance ratio according to Figure 2.

**3.2. Quenching schedule.** A typical quenching schedule in simulated annealing calculations involves the following steps (Vlachos et al., 1992; Chelikowsky and Phillips, 1990):

(i) Choice of an initial configuration of a cluster at an initial temperature (usually $T_i > T_{melt}$). The initial configuration of a cluster is randomly generated. Atoms are placed in a cube so that each atom is at least 0.35 nm from its neighbors and from the walls.

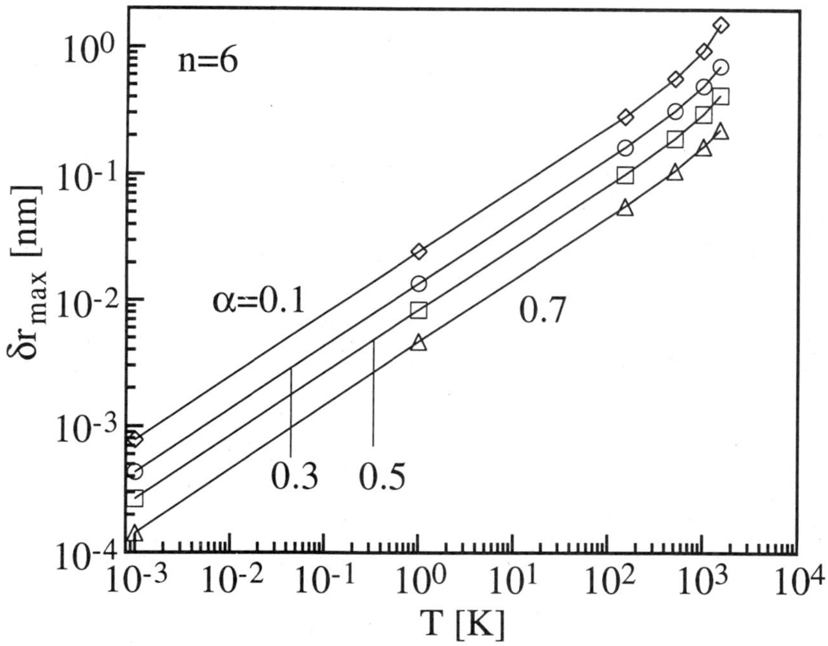


FIGURE 2. Maximum atom displacement in nm versus temperature at different acceptance ratios for a cluster with $n = 6$ atoms using the LJ potential.

(ii) Equilibration of a cluster at a specified temperature T using the Metropolis walk described above.
(iii) Reduction of temperature (quenching) (Chelikowsky and Phillips, 1990; Vlachos et al., 1992):

$$T_{new} = \gamma T_{old}, \tag{3.1}$$

(iv) Repetition of steps ii-iii until a desired low temperature is reached.

Typical parameters encountered in the above quenching algorithm are:

(i) Quenching factor, $\gamma$;
(ii) Initial configuration (initial atom coordinates), $\mathbf{r}^o$;
(iii) Number of MCS at each temperature;
(iv) Acceptance ratio, $\alpha$;
(v) Initial temperature, $T_i$;
(vi) Final temperature, $T_f$;
(vii) Size of the cluster, i.e., number of atoms, n.

When quenching is slow, the probability of formation of an isomer depends on the probability that the initial configuration corresponds to a certain attractor and the probability of isomerization between isomers. In one dimension, the probability that an initial configuration is in the basin of a certain attractor is proportional to the width of the basin of attraction. By an instantaneous quenching of the initial configuration to a very low temperature (e.g., 0 K), the basin of attraction (minimum) in which the cluster started can be determined.

**3.3. Quenching variables and relation to experimental conditions.** The initial temperature can vary from about 2000 K when clusters form in flames such as in the formation of $SiO_2$ to below room temperature in supersonic jets. In most systems, temperature is spatially nonuniform, i.e., clusters form and are transported along a temperature gradient. This gradient may be quite large as happens in flames where temperatures may change by two orders of magnitude in a few millimeters.

The quenching factor $\gamma$ and the number of MCS determine the rates of quenching and isomerization, respectively, and are related to the actual quenching rate in solidification processes. Due to the short time scales studied in molecular simulations (e.g., in molecular dynamics), the slowest quenching on a fast Supercomputer is at present still fast compared to experimental conditions. However, the qualitative characteristics of the solidification process in cluster morphology can still be understood as a function of quenching rate.

The initial configuration can be a structure of known symmetry such as face-centered cubic or icosahedral. Such an initial nucleus may bias the final structure, especially when quenching is fast, and result in a cluster trapped into the initial basin. Here we are interested in simulating conditions where atoms stick rapidly together and there is not sufficient time for isomerization or the experiments are done at high temperatures in a liquid phase, so that a random aggregation of atoms is taken as the initial nucleus. Thus, the initial configuration of a cluster is randomly generated.

**3.4. Statistics.** The MC simulated annealing algorithm is a stochastic method, and there is a probability of formation of various isomers. Simulations have run either with the same initial configuration but different seeds of the random generator

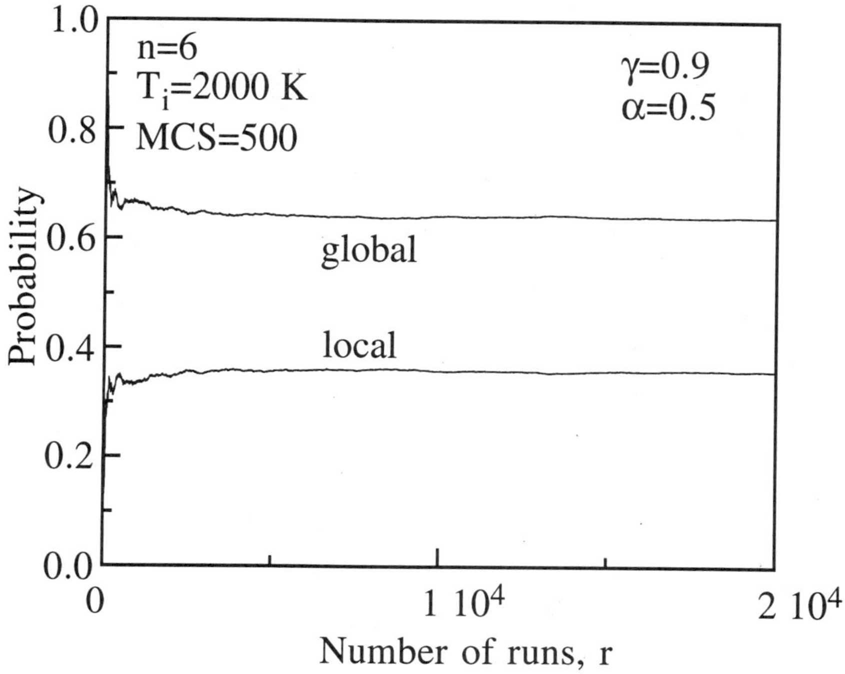


FIGURE 3. Frequency of occurrence of the local and global isomers of $n = 6$ in r runs versus the number of runs for parameters indicated. Asymptotes are observed after a large number of runs which are the average formation probabilities of the two minima.

or different initial configurations (randomly generated) and different sequences of random numbers. Thus, the MC method is typically run thousands of times for every specific set of parameters, and the local, $p_l$, and global, $p_g$, probabilities calculated during each run are averages over r independent runs.

To illustrate the algorithm, the LJ cluster with $n = 6$ atoms with two minima is considered. Figure 3 shows the frequency of occurrence of the local and global isomers of $n = 6$ in r runs versus r for up to 20,000 total runs with parameters indicated and the same initial configuration in all runs. An asymptote is observed with small fluctuations after a sufficiently large number of runs, which is the probability, on the average, of obtaining a particular structure. Figure 3 indicates that thousands of runs are often required to obtain good statistics. More runs are usually needed for intermediate values of $\gamma$ where larger fluctuations are observed.

The probabilities reported below are averages over many thousands runs, and correspond to an asymptotic limit:

$$p_i(\gamma, \mathbf{r}^o, MCS, \alpha, T_i, T_f, n) = \lim_{r\to\infty} \frac{1}{r} N_i(\gamma, \mathbf{r}^o, MCS, \alpha, T_i, T_f, n), \tag{3.2}$$

where $N_i$ is the frequency of occurrence of isomer i in r runs and $\sum_i N_i = r$.

MC simulated annealing calculations are often very intensive. As an example, quenching of the $n = 19$ atom cluster from $T_i = 2000K$ to $T_f = 150K$ requires 120 CPU hours for $\gamma = 0.99$ with 4000 runs on a DEC 3000/600 workstation using the

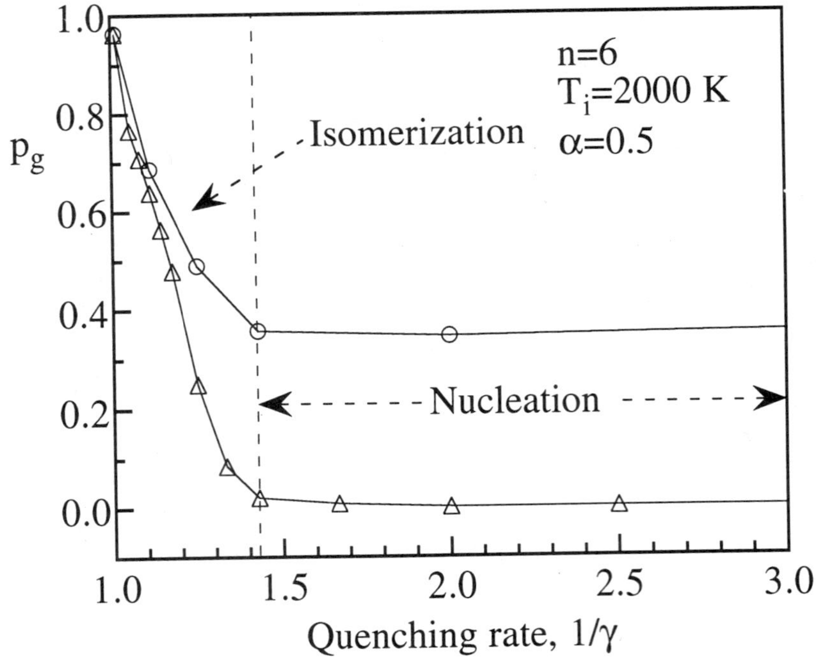


FIGURE 4. Formation probabilities of the global and local minima versus the quenching rate $1/\gamma$ using a single local (open circles) and many randomly generated (filled circles) initial configurations. An isomerization- and a nucleation-controlled regime are found.

LJ potential. Many of these runs have been carried out. Use of a more complex objective function such as that described by Equations (2.2)-(2.4) can result in an order of magnitude more intensive simulations.

## 4. Role of quenching parameters in global optimization

The roles of quenching rate, initial configuration, acceptance ratio, and initial and final temperature in the probability of accessing different minima of the potential hypersurface are examined for the cluster with $n = 6$ atoms. The LJ potential is first discussed. The results described below are obtained with $T_f = 150K$ and subsequent equilibration at 1 K, except otherwise stated.

**4.1. Quenching factor.** The formation probabilities of the global and local isomers are plotted versus the quenching rate $1/\gamma$ in Figure 4 (open circles) with an acceptance ratio of $\alpha = 0.5$ and an initial temperature of $T_i = 2000K$. A single initial configuration is used in all runs which actually corresponds to the local minimum energy.

At high values of $1/\gamma$, quenching is fast, and there is no sufficient time for cluster isomerization. Consequently, the cluster is trapped into the basin of attraction of the initial configuration, i.e., in this case the local minimum. The simulated annealing operates then as a downhill minimization algorithm. However, at sufficiently low values of $1/\gamma$, the cluster can isomerize during quenching, and the probability of getting the global structure increases sharply with decreasing $1/\gamma$.

**4.2. Initial coordinates (nucleus) and potential hypersurface.** The results discussed above are based on a single initial configuration. However, for certain parameter values, the minimum obtained depends considerably on the initial coordinates. In order to calculate the formation probabilities independently of the initial configuration, runs were performed using thousands of randomly generated initial configurations. Asymptotes were obtained similarly to those shown in Figure 3. The filled circles in Figure 4 show the formation probabilities of the global and local isomers versus the quenching rate $1/\gamma$.

Figure 4 shows that when quenching is fast, the formation probabilities depend significantly on the initial conditions. The dependence of the formation probability on the initial coordinates decreases continuously as quenching becomes slower. At sufficiently high values of $\gamma$, the curves obtained with a single local initial configuration (dashed lines) merge with the curves obtained with many random initial configurations (solid lines) as shown in Figure 4. Thus, when quenching is slow, it is very likely to obtain the global structure independently of the initial configuration, i.e., when quenching is slow the cluster looses its memory, and the final structure is almost independent of the initial conditions.

Surprisingly, the global isomer forms in a significant fraction, even at very high quenching rates (low values of $\gamma$). To understand better this behavior, the initial clusters were relaxed using a downhill minimization algorithm where only trials decreasing the energy were accepted. These simulations were performed at a fixed $\delta r_{max}$ which corresponds to a certain temperature. Figure 5 summarizes the results. The open circles show the effect of $\delta r_{max}$ on $p_g$ for thousands of randomly generated configurations. For $\delta r_{max} < 10^{-5}$, $p_g$ is almost zero. For low $\delta r_{max}$, atoms are perturbed with infinitesimally small displacements, and the cluster freezes into the initial basin. This indicates that all initial randomly generated configurations were glassy, and when relaxed sufficiently slowly belonged to the local minimum basin. However, for $\delta r_{max} \subset (10^{-4}, 10^{-1})$, the probability of obtaining the global isomer is almost constant and about 35%. These values of $\delta r_{max}$ correspond to temperatures considerably lower than room temperature. Figure 5 indicates that many initial configurations isomerize to the global isomer, even at very low temperatures. Presumably, these configurations were close to the hypersurface separating the local from the global minimum, and finite temperatures converted these local minima to the most stable isomer. In other words, some nuclei need 'weak' annealing to isomerize into the global structure, and fast quenching converts them into the most stable isomer.

**4.3. Time for isomerization.** The number of MCS is not directly related to real time but expresses the length of sampling the objective function in phase space. As the number of MCS decreases, less sampling occurs, and the cluster does not reach an equilibrium state. In that context, we define the time for isomerization at each temperature, and we adjust it through the number of MCS. For the same value of $\gamma$, we have found that as the number of MCS increases the probability of obtaining the global isomer increases. This is expected because as the number of MCS increases, the time for isomerization increases compared to the quenching time. Since different lengths of equilibration correspond to a different computational cost, Figure 6 shows

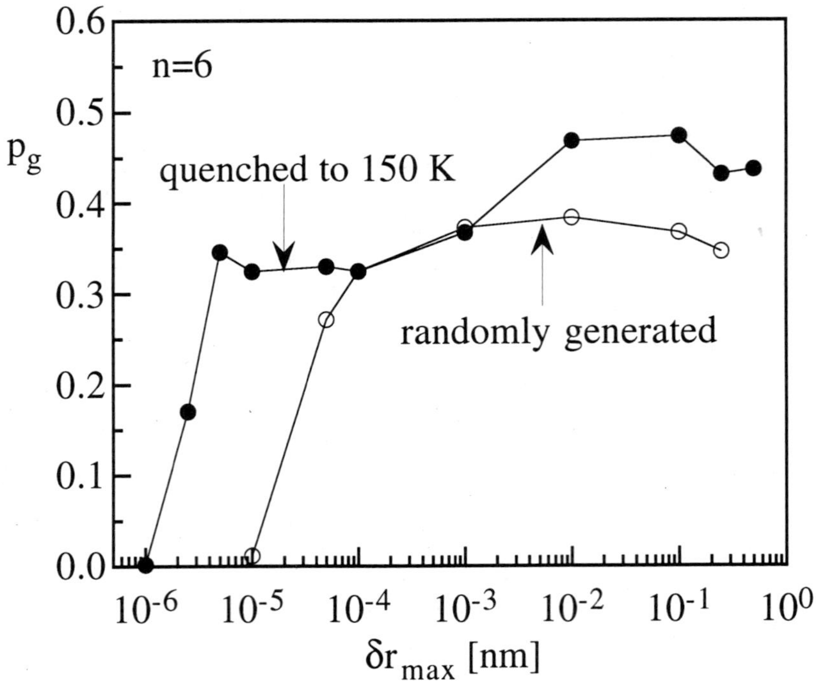


FIGURE 5. Global probability versus the maximum atom displacement for instantaneous quenching of many initial configurations (open circles), and of structures obtained at $T_f = 150K$ with $\gamma = 0.7$, $\alpha = 0.3$, and $T_i = 2000K$ (filled circles).

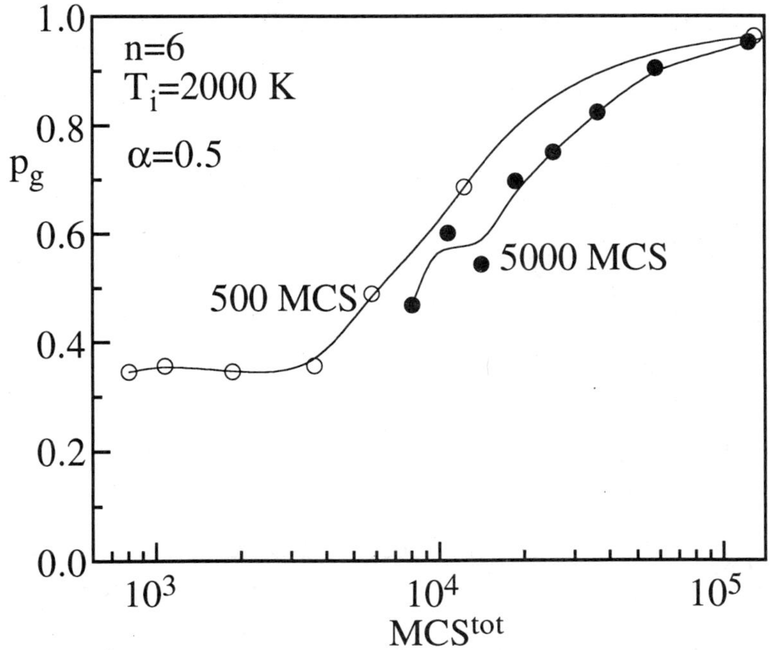


FIGURE 6. Global probability at different equilibration lengths versus the total MCS. For the same computational cost, a higher probability is achieved for fewer MCS.

$p_g$ versus the total number of MCS (computational cost) for a fixed equilibration length (MCS). It should be noted that the same number of total MCS

$$MCS^{tot} = [1 + ln(T_f/T_i)/ln(\gamma)]MCS, \tag{4.1}$$

is achieved for different values of $\gamma$ in the two simulations.

Figure 6 shows that $p_g$ is slightly higher when 500 MCS are used as compared to 5000 MCS. This difference becomes smaller as the number of total MCS increases because the cluster looses memory effects and is quenched through quasi-equilibrium conditions. Figure 6 indicates that more intermediate steps (a higher $\gamma$) and smaller equilibration length (fewer MCS) result in a higher formation probability of the global isomer for the same computational cost. However, the qualitative behavior of $p_g$ does not change at different MCS, and all results below are obtained with 500 MCS.

**4.4. Fraction of successful Monte Carlo trials.** In order to investigate how the formation probability of the global isomer $p_g$ depends on the acceptance ratio, runs have been performed at fixed values of $\gamma$ for various values of $\alpha$. The results are shown in Figure 7. In the limit of very small $\alpha$, displacements are larger than the dimensions of the simulated cube in which the atoms are placed, and atoms essentially evaporate. This behavior determines a lower bound in the simulations.

The probability of obtaining the global structure is higher the slower the quenching rate is, in agreement with Figure 4. A maximum appears at intermediate values

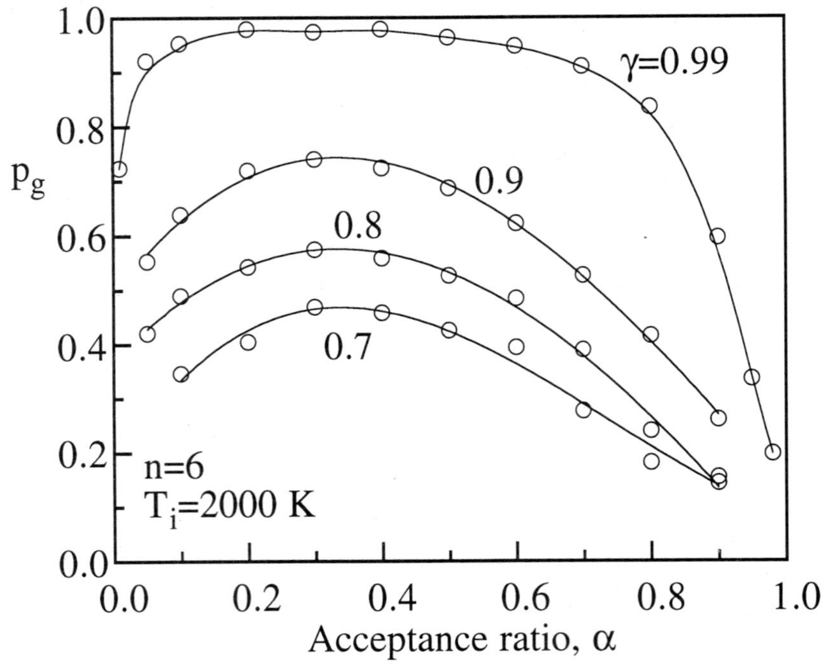


FIGURE 7. Global probability versus acceptance ratio at quenching rates indicated. An optimum acceptance ratio of about 0.3 is found when quenching is fast. A plateau is observed when quenching is slow.

of $\gamma$ indicating that there is an optimum acceptance ratio of about 0.3 where the formation probability of the global isomer is maximum. At sufficiently high values of $\gamma$, a plateau is found for $\alpha \subset (0.2, 0.6)$ indicating that the selection of the acceptance ratio is not significant.

Our results indicate that the dependence of the global probability on the acceptance ratio at various quenching rates is not symmetric around 0.5, i.e., the 50% acceptance ratio is not an optimum value for small molecular clusters as was suggested before. A decrease in the formation probability of the global structure is found at low and high acceptance ratios. In the former case, few but large displacements of atoms are carried out which favor isomerization from the global isomer to the local isomer or vice versa. In the latter case, small movements of atoms are carried out, and the cluster gets frozen in the basin of the initial configuration.

**4.5. Temperature limits.** Figure 8 illustrates the role of the initial temperature in the formation probability of the global isomer for various values of $\alpha$. At sufficiently high temperatures, the effect of temperature on $p_g$ is weak. An increase in the initial temperature results in more intensive simulations without affecting appreciably the probability of formation of the global structure. However, the global probability is considerably reduced as the initial temperature decreases below some value. At sufficiently low temperatures the ergodicity may break down, and the cluster is trapped in the initial basin of attraction. Our simulations indicate that

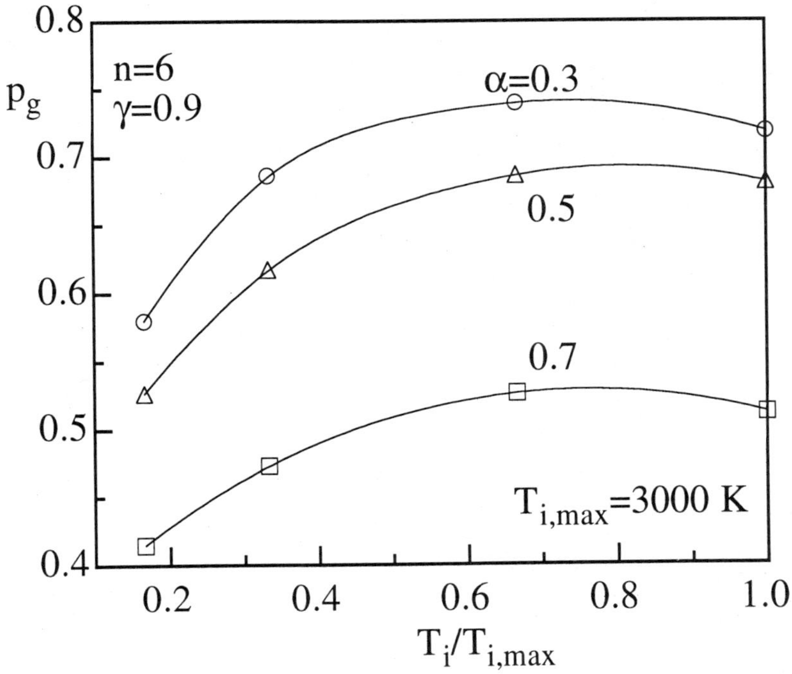


FIGURE 8. Global probability versus initial temperature for various values of acceptance ratio. Low initial temperatures reduce the probability of formation of the global minimum.

the selection of the initial temperature is not very important as far as sufficient time for isomerization occurs.

Simulated annealing calculations are intensive because the temperature is often reduced. slowly by several orders of magnitude. To reduce computational cost, the final temperature is usually considerably higher than 0 K. This is in general justified since isomerization at low temperatures is very slow. In our simulations, the final temperature has been taken to be $T_f = 150K$, and clusters were subsequently equilibrated at 1 K to obtain the cluster energy. At low values of $\gamma$, structures obtained may be glassy-like, and it is expected that the last steps of the quenching affect the formation probabilities.

In order to examine how the final temperature affects the formation probability of the global isomer, structures obtained at $T_f = 150K$, with $\gamma = 0.7$ were relaxed using the downhill minimization algorithm described above. The filled circles in Figure 5 show $p_g$ versus $\delta r_{max}$ for fast quenched clusters. When the clusters are quenched slowly (a low $\delta r_{max}$), only the local minimum is found, i.e., fast quenched clusters are glassy-like. The global probability changes slightly with $\delta r_{max}$ for $\delta r_{max} > 510^{-6}$. Thus, very low but finite temperatures affect strongly isomerization of fast quenched clusters. In contrast, for high values of $\gamma$, the structures obtained at low temperatures are equilibrated, and the final temperature does not affect the results.

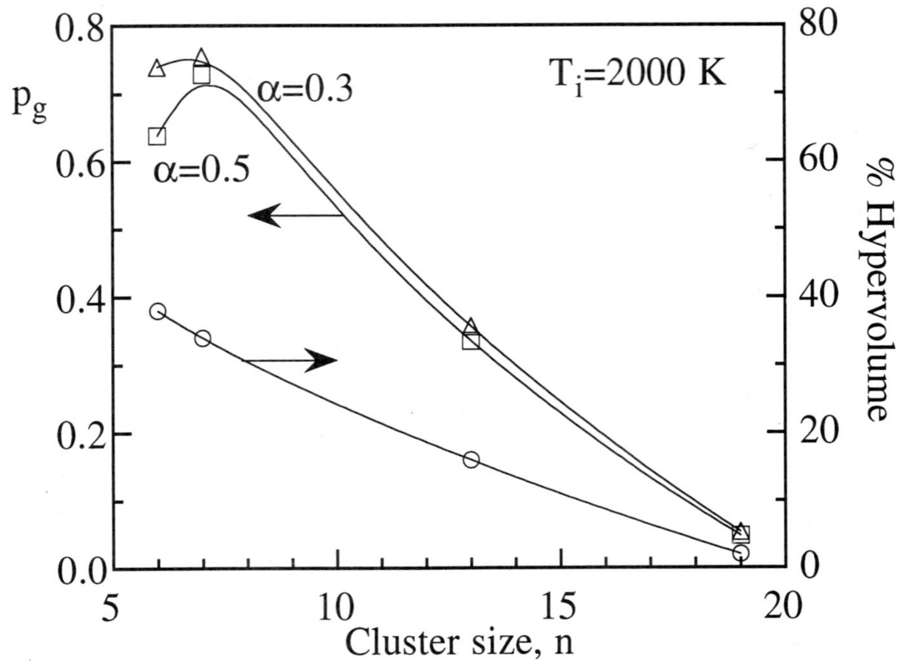


FIGURE 9. Global probability versus cluster size for two acceptance ratios, $n = 6$, and $\gamma = 0.9$. The hypervolume occupied by the global minimum energy isomer (probability of global isomer with $\delta r_{max} = 10^{-3}$) versus size is also shown.

## 5. Role of the number of minimization parameters (cluster size)

Here we investigate the effect of cluster size on the results of simulated annealing by examining clusters with 7, 13, and 19 atoms. For the $n = 7$ atoms cluster four isomers exist. The number of isomers for clusters with 13 and 19 atoms is not definite. For the thirteen icosahedron, 988 possible isomers have been reported many years ago by Hoare and McInnes (1983). With the exception of the $n = 6$ cluster, the most stable structure of the clusters in this size range is noncrystalline.

Figure 10 shows the global probability of formation versus cluster size for two values of acceptance ratio. The global probability exhibits a local maximum at $n = 7$, probably due to the enhanced stability of this cluster (a magic number). The global probability decreases with cluster size almost linearly for $n > 7$. Thus, it is expected that for a certain set of quenching conditions, large clusters will be local minima and small clusters will be global minima. This may be due in part to the increasing isomerization times with increasing cluster size.

In order to obtain an estimate of the width of the basin of attraction, downhill minimization with $\delta r_{max} = 10^{-3}$ was performed for many initial configurations. The results are shown in Figure 9. The hypervolume occupied by the global minimum (percentage of initial configurations resulting in the most stable structure upon instantaneous quenching) decreases as the cluster size increases from about 38% for $n = 6$ to about 2% for $n = 19$. This is probably due to the fact that the

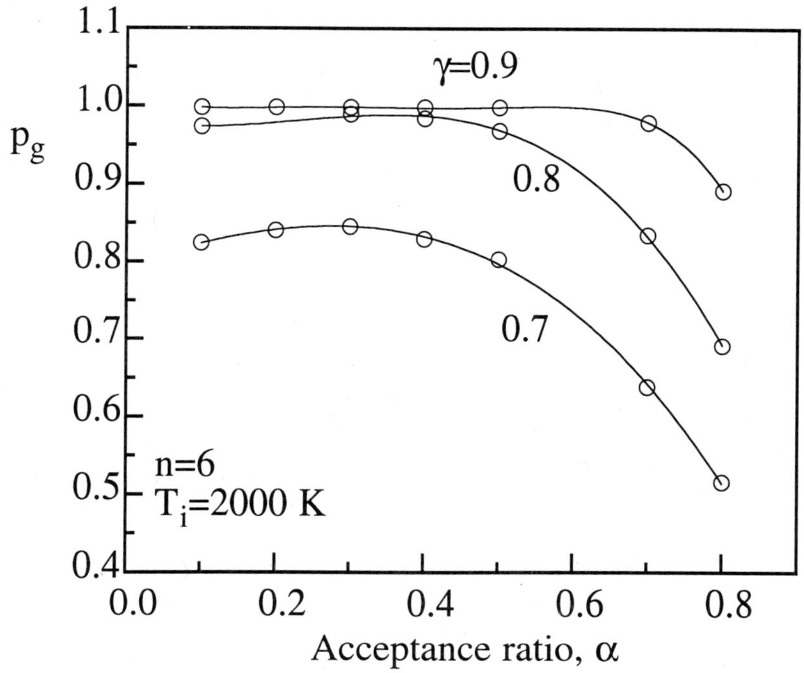


FIGURE 10. Global probability versus acceptance ratio for $n = 6$ at various quenching rates using randomly generated initial configurations and the embedded-atom potential. An optimum acceptance ratio, of about 0.3, is found.

number of isomers increases with the number of atoms n as $exp(n^2)$ (Hoare and McInnes, 1983). As a result, the hypervolume of the global isomer shrinks.

## 6. ROLE OF INTERATOMIC POTENTIAL

In order to access how the results depend on the interatomic potential, runs have been performed using an embedded-atom potential discussed above. Figure 10 shows the formation probability of the global isomer versus $\alpha$ for $n = 6$ atoms at different quenching rates. We obtain similar results as for the LJ clusters concerning the optimum acceptance ratio. However, for the same quenching rates the formation probability of the global isomer is higher. As a result, a plateau is already found at $\gamma = 0.8$. The many body effects favor the formation of the most stable structure as compared to the LJ clusters. At present, it is not clear which characteristics of the hypersurface create this behavior. Our simulations indicate that the qualitative features of the simulated annealing are not affected by the details of the interatomic potential, and thus, are expected to be valid for other materials as well.

## 7. ERGODICITY IN MOLECULAR SIMULATIONS AND NUCLEATION

Next we perform simulated annealing calculations for clusters with $n > 30$ atoms whose most stable structure is quasicrystalline. Figure 11 shows the probability of having 90 degrees bond angles, $p_{90}$, a structural property, versus temperature for

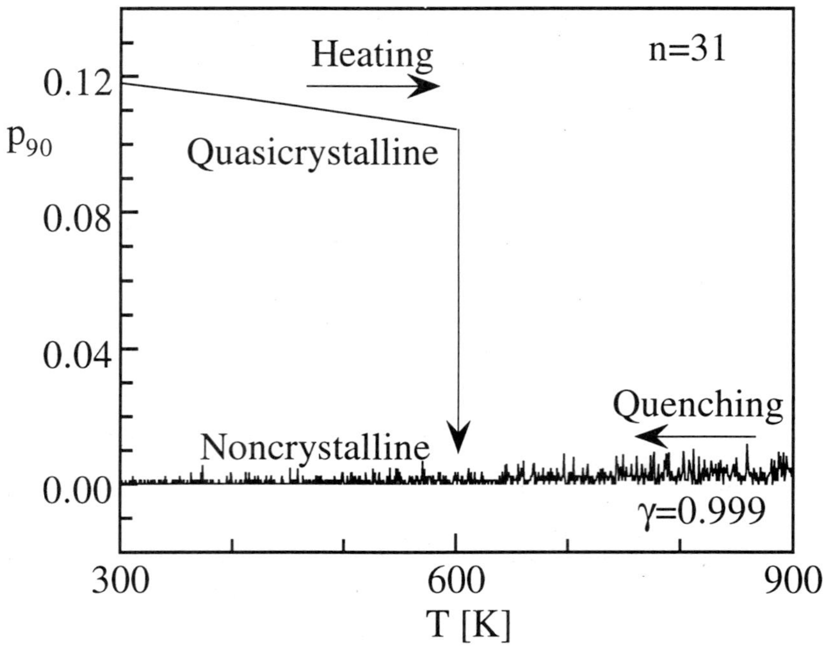


FIGURE 11. Hysteresis upon quenching of $n = 31$ cluster. Slow quenching fails to form the global minimum energy isomer at low temperatures.

$n = 31$ atoms. Upon heating, an abrupt first order-like phase transition occurs at about 600 K. The most stable structure at low temperatures has squares in the structure. At high temperatures, the most stable structure is noncrystalline without any 90 degrees. Upon quenching, hysteresis is found, and the most stable structure does not form. Instead, a structure with $p_{90} = 0$ is found. Slow quenching failed to form the thermodynamically most stable structure at low temperatures. This implies that the ergodicity of the MC algorithm breaks down for this cluster size, and the global minimum energy isomer cannot be obtained using molecular simulations.

From a physical point of view, it is remarkable that an abrupt transition has been found for such a small ensemble of atoms. These abrupt transitions are expected only in bulk-like systems. The abrupt change in structure can imply a considerable change in properties of clusters. For example, surface reactions which need fourfold symmetry may be completely or partially ignited or extinguished at this critical temperature. These ignition/extinction phenomena are caused by a change in catalyst cluster structure.

To investigate whether the lack of ergodicity is generic for all large clusters, simulations have been performed for several larger clusters. Results for $n = 47$ are shown in Figure 12. The circles are ensemble averages from heating simulations, and the noisy solid line is obtained from a quenching run with $\gamma = 0.9999$. Surprisingly, the most stable structure forms upon quenching. This has been possible for all clusters studied with $n > 36$ atoms up to $n = 55$ atoms. This indicates that only

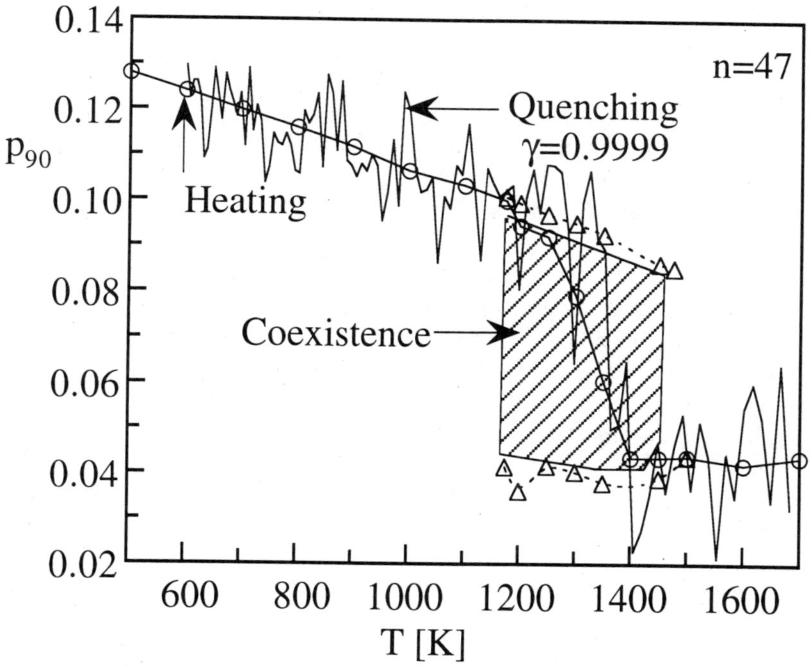


FIGURE 12. The global minimum energy isomer at low temperatures forms upon quenching for $n = 47$. The coexistence regime affects strongly the success of the simulated annealing algorithm.

for clusters with $n = 31$ to $n = 36$ atoms, the ergodicity of the MC simulations breaks down. Besides the quenching process, our recent simulations on growth of clusters by monomer addition revealed hysteresis and failure of formation of quasicrystalline structures at these cluster sizes. This failure of the MC algorithm is caused by a difficulty in the nucleation of crystallinity in clusters. When the number of atoms is sufficiently large, synergetic effects allow nucleation of the new phase.

Thermodynamic considerations revealed that for clusters with $n > 36$ atoms, there is a temperature coexistence regime between the two families of clusters shown by the shaded area in Figure 12. At each temperature within the coexistence regime, oscillations between the two phases are found. The triangles in Figure 12 show the average values for each of the two phases. Outside the coexistence regime, only one phase is found. As the temperature is reduced, the global minimum free energy cluster shifts from a noncrystalline phase at high temperatures to a crystalline phase at low temperatures with a coexistence regime in between. As a result, when quenching in the coexistence regime is sufficiently slow compared to the characteristic time for oscillations, there is a probability for formation of the global minimum energy. The initial conditions are not crucial in affecting global optimization as far as the above criterion is met. For clusters with $n = 31 - 35$ atoms, a coexistence regime is not observed and causes lack of ergodicity of MC simulations.

## 8. Conclusions

The MC simulated annealing method has been employed to minimize the free energy of small clusters. The main thrusts of this work can be outlined as follows:

- The simulated annealing MC method provides global and local isomers at non zero temperatures;
- For high quenching rates, the initial nucleus (atom coordinates) determines the final structure;
- For low quenching rates, cluster isomerization alters the cluster structure;
- The optimum acceptance ratio is about 0.3;
- The probability of forming the most stable size decreases linearly with cluster size and exhibits local maxima due to magic numbers;
- The existence of a critical nucleus size can result in a lack of ergodicity of the conventional Metropolis walk Monte Carlo algorithm due to nucleation barriers;
- Results are qualitative applicable to nanophase clusters of various materials.

## 9. References

M. P. Allen and D. J. Tildesley, *Computer simulation of liquids*, Oxford Science Publications, 1989.

P. A. Braier, R. S. Berry, and D. J. Wales, *How the range of pair interactions governs features of multidimensional potentials*, J. Chem. Phys. **93** (1990), 8745-8756.

X.-Y. Chang and R. S. Berry, *Geometry, interaction range, and annealing*, J. Chem. Phys. **97** (1992), 3573-3575.

J. R. Chelikowsky and J. C. Phillips, *Surface and thermodynamic interatomic force fields for silicon clusters and bulk phases*, Phys. Rev. B **41** (1990), 5735-5745.

E. Clementi and C. Roetti, *Atomic Data and Nuclear Data Tables*, Academic, 1974.

M. S. Daw and M. I. Baskes, *Embedded-atom method: Derivation and application to impurities, surfaces, and other defects in metals*, Phys. Rev. B **29** (1984), 6443-6453.

M. R. Hoare and J. A. McInnes, *Morphology and statistical statics of simple microclusters*, Adv. Phys. **32** (1983), 791-821.

M. R. Hoare and P. Pal, *Physical cluster mechanics: Statics and energy surfaces for monatomic systems*, Adv. Phys. **20** (1971), 161-196.

M. R. Hoare and P. Pal, *Statistics and stability of small assemblies of atoms*, J. Cryst. Growth **17** (1972), 77-96.

S. Kirkpatrick, C. D. Gelatt, Jr., and M. P. Vecchi, *Optimization by simulated annealing*, Science **220** (1983), 671-680.

C. Kittel, *Introduction to Solid State Physics*, Wiley, 1986.

R. S. Maier, J. B. Rosen, and G. L. Xue, *A discrete-continuous algorithm for molecular energy minimization*, AHPCRC Report No. 92-031, Minneapolis (1992), 1-15.

C. D. Maranas and C. A. Floudas, *A global optimization approach for Lennard-Jones microclusters*, J. Chem. Phys. **97** (1992), 7667-7678.

N. Metropolis, A. W. Rosenbluth, M. N. Rosenbluth, A. H. Teller, and E. Teller, *Equation of state calculations by fast computing machines*, J. Chem. Phys. **21** (1953), 1087-1092.

J. A. Northby, *Structure and binding of Lennard-Jones clusters:* $13 < N < 147$, J. Chem. Phys. **87** (1987), 6166-6177.

P. M. Pardalos, S. D., and G. Xue, *Optimization Methods for Computing Global Minima of Nonconvex Potential Energy Functions*, J. Global Optimization **42** (1994), 117-133.

W. H. Press, B. P. Flannery, S. A. Teukolsky, and W. T. Vetterling, *Numerical Recipes*, Cambridge University Press, 1986.

J. P. Rose and R. S. Berry, $(KCl)_{32}$ *and the possibilities for glassy clusters*, J. Chem. Phys. **98** (1993), 3262-3274.

F. H. Stillinger and D. K. Stillinger, *Cluster optimization simplified by interaction modification*, J. Chem. Phys. **93** (1990), 6106-6107.

F. H. Stillinger and T. A. Weber, *Hidden structure of liquids*, Phys. Rev. A **25** (1982), 978-989.

F. H. Stillinger and T. A. Weber, *Dynamics of structural transitions in liquids*, Phys. Rev. A **28** (1983), 2408-2416.

F. H. Stillinger and T. A. Weber, *Nonlinear optimization simplified by hypersurface deformation*, J. Stat. Phys. **52** (1988), 1429-1445.

J. Uppenbrink and D. J. Wales, *Packing schemes for Lennard-Jones clusters of* $13 - 150$ *atoms: Minima, transition states, and rearrangement mechanisms*, J. Chem. Soc. Faraday Trans. **87** (1991), 215.

D. G. Vlachos, L. D. Schmidt, and R. Aris, *Structures of small metal clusters: I Low temperature behavior*, J. Chem. Phys. **96** (1992), 6880-6890.

L. T. Wille, *Minimum-energy configurations of atomic clusters: New results obtained by simulated annealing*, Chem. Phys. Letters **133** (1987), 405-410.

G. L. Xue, R. S. Maier, and J. B. Rosen, *Minimizing the Lennard-Jones potential function on a massively parallel computer*, UMSI Report No. 91/309, Minneapolis (1991), 1-18.

Department of Chemical Engineering, University of Massachusetts, Amherst, Massachusetts 01003

*E-mail address*, D. G. Vlachos: vlachos@ecs.umass.edu